Condensed Matter Physics

Springer

Berlin
Heidelberg
New York
Hong Kong
London
Milan
Paris
Tokyo

Gert Strobl

Condensed Matter Physics

Crystals, Liquids, Liquid Crystals, and Polymers

With 169 Figures

Translation of the Original German Version
by Steven P. Brown

Springer

Professor Gert Strobl

Physikalisches Institut
Albert-Ludwigs-Universität Freiburg
79104 Freiburg
Germany
e-mail: gert.strobl@physik.uni-freiburg.de

Translator:
Dr. Steven P. Brown

Department of Physics
University of Warwick
Coventry CV4 7AL
United Kingdom

Catalog-in-Publication Data applied for

A catalog record for this book is available from the Library of Congress.

Bibliographic information published by Die Deutsche Bibliothek Die Deutsche Bibliothek lists this publication in the Deutsche Nationalbibliografie; detailed bibliographic data is available in the Internet at <http://dnb.ddb.de>.

ISBN 3-540-00353-3 Springer-Verlag Berlin Heidelberg New York

Springer-Verlag Berlin Heidelberg New York
a member of BertelsmannSpringer Science+Business Media GmbH

http://www.springer.de

© Springer-Verlag Berlin Heidelberg 2004
Printed in Germany

Typesetting and production: LE-TEX Jelonek, Schmidt & Vöckler GbR, Leipzig
Cover design: *design & production*, Heidelberg

Printed on acid-free paper   2/3141/YL   5 4 3 2 1 0

Preface

While an introduction to solid-state physics always constitutes an important part of a physics degree course, the students usually learn very little, if anything, about the physics of liquids and non-crystalline solids. This is difficult to understand given the importance of these materials in today's world. Clearly crystalline solids occupy the central position in commercial technical applications but other materials, in particular polymers and liquid crystals, have also become really indispensable. Their physical properties can be so well controlled that completely new possibilities have opened up. It could perhaps be thought that the reason for the deficiencies in teaching in this area is a lack of fundamental understanding of the physical properties of these systems – this, however, is not correct. Actually, there exists for liquid crystals and polymers a well-rounded basic physical knowledge based on self-contained concepts which have stood the test of time. As a reason for the neglect of these areas in the education there remains, therefore, only something very simple: The physics of polymers and liquid crystals was developed only late on, indeed too late for it to have made a substantial entry into physics faculties. Freiburg is one of a few exceptions. Here, one finds in addition to research in solid-state physics, as carried out at the local Fraunhofer Institute for Applied Solid-State Physics, also research in polymer physics, conducted in my group in close cooperation with the renowned Institute of Macromolecular Chemistry founded by Staudinger, as well as research in the field of liquid crystals by Professor Finkelmann and his coworkers. As a consequence of this, a lecture course entitled "Physics of condensed matter", which teaches in equal measure solid-state physics and the physics of liquids, liquid crystals and polymers, has for almost twenty years been a core element of our curriculum for physics students. This textbook which I have now written is based on this lecture course. As in the lecture course, about half of the book is devoted to the physics of crystalline solids, with the other materials, liquids, liquid crystals and polymers, being described in the other half. My experience from giving the lecture course has shown me that the two halves are not separated from each other, but rather a common treatment can be adopted. With this book, I am attempting to demonstrate this to a wider readership.

Like the lecture course, this book is aimed at physics students at the beginning of their graduate studies. As regards its level and assumption of previous knowledge, it is comparable to standard textbooks like the "Kittel". I always dictated the text after having given a lecture, and so the book sometimes contains "as-spoken" expressions. The intention of the lecture course and the book is a coherent representation of the ideas and concepts which are introduced when describing the properties of condensed matter, this being achieved at the expense of going into great detail. The aim is, thus, to create a broad fundamental understanding, from which a simple progression into the different fields of condensed matter physics is made possible. As far as its content is concerned, there is nothing in the book which has not already been treated in individual texts, what is new is solely the combination with an emphasis on the overall interrelations. Perhaps such a common treatment opens now a way for more students to be introduced to the physics of non-crystalline solids, i.e., polymers, and the physics of liquids, and there in particular, of liquid crystals.

This book, like many other 'monographs' is, in truth, a product of teamwork:

Without my secretary, Christina Skorek, who produced the LaTeX file from dictated text and equations in what was for me always an unbelievably short time and furthermore prepared almost all the figures, this book would not have come into being. Without the input of my assistant Dr. Werner Stille and Dr. Thomas Thurn-Albrecht, now professor at the University of Halle, would have been neither exercises nor solutions, while some of the presentations would have remained misunderstandable and erroneous. Without the explanations and advice of Professor Jürgen Heinze, the section about electrolytes would not have been written.

In addition, my translator, Dr. Steven P. Brown, would like to thank Martina Zimmermann for spending many hours checking the English translation with the German, so ensuring a faithful representation of the original text.

To all of them I express here my gratitude for the naturally provided, steady, effective support which I value very much.

Freiburg, July 2003   *Gert Strobl*

Contents

1

Structures

What is 'condensed matter'? Where does the name come from? What characterises this state? Answers are easy to give: Condensed matter is the general description for atomic and molecular substances in the liquid and solid state, and, as the name suggests, this is the state which forms from the condensation of a gas. Such a condensation process leads to an increase of several orders of magnitude in the density. Moreover, in contrast to the case of a gas, where densities can be arbitrarily small, the densities encountered in condensed phases are restricted in range and exhibit only small changes as a function of pressure and temperature. The transition from the gas phase results in a complete change in the motion of the atoms or molecules on a microscopic scale. In gases, motions occur largely unhindered, with the translational movement of the molecules only being altered by collisions. In condensed phases, attractive interactions become active because of the much higher densities. This causes any motion to be severely restricted and leads to the stabilisation of states, which 'hold themselves together'. Two different fundamental states, **solid** and **liquid**, are to be distinguished. Solids have, on their own, a definite form, while liquids take up the form of their container. In between states also exist, for example, highly viscous liquids, which can maintain their form for a certain time, as well as very malleable solids, whose shape changes upon the application of only small external forces. How do liquids and solids differ in their microscopic structure and dynamics? First, it is to be stated that structure and local motion are closely coupled. In a crystalline solid, where there is order over macroscopic distances, motion is restricted to small oscillations of the atoms about their equilibrium positions in the lattice. In contrast, the much greater, for long times even unrestricted motion in a liquid allows only a minor local ordering, which is constantly changing. In a liquid, the interatomic interactions are strong enough such that the molecules are attracted together by the action of a commonly created **molecular field**, but they are not sufficient to force the molecules to occupy fixed positions.

Besides **simple liquids** made up of atoms or quasi-spherical molecules, there exists an additional important group, namely **complex fluids**. The latter form from structured molecules, which, depending on their inner conformational dynamics, can be either rigid or flexible. Two examples, polymers and liquid crystals, are of much importance because of the many possible applications arising from their specific properties. **Liquid crystals** are indispensable components of display devices. As the name suggests, they are in a state in between the crystalline and the liquid phase. It is often formed from rod-like molecules, which, while possessing a liquid-like centre of mass mobility, retain a collective preferred orientation. As a consequence, liquid crystals are optically anisotropic. **Polymers** are materials with unusual mechanical properties, which can be adjusted in a controlled fashion over a wide range. They are processed in the liquid state at high temperatures. In this state, long-range couplings due to the chain character of the molecules strongly influence the flow properties and even lead to the temporary presence of elastic restoring forces. Upon cooling, a solidification process occurs, by means of a transition to either a glass or to a semi-crystalline state, in which crystalline and liquid regions co-exist.

In this chapter, we will consider the main features of the structure of crystals, simple liquids, liquid crystals, and polymers. There are many methods, by which the structure of condensed phases can be analysed, with the investigated length-scales ranging from Å to mm. The most important method, which is generally applicable and does not alter the sample, is scattering experiments. Such experiments can be performed using electromagnetic radiation, in particular X-rays (ideally from a synchrotron source), and also neutrons. The final section of this chapter presents the general principles of this technique.

1.1 Crystals

1.1.1 Crystal Structures

Crystals possess a spatially periodic structure. The repeating unit is termed the **unit cell**. The unit cell is found with the same dimensions and the same arrangement of the same constituent atoms throughout the crystal. The following diagrams show examples of unit cells, and we will use them to introduce several terms and concepts.

The unit cell of caesium chloride is shown in Fig. 1.1a. It is **cubic** and comprises a single caesium and a single chlorine atom. Both atoms are separated by a distance corresponding to half the cube diagonal. If both positions, i.e., the corner and the centre of the cube, are occupied by the same type of atom, the resulting unit cell is termed **body-centred cubic**. This is represented in Fig. 1.1b. Figure 1.1c shows a third type of a cubic unit cell, which is termed **face-centred cubic**, where four atoms of the same type are positioned in a corner and the middle of three outer surfaces of the cube. Atoms are of course also present at the other three outer surfaces of the cube, however, these are counted as belonging to the neighbouring unit cells.

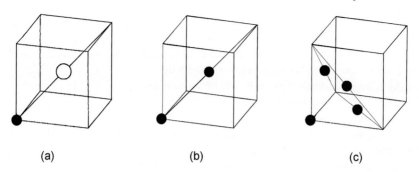

(a)   (b)   (c)

Fig. 1.1. Three cubic unit cells. (**a**) The caesium chloride unit cell, which is made up of a caesium and a chlorine atom. (**b**) The body-centred cubic unit cell. (**c**) The face-centred cubic unit cell – the atoms are hexagonally packed in the shown layer.

The two and four atoms which occupy the body-centred cubic and the face-centred cubic unit cells, respectively, all occupy equivalent positions; i.e., each of them could be chosen to be the corner atom of a cube. This is not the case for the caesium chloride structure, where the relative arrangement of the two atoms is viewed differently from all points in the unit cell. Cells of this latter type are referred to as **primitive**.

The face-centred cubic unit cell possesses a special property: The atoms in the shown plane in Fig. 1.1c are **hexagonally-close** packed. This arrangement is illustrated in Fig. 1.2, which shows the view perpendicular to this plane, with the atoms having been drawn larger to make it clear that a close packing is involved here. The diagram also shows how the planes are arranged with respect to each other – there are three different relative positions of the planes, which are labelled A, B, and C. In a face-centred cubic structure, there is always a regular ABCABC pattern. Hexagonally-close packed structures are often found in the crystals formed by atoms, e.g., in many metals. The stacking order is not necessarily that of the face-centred cubic structure, it could, for example, be ABAB, in which case the unit cell is no longer cubic but rather takes the form of a column with an hexagonal base.

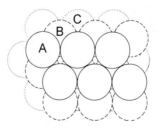

Fig. 1.2. The view perpendicular to the marked plane in the face-centred cubic unit cell of Fig. 1.1c. The planes are arranged according to an ABCABC stacking order.

As the next example, Fig. 1.3a shows the unit cell of diamond. It is also cubic, but is occupied in a special way by eight atoms in equivalent positions. As can be seen, each carbon atom is in a **tetrahedral** environment and is covalently bonded to its four nearest neighbours.

As a final example, the unit cell of sodium chloride (salt) is presented in Fig. 1.3b. It is also face-centred cubic and is occupied by four sodium and four chlorine atoms.

The atoms, which belong to a unit cell, form the **basis** of the unit cell. The basis consists of four sodium and four chlorine atoms, eight carbon atoms, and one caesium and one chlorine in the structure of salt, diamond, and caesium chloride, respectively.

How can the periodic structure of a crystal be formally described? It is obvious that a property is required which changes regularly within the crystal and at the same time describes well the structure within the unit cell. The microscopic electron density fulfils these requirements since it varies, first, within each atom because of the distribution within the electronic shells and, then, correspondingly within each unit cell and, finally, periodically through the crystal. Geometrically speaking, the unit cell is a parallelepiped, which is fully defined by three basis vectors, which are denoted here $\boldsymbol{a}_1, \boldsymbol{a}_2, \boldsymbol{a}_3$. By putting the unit cells together, a **lattice** is formed from the corners of the individual unit cells, with the lattice points, $\boldsymbol{R}_{uvw}$, being given by

$$\boldsymbol{R}_{uvw} = u\boldsymbol{a}_1 + v\boldsymbol{a}_2 + w\boldsymbol{a}_3 \ . \tag{1.1}$$

The coefficients u, v and w take integer values, and each combination fixes a particular lattice point. The periodicity of the lattice leads to the electron density, $\rho_e(\boldsymbol{r})$, possessing the following property:

$$\rho_e(\boldsymbol{r}) = \rho_e(\boldsymbol{r} + \boldsymbol{R}_{uvw}) \ . \tag{1.2}$$

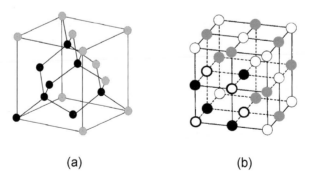

(a)   (b)

Fig. 1.3. Two further cubic unit cells: (**a**) The diamond structure with its tetrahedral connectivities – this structure can be considered as two interlocked face-centred cubic lattices. (**b**) The unit cell of sodium chloride is face-centred cubic and is occupied by four sodium and four chlorine atoms.

The periodic structure of a crystal is described by means of this lattice, i.e., by the set of lattice points $\boldsymbol{R}_{uvw}$ defined according to Eq. (1.1). This lattice is referred to as the **Bravais lattice**. Each crystal is assigned a Bravais lattice according to its spatial periodicity. It expresses the **translational symmetry** of the crystal. In the following, we will discuss in detail how the symmetry of objects can be characterised by the identification of the **symmetry operations**. The Bravais lattice is the first example. It consists of the collection of translation vectors, the action of which lead to the crystal being superimposed upon itself. Strictly speaking, this applies only to an infinitely extended crystal, and indeed the definition is based upon this case.

After the Bravais lattice of a crystal has been established, in order to fully describe the structure, it is necessary to specify the arrangement of the atoms in the unit cell. This is usually achieved in the way

$$r_j = x_{1j}\boldsymbol{a}_1 + x_{2j}\boldsymbol{a}_2 + x_{3j}\boldsymbol{a}_3 \quad , \tag{1.3}$$

i.e., the positions are defined with reference to the basis vectors, $\boldsymbol{a}_1, \boldsymbol{a}_2, \boldsymbol{a}_3$, by means of the relative coordinates, x_{1j}, x_{2j}, x_{3j}. The caesium chloride and sodium chloride lattices are thus described as

$$|\boldsymbol{a}_1| = |\boldsymbol{a}_2| = |\boldsymbol{a}_3| = 4.1 \text{ Å} \qquad \text{(cubic)} \quad ,$$

$$\text{Cl} : (0; 0; 0) \quad , \quad \text{Cs} : (0.5; 0.5; 0.5)$$

and

$$|\boldsymbol{a}_1| = |\boldsymbol{a}_2| = |\boldsymbol{a}_3| = 5.6 \text{ Å} \qquad \text{(cubic)} \quad ,$$

$$\text{Na} : (0; 0; 0)(0.5; 0.5; 0)(0.5; 0; 0.5)(0; 0.5; 0.5) \quad ,$$

$$\text{Cl} : (0; 0; 0.5)(0.5; 0.5; 0.5)(0; 0.5; 0)(0.5; 0; 0) \quad .$$

For non-cubic crystals, it is additionally necessary to specify the angles enclosed by the basis vectors $\boldsymbol{a}_1, \boldsymbol{a}_2, \boldsymbol{a}_3$.

Classification according to Symmetry. A characteristic which is common to all crystals is their translational symmetry. As described above, this is specified for each crystal by means of its Bravais lattice, i.e., the collection of translations by which the crystal is superimposed upon itself. However, translational symmetry represents only a part of the symmetry properties of a crystal. That this is so becomes clear upon thinking about the many and different symmetrically formed natural minerals. The way in which a handle can be put upon the symmetry of an object became apparent in the introduction to translational symmetry: It is necessary to identify the symmetry operations by which the crystal is superimposed upon itself. As well as translations, **rotations, reflections** and combinations of such elements come into play. Crystals can be classified according to a consideration of their symmetry. Indeed, such a classification is not solely a cataloguing process – all important properties of a crystal are influenced by its symmetry, and thus a knowledge of the symmetry and, in particular, a knowledge as to which 'symmetry class' a crystal belongs allows important conclusions to be drawn.

We will begin our consideration of symmetry not with crystals, but rather we look first at the symmetry of molecules, and consider some examples. Figure 1.4 shows a water molecule together with all of its symmetry elements. Firstly, there is a two-fold **rotation axis**, which is denoted C_2. This means that a rotation of 180° degrees, and also trivially a rotation of 360°, causes the molecule to be superimposed upon itself. In addition, there are two **mirror planes**, which are denoted $\sigma^{(1)}$ and $\sigma^{(2)}$. The reflection $\sigma^{(1)}$ leaves the atoms in the same positions, while $\sigma^{(2)}$ swaps the two hydrogen atoms. It is important to note that these four symmetry operations together form a closed group. Obviously, the combination of two elements, which is defined as the successive operation of the two elements, also constitutes a symmetry operation and corresponds to a further element of the group.

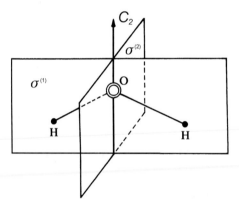

Fig. 1.4. The group of symmetry elements of the water molecule.

The symmetry group of the ammonia molecule, which is shown in Fig. 1.5, consists of a three-fold rotation axis, C_3, as well as the three mirror planes, $\sigma^{(1)}$, $\sigma^{(2)}$ and $\sigma^{(3)}$.

A different type of symmetry element appears in the case of the methane molecule, which is shown in Fig. 1.6. This molecule possesses a four-fold **rotation-reflection axis**, which is labelled by the symbol S_4. As is illustrated in the figure, performing a single rotation-reflection operation consists of a rotation of 90°, followed immediately by a reflection about a plane perpendicular to the rotation-reflection axis. Performing this operation twice is equivalent to a rotation by 180°.

The symmetry of a molecule is completely described by specifying its symmetry group. These groups are named **point groups**, since at least one point remains unchanged for all symmetry operations. Out of which operations do

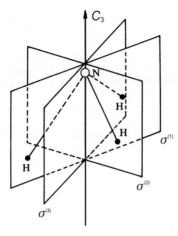

Fig. 1.5. The elements of the symmetry group of the ammonia molecule.

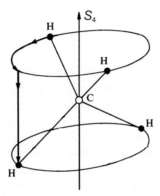

Fig. 1.6. The four-fold rotation-reflection axis of a methane molecule.

the elements of the point groups of molecules belong? The answer is that point groups are made up exclusively of rotations and rotation-reflections. In addition, the identity element is always present. The mirror planes, $\sigma^{(i)}$, in the previous figures correspond to a one-fold rotation-reflection axis S_1, while the **inversion** operation, which is a further commonly encountered symmetry element, is equivalent to a rotation-reflection with a rotation angle of 180°.

Following this short discussion of the symmetry properties of molecules, the nature of the symmetry groups for crystals should be evident: They contain all the translations, rotations, and rotation-reflections, which cause a given crystal to be superimposed upon itself. Actually, this is not completely correct, since two special symmetry elements, known as **screw axes** and **glide planes** also need to be included. They are represented in Fig. 1.7, where a part of a crystal is illustrated. The symmetry operation associated with a screw

Fig. 1.7. A screw axis (*left*) and a glide plane (*right*).

axis involves a translation over a distance corresponding to a whole-number fraction of a lattice constant – in Fig. 1.7, it is half the lattice constant a – followed by a rotation about the screw axis by a whole-number fraction of $360°$ – in the example, $180°$. For a glide plane, the translation is accompanied by a subsequent reflection in the plane. The special nature of these operations is apparent: A rotation of $180°$ or a reflection would not cause the crystal to be superimposed upon itself. This can only be achieved by coupling the operation with a translation, which is itself not a translation vector of the Bravais lattice. The group of all operations which cause a crystal to be superimposed upon itself is termed its **space group**, and consists, in general, of translations, rotations, rotation-reflections, screw axes, and glide planes. The symmetry of a crystal is completely defined upon specifying its space group.

How many possible space groups are there? The answer is that there is not an infinite number, with there being exactly 230 possible space groups. An important reason as to why there is a finite number is that only two-, three-, four-, and six-fold rotation axes can exist in a crystal, which is in contrast to the case of molecules where any rotation axis can, in principle, be found. That such a restriction should exist becomes clear when it is recognised that while regular triangles, rectangles, and hexagons can be arranged to completely cover a surface, i.e., there are no gaps, this is not possible with pentagons. Considering all possible symmetries, there are 230 different crystal types. In fact, in nature only a fraction of these is actually found to exist.

The fundamental property of a crystal is its translational symmetry – it can then be asked: Is it possible to distinguish the different Bravais lattices on the basis of their symmetry? To do this, it is necessary to identify the rotations and rotation-reflections which cause a given Bravais lattice to be

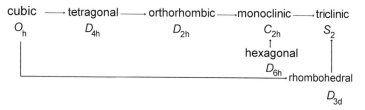

Fig. 1.8. The hierarchy of the 7 crystal systems with the accompanying point groups of the Bravais lattices.

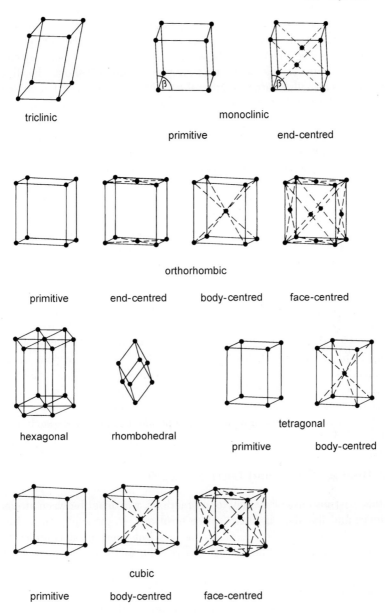

Fig. 1.9. The 14 different Bravais lattices.

superimposed upon itself. The theoretical analysis shows that there are seven different point groups, and there are, thus, by considering the Bravais lattices, seven different **crystal systems**. The different types, from **cubic** to **triclinic**, are given in their hierarchical order in Fig. 1.8. The symbol (O_h, D_{4h}, etc.)

is the **Schönflies symbol** of the point group of the crystal system, and the degree of symmetry decreases on going from left to right. Hierarchical order means here that the direction of the arrow indicates that the point group gets smaller, i.e., symmetry elements are removed. Figure 1.9 shows the different unit cells, and also illustrates a further differentiating property: For cubic crystal systems, three different unit cells are shown, which, although having the same symmetry, differ in their **type**. In addition to the primitive type, where there is one lattice point per cell, for the cubic, tetragonal, orthorhombic, and monoclinic systems, different types arise by means of one face- ('end'), all face- ('face') and body-centring. In total, there are fourteen different Bravais lattices.

The third and final concept by which the symmetry properties of a crystal can be described is termed the **crystal class**. This describes the symmetry of the direction vectors. The description of this symmetry is again achieved by means of a point group comprised of the elements, which transform all direction vectors into equivalent vectors. It is evident which type of elements are involved here – it is the rotations and rotation-reflections of the space group together with the rotation and reflection parts of any screw axes and glide planes, respectively. Since we are only interested here in processes which cause equivalent direction vectors to be superimposed, translations are not relevant. The theoretical analysis shows that there are 32 different crystal classes. In fact, the crystal class is decisive as far as many physical properties are concerned, in particular, it defines the symmetry of all tensorial properties, e.g., of the magnetic susceptibility or the electrical conductivity. What is important here is that it is possible merely from the symmetry, i.e., the crystal class, to determine how many independent components such a tensor has.

1.1.2 Binding Forces and Lattice Energies

Crystals exist because of attractive interactions, which act between atoms and molecules and fix them in their lattice positions. In principle, these binding forces are always electromagnetic in nature, but there exist completely different types with very different strengths. The following ones are found in crystals:

- ionic interactions
- van der Waals forces between neutral atoms and molecules
- valence forces, acting along covalent bonds
- the interactions by means of hydrogen bonds
- the interactions which are active in metals and hold together the metallic ions and the electron gas.

Van der Waals forces are the weakest, while covalent and ionic interactions are the strongest. For all the types of interactions, there are example materials: The alkali halides are typical ionic crystals, crystals of organic molecules or noble gases are largely held together by van der Waals forces, diamond has covalent bonds, hydrogen bonds are important in ice, and all metallic conductors use electron mediated interactions.

Metals will be discussed later in Sect. 4.1, and we begin here with a short sketch of the nature of valence forces and hydrogen-bonding interactions. Covalent bonds generally arise from the overlap of two orbitals, which are both occupied by a single electron, with the two spins being orientated opposite to each other. The quantum-mechanical exchange energy leads in this situation to a lowering of the energy and thus to the formation of a chemical bond. Figure 1.10 shows, for a carbon atom, how four orbitals with the characteristic tetrahedral arrangement, which is found in the diamond structure, form. On the left-hand side of the diagram, the four basis second-shell orbitals (2s, $2p_x$, $2p_y$, $2p_z$) are represented. Since these four states are energetically degenerate, they can be superimposed, with the result being shown on the right-hand side. The lobe with the plus sign arises from a sum of all four orbitals; the other three lobes correspond to the other three independent linear combinations, this time involving subtractions. The name for these orbitals which are created by linear combinations is 'sp^3 hybrid'. Each carbon atom can, by means of the four sp^3 hybrids, form four tetrahedrally orientated bonds, each to a neighbouring atom, which possesses an orbital of the opposite sign also directed along the bond. The four covalent bonds formed in this way by each carbon have fixed bond angles and distances relative to the neighbouring atoms.

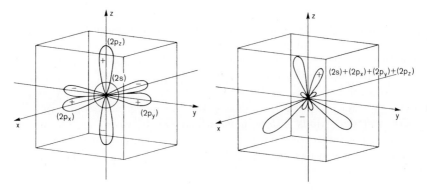

Fig. 1.10. The tetrahedral covalent-bonding structure in the diamond lattice: The build-up through hybridisation (*right*) from a 2s and three 2p orbitals (*left*).

Figure 1.11 shows the crystal structure of ice. In the vertical direction water molecules are grouped in rows. The molecules are positioned such that tetrahedral bonding arrangements form between water molecules in neighbouring rows. Each oxygen atom can form hydrogen bonds to its four nearest neighbours, with a proton moving back and forth along the bond vector,

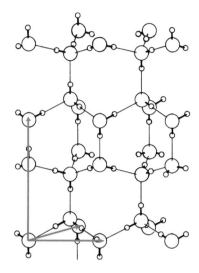

Fig. 1.11. Hydrogen bonding interactions in ice.

changing its position from being near one oxygen atom to a position near the other. The pre-requisite for the formation of the binding interactions is the donation of the hydrogen's electron to an electronegative partner such as oxygen. After the donation, there exist attractive interactions between the neighbouring oxygens mediated by the so-formed proton. In a similar way as in ice, hydrogen bonds influence the structure of other crystals and also, in a very important and function-decisive way, structures in quite different surroundings – in the cells of plants and animals. Here biopolymers such as DNA and proteins play decisive roles. For their biological activity well-defined conformations are necessary, and these are selected and stabilised by hydrogen bonds.

For van der Waals interactions, which act between neutral atoms or molecules, the interaction energy depends in a characteristic way upon the distance. For a distance r_{jk} between two particles j and k, the interaction energy is determined according to the following equation:

$$u(r_{jk}) = 4u_L \left(\left(\frac{r_L}{r_{jk}} \right)^{12} - \left(\frac{r_L}{r_{jk}} \right)^{6} \right) \; , \tag{1.4}$$

where u_L and r_L are two parameters. This is referred to as the **Lennard–Jones potential**. The interaction energy contains, in addition to the attractive van der Waals part, a repulsive term. The inverse dependence to the sixth power upon the distance between the atoms is a characteristic feature of the van der Waals energy. What is the physical origin of the attraction? Neutral atoms, for example argon, cannot, in the first instance, exert electrical forces upon each other. However, this is only valid so far as no excitation of the electrons is possible. Changes in the electronic states, even when they only occur for a short time, can lead to the generation of electric dipole moments. As soon as such a dipole moment forms, even if only for a short time, an interaction with the other atom exists because of the electric field associated with the dipole moment. The field can then cause a polarisation of the second atom, i.e., generate another dipole moment. Through this mutual induction of dipole moments, an attractive force arises. The attractive interaction is proportional to the polarisability of both partners and is found to fall off with the sixth power of the distance.

The origin of the first, repulsive term in Eq. (1.4) is, generally speaking, the Pauli principle. It works against an overlap of the electron shells of two atoms which would occur upon an approach to a sufficiently close distance. If no chemical bond can arise, and that is only so in special cases, an overlap causes, because of the Pauli principle, an energy increase of one of the electrons, and this leads to a repulsive force. The representation of the repulsive energy by a steeply rising potential function of order 12 is purely empirical, but experience shows that it provides a good description of reality.

A very important property which is common to covalently bonded crystals, hydrogen-bonded systems, van der Waals crystals, and the yet to be discussed ionic crystals is that the complete **lattice energy**, which is stored in the crystal as potential energy, corresponds to the sum of the bond energies between pairs of building blocks. The lattice energy U_b can be expressed as

$$U_{\mathrm{b}} = \frac{1}{2} \sum_{j \neq k} u(r_{jk}) \ . \tag{1.5}$$

The factor of $1/2$ is included since otherwise the sum would count each interaction twice. In order to calculate the interaction energy of a van der Waals crystal, Eq. (1.4), which describes the Lennard–Jones potential, has to be used. As a basic constant, we introduce the distance a between two neighbours and write

$$r_{jk} = a s_{jk} \ . \tag{1.6}$$

The coefficients s_{jk} represent relative distances with respect to the basic distance a. The energy U_b of a crystal comprising $\mathcal{N}$ atoms or molecules is then given by

$$U_{\mathrm{b}} = 4 u_{\mathrm{L}} \frac{\mathcal{N}}{2} \left[\left(\frac{r_{\mathrm{L}}}{a} \right)^{12} \sum_{k \neq 1} \left(\frac{1}{s_{1k}} \right)^{12} - \left(\frac{r_{\mathrm{L}}}{a} \right)^{6} \sum_{k \neq 1} \left(\frac{1}{s_{1k}} \right)^{6} \right] \ . \tag{1.7}$$

The advantage of this notation is that the result of both sums only depends on the type of the crystal lattice and no longer on the lattice constants. For example, for a face-centred cubic structure, the first and second sums equal 12.1 and 14.5, respectively.

In Eq. (1.7), the basic distance a appears as a variable, while the parameters u_L and r_L from Eq. (1.4) are fixed coefficients. The knowledge of the dependence of the lattice energy upon the basic distance a allows the equilibrium distance between nearest neighbours in the crystal to be calculated: It is that where U_b is a minimum and is, therefore, given by

$$\left.\frac{\partial U_b}{\partial a}\right|_{eq} = 0 \ . \tag{1.8}$$

In the case of a face-centred cubic lattice, this leads to

$$a = 1.09 r_L \ . \tag{1.9}$$

An analogous approach can be employed for ionic crystals. We choose here a different, likewise frequently employed empirical description of the repulsive term known as the Buckingham potential. Equation (1.10) describes the interaction energy between nearest neighbours in an ionic crystal – the first term is the Buckingham potential, which is a negative exponential term parameterised by u_B and r_B, and the second term is the attractive Coulomb energy

$$u(r_{jk} = a) = u_B \exp\left(-\frac{a}{r_B}\right) - \frac{Q^2}{4\pi\varepsilon_o a} \ . \tag{1.10}$$

The repulsive term is neglected for pairs of atoms which are not nearest neighbours, i.e.,

$$u(r_{jk} > a) = \pm\frac{Q^2}{4\pi\varepsilon_o a}\frac{1}{s_{jk}} \ . \tag{1.11}$$

Both Eqs. (1.10) and (1.11) contain the product of the charges of the interacting ions, Q^2. Depending on the particular charges, this product can be positive or negative, and this is to be taken into account when the lattice energy is calculated. The result can be written as

$$U_b = \frac{\mathcal{N}}{2}\sum_{k\neq 1} u(r_{1k}) = \frac{\mathcal{N}}{2}\left[z u_B \exp\left(-\frac{a}{r_B}\right) - \beta_M\frac{Q^2}{4\pi\varepsilon_o a}\right] \ . \tag{1.12}$$

The first term contains the number z of nearest neighbours, which determines the total repulsive term. The second term includes the parameter β_M. It is known as the **Madelung constant** and corresponds to the sum

$$\beta_M = \sum_{k\neq 1} -\frac{\text{sign}(Q_1 Q_k)}{s_{1k}} \ . \tag{1.13}$$

The Madelung constant again depends solely on the crystal type, e.g., for sodium chloride, $\beta_M = 1.75$. As above, the basic distance a can be determined by applying the minimum condition of Eq. (1.8).

Generally, lattice energies are determined by thermodynamic methods, i.e., measurements of specific heats as well as the latent heats associated with melting and evaporation. The thermodynamic **cohesive energy**, denoted here as $\mathcal{U}_b$, is defined as the difference between the internal energy $\mathcal{U}$ of a crystal and that of an ideal gas of the same atoms or molecules, $\mathcal{U}_{ig}$:

$$\mathcal{U}_b = \mathcal{U} - \mathcal{U}_{ig} \quad . \tag{1.14}$$

Note that $\mathcal{U}_b$ depends, together with $\mathcal{U}_{ig}$ and $\mathcal{U}$, on the temperature. The largest part of $\mathcal{U}_b$ corresponds to the lattice energy U_b, which is calculated using Eq. (1.7) or (1.12) using the appropriate basic distance. There are additional, although usually small, contributions, for example, if the kinetic energy in the lattice is less than in an ideal gas because of quantum effects. When the aim of an investigation is the exact determination of fundamental interaction energies, such effects must be considered.

Thermal Expansion. It was stated above that the basic distance a in a crystal can be determined from the minimum condition in Eq. (1.8). Actually, this is not completely true, since the statement is only strictly valid at a temperature of absolute zero. It is well known that all crystals exhibit a thermal expansion and this was not considered above. It is still the case at a finite temperature that the lattice constants can be determined from a minimum condition, however it is not the lattice energy, but rather the Helmholtz free energy of the crystal which is seeking to reach a minimum at thermal equilibrium. Under isothermal conditions, the following general equation is valid:

$$\left. \frac{\partial \mathcal{F}}{\partial \mathcal{V}} \right|_T = -p \tag{1.15}$$

(see Eq. (A.6) of Appendix A), i.e., under zero external pressure

$$\left. \frac{\partial \mathcal{F}}{\partial a} \right|_{eq} = 0 \quad . \tag{1.16}$$

There are two contributions to the free energy of a crystal. In addition to the lattice energy U_b, there is a second term which is due to thermal motion, i.e., lattice vibrations. We denote this second term as $\mathcal{F}_{vib}$, and the Helmholtz free energy of a crystal is then given as

$$\mathcal{F} = U_b + \mathcal{F}_{vib} \quad . \tag{1.17}$$

Statistical thermodynamics provides a mathematical description for $\mathcal{F}_{vib}$. Each lattice vibration corresponds to a single harmonic oscillator, and $\mathcal{F}_{vib}$ is then calculated as the sum over all contributions $\mathcal{F}_j$ of all lattice vibrations. As discussed in detail later in Sect. 5.1.1, the calculation must take into account that the energy uptake of an harmonic oscillator is not continuous but quantised with quanta of magnitude $\hbar\omega_j$. This means that in deriving the contribution $\mathcal{F}_j$ of a lattice vibration from the associated partition function $\mathcal{Z}$ using

$$\mathcal{F}_j = -k_B T \ln \mathcal{Z} \tag{1.18}$$

it is necessary to calculate the sum over all eigenstates with energies ϵ_k:

$$\mathcal{Z} = \sum_k \exp -\frac{\epsilon_k}{k_B T} \quad . \tag{1.19}$$

This leads to (see Eq. (5.44))

$$\mathcal{Z} = \exp -\frac{\hbar\omega_j}{2k_B T} \cdot \left(1 - \exp -\frac{\hbar\omega_j}{k_B T}\right)^{-1} \tag{1.20}$$

and thus to the following contribution to the free energy

$$\mathcal{F}_j = k_B T \left[\frac{\hbar\omega_j}{2k_B T} + \ln\left(1 - \exp -\frac{\hbar\omega_j}{k_B T}\right)\right] \quad . \tag{1.21}$$

The free energy of all lattice vibrations in a crystal is therefore given as

$$\mathcal{F}_{vib} = k_B T \sum_j \left[\frac{\hbar\omega_j}{2k_B T} + \ln\left(1 - \exp -\frac{\hbar\omega_j}{k_B T}\right)\right] \quad , \tag{1.22}$$

where ω_j is the frequency of the lattice vibration j.

How does the Helmholtz free energy of the lattice vibrations change upon changing the lattice constant a? For a crystal with pure harmonic interaction potentials, there would be no change. The reality is, however, different, since real interaction potentials always contain anharmonic contributions, which generally cause a reduction in the restoring forces and thus the vibration frequencies for increasing lattice constants. This means that, in contrast to the interaction energy U_b which increases if the lattice is extended, the following equation applies for the lattice vibration part of the Helmholtz free energy:

$$\frac{\partial \mathcal{F}_{vib}}{\partial a} < 0 \quad . \tag{1.23}$$

We now have gathered together all the equations which are required in order to describe the thermal extension of a crystal. In the general case, considering the external pressure, the following is valid:

$$p(T, V) = -\left.\frac{\partial \mathcal{F}}{\partial V}\right|_T = -\left.\frac{\partial U_b}{\partial V}\right|_T - \left.\frac{\partial \mathcal{F}_{vib}}{\partial V}\right|_T \quad . \tag{1.24}$$

The pressure applied outwardly by a crystal in order to stay in equilibrium with the external pressure consists of two contributions:

$$p = p_{el} + p_{th} \quad . \tag{1.25}$$

The first **elastic** term originates directly from the bonding interactions in the lattice, while the second **thermal** term comes from the lattice vibrations. When there is no external pressure, the following is true for all temperatures:

$$-p_{\text{el}} = p_{\text{th}} \ \cdot \tag{1.26}$$

This equation means that an equilibrium is maintained between the thermal pressure and the elastic part which act to cause expansion and contraction, respectively. With increasing temperature, the thermal pressure increases and there is a corresponding expansion of the crystal. The magnitude of the thermal pressure depends on the change of the frequencies of the lattice vibrations with the crystal volume according to the derived equations. To specify this change a **Grüneisen constant** is introduced for each lattice vibration; it is

$$\frac{\partial \ln \omega_j}{\partial \ln V} \ \cdot \tag{1.27}$$

Crystals with large thermal expansion coefficients have large Grüneisen constants, and it is sufficient when that is only the case for some specific lattice vibrations.

1.2 Liquids

1.2.1 When can a Liquid exist?

If a crystal is heated, a transition into the liquid phase occurs at the melting point. Even though the density usually only changes slightly upon this transition, the structure undergoes a complete transformation. The periodic structure of a crystal with its **long-range ordering** is lost, and a new state arises in which only a **local order** exists.

It is not always the case that a crystal can be transformed into the liquid phase. The range of conditions for which a liquid can exist is limited. As an example Fig. 1.12 illustrates the phase behaviour of argon by means of a p, T **phase diagram**. It shows which phase is stable at a given temperature and pressure. The points along the given phase boundaries represent those pairs of pressure and temperature under which two phases, solid and liquid, liquid and gas, or solid and gas, can exist together. If the pressure or temperature deviates from the boundary line, such a co-existence is no longer possible, which implies that upon crossing a boundary, a phase change takes place. As an example, the dashed line refers to the changes observed upon increasing the temperature at a constant pressure of 1 bar. At very low temperatures, a crystal exists. At the melting point T_{m} the transition into the liquid phase occurs, followed by that into the gas phase at the boiling point T_{b}. Both transitions involve a jump in the volume V, together with an associated change in the internal energy U and entropy S, although it is to be noted that the change upon melting is always small compared to that upon evaporation of

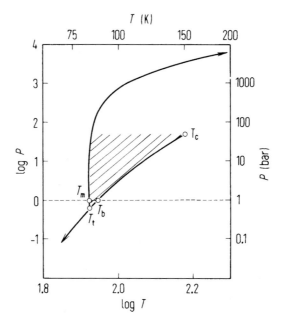

Fig. 1.12. The p, T phase diagram of argon [1].

the liquid. $\mathcal{U}$ and $\mathcal{S}$ always increase upon melting. Generally, $\mathcal{V}$ also increases, although there are exceptions. The best known anomalous case is H_2O, where the crystal structure of ice is such that the hydrogen bonding interactions are maximised, with the result that there is an 'open' crystal with a comparatively low density is formed. The hatched region shows that the range of conditions under which a liquid can exist is limited. Two characteristic points fix the limits, namely the **triple point** at the temperature T_t and the critical point at the temperature T_c. The upper end of the hatched region at the height of the critical point indicates that it is no longer possible, for higher temperatures, to distinguish between the gas and liquid phase. In this **supercritical** pressure region, the transition from the liquid to the gas phase occurs continuously without a jump. The line of co-existence of gas and liquid phases ends at T_c. The lower limit for the existence of a liquid is given by the triple point at the temperature T_t. The name triple point indicates that here all three phases, i.e., the crystalline, liquid, and gaseous phase, coexist. At lower pressures a liquid phase does not form. A crystalline and a gas phase then coexist along the phase boundary. Crossing the boundary corresponds to the sublimation of the solid or the condensation of the gas. If, at some temperature below T_t, a solid is placed in a closed space from which the air has been removed, the vapour pressure fixed by the phase boundary builds up itself by sublimation. The phase boundary which separates the crystalline phase from a phase of lower order, be it liquid or gaseous, has no ends and continues to run, as indicated by the arrows, in the two directions. The reason for this is clear: It is not possible that the transition from a crystalline structure which possesses long-range order to a phase which, if at all, possesses only a local order, occurs in a continuous way. In terms of symmetry there can be no continuity, as a system either possesses a definite symmetry or none at all.

The course of the phase boundaries is fixed by the criteria of thermodynamics: Coexistence can only occur when the chemical potentials of both phases are equal. For the three phase boundaries of the phase diagram the following equations are true:

$$\tilde{\mu}_s = \tilde{\mu}_g \quad ; \quad \tilde{\mu}_s = \tilde{\mu}_l \quad ; \quad \tilde{\mu}_g = \tilde{\mu}_l \quad . \tag{1.28}$$

The subscripts on the symbol for the molar chemical potential $\tilde{\mu}$ refer to *solid, liquid*, and *gas*. At the triple point, the condition

$$\tilde{\mu}_l = \tilde{\mu}_g = \tilde{\mu}_s \tag{1.29}$$

is met.

It is possible to describe the slopes of the different co-existence lines, i.e., the changes of the transition temperatures with respect to pressure, in a straightforward manner by means of the **Clausius–Clapeyron** equation. Consider, for example, the melting point T_m. Along the corresponding phase boundary, the following is true for the changes of both chemical potentials:

$$d\tilde{\mu}_l = d\tilde{\mu}_s \quad . \tag{1.30}$$

This means that

$$\left. \frac{\partial \tilde{\mu}_l}{\partial T} \right|_p dT + \left. \frac{\partial \tilde{\mu}_l}{\partial p} \right|_T dp = \left. \frac{\partial \tilde{\mu}_s}{\partial T} \right|_p dT + \left. \frac{\partial \tilde{\mu}_s}{\partial p} \right|_T dp \tag{1.31}$$

and, thus, by using fundamental equations of thermodynamics (see Eq. (A.10) in Appendix A)

$$-\tilde{s}_l dT + \tilde{v}_l dp = -\tilde{s}_s dT + \tilde{v}_s dp \quad , \tag{1.32}$$

where $\tilde{s}_l, \tilde{s}_s$ and $\tilde{v}_l, \tilde{v}_s$ represent the molar entropies and volumes of the two phases. The change in the melting temperature for a given pressure change is, thus, given by

$$\left. \frac{dT}{dp} \right|_{coex} = \frac{\tilde{v}_l - \tilde{v}_s}{\tilde{s}_l - \tilde{s}_s} \quad . \tag{1.33}$$

The entropy change upon melting can be determined from the increase in the enthalpy, $\tilde{h}_l - \tilde{h}_s$, i.e., the heat of fusion, since the following equation can be derived from $\tilde{\mu}_l = \tilde{\mu}_s$:

$$\tilde{s}_l - \tilde{s}_s = \frac{\tilde{h}_l - \tilde{h}_s}{T_m} \quad . \tag{1.34}$$

1.2.2 Local Order and the Pair Distribution Function

How can the microscopic structure of a liquid, which is constantly changing and yet still has characteristic properties, be described? The answer is that it is necessary to use statistical concepts. Snapshots of the liquid would show that some order in the structure exists, but it is only temporary and on an Å to nm length-scale. The **pair distribution function** is a suitable way to characterise such a local ordering. It describes the average structure in the region around a chosen atom or molecule, and is defined in the following way:

$$g_2(\boldsymbol{r})\mathrm{d}^3\boldsymbol{r}$$

specifies the number of atoms or molecules whose centre of gravity is, on average, in a volume element of size $\mathrm{d}^3\boldsymbol{r}$ at a distance $\boldsymbol{r}$ from the chosen atom. The pair distribution function g_2, thus, describes a particle density averaged over time for a particular location away from the reference atom. Since a liquid does not possess any specific orientation, the pair distribution is isotropic:

$$g_2(\boldsymbol{r}) = g_2(|\boldsymbol{r}| = r) \quad . \tag{1.35}$$

Figure 1.13 shows how a pair distribution function typically appears. On account of the finite extension of an atom, there is a lower limit for the centre-to-centre distance. Upon passing this limit, g_2 rises rapidly to a maximum. The presence of a maximum shows that a shell of atoms forms around the reference atom. The additional maxima in g_2 indicate that there are further shells of atoms at larger distances. If the local ordering dies away after some nanometres, the averaged particle density ρ is reached as a limiting final value.

The pair distribution function is of central importance for the physics of liquids. The simplest situation is found for liquids comprising neutral atoms such as argon or spherical molecules such as CCl_4, where the properties and also the pair distribution function are fully determined by a pair interaction

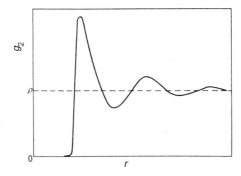

Fig. 1.13. Appearance of the pair distribution function of an atomic liquid.

potential $u(r)$ of the Lennard–Jones type (see Eq. (1.4)). The attractive van der Waals term generates the field which holds the liquid together, while it is the repulsive term which largely determines the local structure, i.e., that which is described by $g_2(r)$. It is harder to describe other types of liquid. For ionic melts, for example molten NaCl, the long-range Coulomb potential appears in place of the attractive van der Waals term, and three pair distribution functions are required, i.e., for Na-Na, Cl-Cl, and Na-Cl pairs. For liquids made up of non-spherical molecules, the pair interaction energy doesn't depend solely on the distance between the centres of gravity, but also upon the orientation of both molecules. It is then necessary to generalise the expression for the pair distribution function, i.e., in addition to $\boldsymbol{r}$, the orientations are further variables. For monoatomic metallic melts, $g_2(r)$ is again sufficient for a complete statistical description of the structure, but here a pair interaction potential no longer exists. The positive ions are surrounded by electron clouds, which partially shield the charge and so weaken the repulsive Coulombic potential. On the other hand, the electrons altogether provide the sufficiently strongly attractive homogeneous background potential, which is required in the liquid state.

The theory of liquids has as its aim the calculation of the pair distribution function, starting from the interaction forces which are present in the liquid. Finding a solution is a harder challenge as compared to the case of gases or crystals. For gases and crystals, there is the ideal gas and the perfectly ordered crystal with harmonic interaction energies, respectively, whose properties are known exactly and from which real gases and crystals can be handled by considering perturbations to the ideal case. For liquids, there is no comparable ideal system. Even the simplest system which can be envisaged, namely a liquid made up of hard spheres without attractive forces, held together in a given fixed volume, cannot be described analytically. Starting equations for the calculation of the pair distribution function can be formulated in terms of $u(r)$, however, they cannot be exactly solved and only be evaluated by means of iterative approximation procedures. Computer simulations are especially helpful and important in such a situation. The employed methods, known as **molecular dynamics** and **Monte-Carlo** algorithms, are capable of calculating numerically exact pair distribution functions for differently chosen interaction potentials.

If both the interaction potential and the pair distribution function are known, a number of important quantities can be determined. The internal energy $\mathcal{U}$ of a liquid is obviously given by the following equation:

$$\mathcal{U} = \frac{3}{2}\mathcal{N}k_{\mathrm{B}}T + \frac{\mathcal{N}}{2}\int u(r)g_2(r)\mathrm{d}^3\boldsymbol{r} \ . \tag{1.36}$$

The equation separates the internal energy into kinetic and potential energies. It is a fundamental property of a liquid that the kinetic energy is comparable to the negative interaction energy. This distinguishes liquids from gases, where the kinetic energy dominates by far, and also from crystals at low temperatures, where, in the opposite manner, the (negative) potential energy can clearly exceed the kinetic energy.

At the end of this section, two further equations are given without explanation. The pressure can be calculated as

$$p = \rho k_B T \left[1 - \frac{1}{6k_B T} \int r \frac{du}{dr} g_2(r) d^3 r \right] \quad , \tag{1.37}$$

while the compressibility is given as follows:

$$k_B T \frac{\partial \rho}{\partial p} = \int (g_2(r) - \rho) d^3 r + 1 \quad . \tag{1.38}$$

Note that only the pair distribution function appears in Eq. (1.38).

1.3 Liquid Crystals

Liquids are always isotropic, i.e., in contrast to the case of crystals, their properties are not orientation dependent. That this is so is because the thermal motion has no orientational preference. Indeed while locally ordered structures are constantly formed, the alignment of the atoms, which form a shell around a particular atom, is completely arbitrary. There is a special class of molecules, for which an inner ordering spontaneously develops in the liquid phase such that a macroscopic anisotropy arises. Such molecules are termed **nematogens** and the anisotropic state which they form is referred to as **nematic**.

1.3.1 The Nematic Liquid-Crystalline State

A common feature of the majority of molecules which can form such an anisotropic liquid phase is that they have a rod-like form. Table 1.1 gives as examples the chemical structure of three nematogen substances with the abbreviations PAA, MBBA, and 5CB. All three have a rigid core and 'wing' groups, which can be flexible. MBBA and 5CB are crystalline at room temperature, but undergo a transition into a nematic liquid-crystalline state at a temperature of 22 °C. By comparison, for PAA, the transition into the liquid-crystalline state occurs only at a temperature of 118 °C. Figure 1.14 shows the image of the nematic phase of 5CB obtained using a polarising microscope for the case where the polariser and the analyser are perpendicular to each other. Although a liquid, the phase is birefringent and stays like this up to a temperature of 35 °C. Here, at the **clearing point**, the birefringence disappears and a transition into the normal isotropic phase occurs. The characteristic pattern with the dark threads is referred to as a **Schlieren texture**. It also gave the structure its name – the Greek word for threads is νημα.

The sketch in the middle part of Fig. 1.15 illustrates the microscopic structure of the nematic phase. The centres of gravity of the rod-like molecules are

Table 1.1. The chemical structure of the nematogen molecules PAA, MBBA and 5CB. The temperatures (in °C) at which the transitions from the crystalline to the nematic phase (T_{cn}) and from the nematic to the isotropic-liquid phase (T_{ni}) occur are given on the right-hand side.

p-azoxyanisole

PAA

$T_{cn} = 118$ $T_{ni} = 135.5$

N-(p-methoxybenzylidene)p'-butylaniline

MBBA

$T_{cn} = 22$ $T_{ni} = 47$

p-pentyl-p'-cyanobiphenyl

5CB

$T_{cn} = 22.5$ $T_{ni} = 35$

arranged as in a liquid. In contrast to the case of an isotropic liquid, which is sketched on the left-hand side, there is, however, no isotropic distribution of the orientation of the molecular axes. There is an orientation distribution, but with a well-defined preferred overall orientation. As is illustrated by Fig. 1.14, the preferred direction is not constant over the whole sample, but it changes slowly over macroscopic (μm-) length-scales. A liquid-crystal is, therefore, characterised by, on the one hand, the local ordering of the positions of the centres of gravities as is usually the case in a liquid, but, on the other hand, a long-range ordering of the molecular orientations. The preferred orientation can be specified by means of a unit vector, which is referred to as the **nematic director**, and is denoted by n in the figure. Physically, n determines the direction of the optical axis.

Polarisation microscope images such as that in Fig. 1.14 portray the **director field**, i.e., the director orientations adopted in a sample. Figure 1.16 immediately reveals for the case of a different nematic phase how a Schlieren texture forms. In addition to the threads, the director field can be recognised – a suitable staining method has been used which makes it

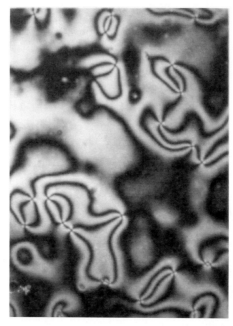

Fig. 1.14. A typical polarisation microscope image (Schlieren texture) of a nematic phase (5CB, from Vertogen and de Jeu [2]).

Fig. 1.15. The isotropic (*left*), nematic (*middle*), and smectic (*right*) phase of a liquid-crystalline substance.

visible. It is interesting that two singularities are observed, i.e., points where the director field is not fixed. As can be seen, these are found exactly at the starting or branching points of the strands. The points are the ends of line-like singularities which form within the sample.

Frank has introduced the term **disclination** for this defect which is characteristic for nematic liquid crystals. As illustrated in Fig. 1.16, there are different types of disclinations – Fig. 1.17 presents in schematic form two further examples. The specification of a particular disclination can be achieved in a simple manner by means of a description of the change in the director

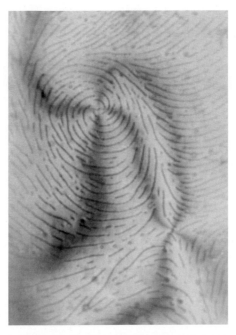

Fig. 1.16. A nematic phase, for which the director field has been made visible by means of a surface staining technique. The structure of two disclinations can be identified (from Cladis et al. in [3]).

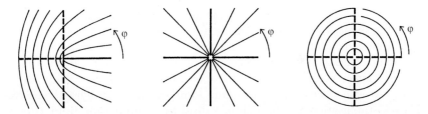

Fig. 1.17. The structure of three different disclinations as described by Eq. (1.39) with strengths $m = 1/2$ (*left*) and $m = 1$ (*middle and right*).

direction around the singularity. If we define the polar angle (see Fig. 1.17) and the director direction by φ and Ψ, respectively, the following is valid for all disclinations:

$$\Psi = m\varphi + \varphi_0 \ . \tag{1.39}$$

The parameter m, which is referred to as the **strength** of the disclination, takes only integer or half-integer values. The three cases shown in Fig. 1.17 correspond to $m = 1/2$, $\varphi_0 = 0$ (left), $m = 1$, $\varphi_0 = 0$ (middle) und $m = 1$, $\varphi_0 = \pi/2$ (right).

Liquid crystals were first observed in 1888 under a microscope by Reinitzer, a biologist at the University of Prague. Immediately afterwards, Lehmann, Professor of Physics at the Universities of Aachen and Karlsruhe, carried out a more detailed examination and, having understood the basic nature of this state, gave it its name. The early investigation of the properties of liquid crystals was of scientific interest, but without technical and economic importance. Today, this has completely changed, with nematogens having become indispensable elements in the construction of screens and display elements. It is clear where the suitability for this type of application comes from: On account of the liquid-like low viscosity, the arrangement of the directors, and, thus, the optical axis, can be altered by means of weak forces. For example, by simply rubbing the surface of a glass substrate, a process which introduces sub-microscopic grooves, it is possible to transform a liquid-crystalline layer, which initially shows a Schlieren texture, into a **planar** single-domain structure with a uniform director direction parallel to the grooves and with, thus, homogeneous optical properties. Starting from such a prepared state, the arrangement of the directors can be altered by the somewhat stronger force of an electric field. Electric fields can exert a torque on the rod-like molecules because of both the anisotropy of the polarisability and the normally found electric dipole moment. A reorientation of the directors changes the degree of transparency with respect to light, and it is exactly this which is relevant for the construction of display elements.

How can the existence of liquid-crystalline phases be understood? In general terms, their appearance is a consequence of the fact that the interaction energy between two partners depends, for rod-like molecules, not only on the distance between the centre of gravities but also upon the relative orientation of the rod axes. A parallel arrangement is energetically preferred regardless of the arrangement of the centres of gravity, and, thus, the formation of **clusters** of molecules is promoted, where there is only a limited variation in the axis direction. If the anisotropy of the interaction forces is sufficiently strong, aggregates with macroscopic dimensions form, which can stabilise themselves. This self-stabilisation can be well described by a **molecular field theory**, and we will come back to this in Sect. 3.3.

A director field with curved field lines and singularities, like that in Fig. 1.16, obviously has a higher potential energy relative to that of the homogeneous case. In Sect. 2.1.2, it will be shown how this can be described. Nevertheless, there are materials for which a deformation of the director field can lead to a lowering in energy: Figure 1.18 shows a twisted equilibrium state. This arises when a chiral molecule, i.e., one which possesses no mirror symmetry, is dissolved in a nematogen. Mostly, this involves a molecule with an asymmetric carbon atom, bonded to four different neighbouring groups or atoms. In this case, there are always two different forms, a **left-handed** and a **right-handed** type. If only one type of the molecule is dissolved in a nematic liquid crystal, the director field adopts, depending on the type, a left- or right-handed screw arrangement. There also exist nematogens which

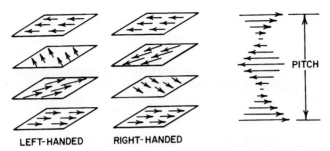

Fig. 1.18. The basic structure of a chiral-nematic (cholesteric) phase. This liquid-crystalline phase in which the director exhibits a periodic twist is formed by chiral nematogens or by dissolving a chiral molecule into a nematic phase. The chirality of the molecule (left- or right-handed) determines which one of the two senses of rotation is found.

contain an asymmetric carbon atom and they, themselves, thus, form a screw-like nematic phase. Cholesterol esters were the first substances for which such a **chiral nematic** phase was observed, and it then led to the commonly used name of **cholesteric liquid crystals**.

The pitch depends on the molecule and changes in the case of a solution not only with the concentration of the chiral component but also with the temperature. It is often the case that the pitch is of about the same size as the wavelength of visible light. This leads to striking colour phenomena, which are due to the reflection of light at the surface, this being selective over the range of frequencies ν

$$\nu = \frac{c_1}{n_\parallel a} \rightarrow \nu = \frac{c_1}{n_\perp a} \quad . \tag{1.40}$$

a is the pitch, c_1 is the velocity of light in a vacuum, $n_\parallel$ and $n_\perp$ refer to the refractive index parallel and perpendicular to the director, respectively. If the reflected light is examined, it is found that it is circularly polarised with the same sense of rotation as that of the screw in the chiral-nematic material. To explain this effect, it is necessary to consider the relevant Maxwell equations. It is then found that a light wave which is circularly polarised in the same sense as the liquid-crystal and has a frequency in the range given by Eq. (1.40) and a wavevector in the direction of the screw axis cannot propagate through the liquid crystal – it is, thus, completely reflected at the surface.

As well as the nematic state, there are other **mesomorphic** phases, i.e., states with a degree of ordering between that of an isotropic liquid and that of a crystal. The right-hand side of Fig. 1.15 shows the arrangement of rod-like molecules in a **smectic** phase – it is apparent that there is a greater ordering as compared to that in a nematic phase. In addition to the long-range orientational order, there is long-range positional order in one of the three spatial dimensions. Smectic liquid-crystals exist as layers, whereby each layer is liquid-like; they can, thus, be considered as two-dimensional liquids. The name is also of Greek origin, coming from $\sigma\mu\eta\gamma\mu\alpha$, which means soap – Friedel found that soap, when mixed with water, also forms such a layer-like structure. Smectic phases are more viscous than nematic phases because of their different nature. In contrast to the case of nematogens, no significant technical use has been found so far for substances in the smectic phase.

1.3.2 Orientational Order and Optical Anisotropy

As is illustrated in the earlier sketch of the nematic liquid-crystal, there is not a strict parallel arrangement of all molecules, but rather a collective preferred orientation. The distribution of orientations covers the whole range of possible angles. The width of the distribution determines the optical anisotropy, and it is, therefore, important to define a parameter which describes in a suitable way the **degree of orientation** of a nematic phase.

In order to describe the distribution of the orientation of the molecular axes, we introduce, in the framework of the spherical coordinates with $\vartheta = 0$ corresponding to the direction of the director, a function $w(\vartheta, \varphi)$, where

$$w(\vartheta, \varphi) \sin \vartheta \mathrm{d}\vartheta \mathrm{d}\varphi$$

gives the fraction of molecules whose long axis lies within the range of angles $\mathrm{d}\vartheta \mathrm{d}\varphi$. In the case here of uniaxial symmetry, w does not depend on φ, and we write, upon introduction of a second distribution function w' which is only dependent on ϑ

$$w(\vartheta, \varphi) = \frac{w'(\vartheta)}{2\pi} \quad . \tag{1.41}$$

The following normalisation equation applies

$$1 = \int_{\vartheta=0}^{\pi} \int_{\varphi=0}^{2\pi} w(\vartheta) \sin \vartheta \mathrm{d}\vartheta \mathrm{d}\varphi = \int_{\vartheta=0}^{\pi} w'(\vartheta) \sin \vartheta \mathrm{d}\vartheta \quad . \tag{1.42}$$

It is a fundamental property of a nematic phase that

$$w'(\vartheta) = w'(\pi - \vartheta) \quad . \tag{1.43}$$

Even in the case that a molecule possesses an electric dipole moment, the nematic phase is always non-polar. This also means that the director $\boldsymbol{n}$ is not uniquely fixed – it can always be chosen as well in the opposite direction.

Orientation distribution functions $w'(\vartheta)$ can be expanded as a series of Legendre polynomials $P_l(\cos \vartheta)$:

$$w'(\vartheta) = \sum_{l=0}^{\infty} \frac{1}{2}(2l + 1)S_l P_l(\cos \vartheta) \quad . \tag{1.44}$$

These polynomials are orthogonal to each other – this is expressed mathematically by

$$\int_{\vartheta=0}^{\pi} P_l P_k \sin\vartheta \, d\vartheta = \frac{2}{2l+1} \delta_{lk} \quad . \tag{1.45}$$

The series coefficients S_l can be calculated according to:

$$S_l = \int_{\vartheta=0}^{\pi} P_l(\cos\vartheta) w'(\vartheta) \sin\vartheta \, d\vartheta = \langle P_l \rangle \quad . \tag{1.46}$$

Regardless of the form of w', it is always true that

$$S_0 = 1 \quad . \tag{1.47}$$

In the given case of a non-polar system, the first-order series coefficient is zero:

$$\langle \cos\vartheta \rangle = 0 \quad . \tag{1.48}$$

The first non-zero contribution after the constant S_0 term is thus the second-order term, whose coefficient is given by

$$S_2 = \left\langle \frac{3\cos^2\vartheta - 1}{2} \right\rangle \quad . \tag{1.49}$$

S_2 characterises the degree of orientation and is suited to take the role of the **nematic order parameter**. There are two limiting cases for S_2:

- for an isotropic distribution, $S_2 = 0$
- for a perfect orientation, $S_2 = 1$.

There is a further reason for the choice of S_2 as the order parameter: The difference in the refractive indexes, $\Delta n = n_\parallel - n_\perp$, which is responsible for birefringence is proportional to S_2. We present here the proof of this and introduce first two coordinate systems. The first, with the coordinates x, y and z, has its z axis aligned parallel to the director, while the second, with the coordinates x', y' and z', is fixed at the molecule. The z' axis corresponds to the long axis of the molecule, such that the matrix representation of the polarisability tensor β' in the second system is diagonal

$$\beta' = \begin{pmatrix} \beta_\perp & 0 & 0 \\ 0 & \beta_\perp & 0 \\ 0 & 0 & \beta_\perp + \Delta\beta \end{pmatrix} \quad . \tag{1.50}$$

We assume here that the molecule has uniaxial symmetry. β' can be transformed into the director-fixed axes system:

$$\beta = \Omega^{-1} \cdot \beta' \cdot \Omega \quad . \tag{1.51}$$

$\mathbf{\Omega}$ is the rotation matrix which accomplishes the transformation. On account of the uniaxial symmetry, all off-diagonal elements of the averaged polarisability tensor disappear in the coordinate system of the director, and it is only necessary to calculate the diagonal elements. The following is obtained for β_{xx}:

$$
\begin{aligned}
\beta_{xx} &= \sum_l \Omega_{xl}^{-1} \beta'_{ll} \Omega_{lx} \\
&= \beta_\perp \cos^2 \theta_{x',x} + \beta_\perp \cos^2 \theta_{y',x} + (\beta_\perp + \Delta\beta) \cos^2 \theta_{z',x} \\
&= \beta_\perp + \Delta\beta \cos^2 \theta_{z',x} \quad,
\end{aligned}
\tag{1.52}
$$

where $\theta_{i',j}$ denotes the angle between the axes i' and j. β_{yy} and β_{zz} are given by

$$
\beta_{yy} = \beta_\perp + \Delta\beta \cos^2 \theta_{z',y}
\tag{1.53}
$$

$$
\beta_{zz} = \beta_\perp + \Delta\beta \cos^2 \theta_{z',z} \quad.
\tag{1.54}
$$

Performing an average over all orientations of the molecule leads to

$$
\langle \beta_{xx} \rangle = \langle \beta_{yy} \rangle = \beta_\perp + \Delta\beta \left\langle \cos^2 \theta_{z',x} \right\rangle
\tag{1.55}
$$

$$
\langle \beta_{zz} \rangle = \beta_\perp + \Delta\beta \left\langle \cos^2 \theta_{z',z} \right\rangle \quad.
\tag{1.56}
$$

Since

$$
\cos^2 \theta_{z',x} + \cos^2 \theta_{z',y} + \cos^2 \theta_{z',z} = 1
\tag{1.57}
$$

it follows that

$$
2 \left\langle \cos^2 \theta_{z',x} \right\rangle = 1 - \left\langle \cos^2 \theta_{z',z} \right\rangle
\tag{1.58}
$$

and we, thus, obtain

$$
\langle \beta_{zz} \rangle - \langle \beta_{xx} \rangle = \Delta\beta \cdot \frac{3 \left\langle \cos^2 \theta_{z',z} \right\rangle - 1}{2} \quad,
\tag{1.59}
$$

or, remembering the definition of S_2 (Eq. (1.49)),

$$
\langle \beta_{zz} \rangle - \langle \beta_{xx} \rangle = \Delta\beta \cdot S_2 \quad.
\tag{1.60}
$$

It is now necessary to calculate the dielectric tensor

$$
\varepsilon = \begin{pmatrix} \varepsilon_\perp & 0 & 0 \\ 0 & \varepsilon_\perp & 0 \\ 0 & 0 & \varepsilon_\parallel \end{pmatrix} \quad.
\tag{1.61}
$$

Applying the Clausius-Mosotti equation (Eq. (2.121) in Sect. 2.2) yields

$$
\varepsilon_\parallel - 1 = (\varepsilon_\parallel + 2) \frac{1}{3\varepsilon_0} \rho \langle \beta_{zz} \rangle \approx (\bar\varepsilon + 2) \frac{1}{3\varepsilon_0} \rho \langle \beta_{zz} \rangle
\tag{1.62}
$$

and

$$\varepsilon_\perp - 1 \approx (\bar{\varepsilon} + 2)\frac{1}{3\varepsilon_0}\rho\langle\beta_{xx}\rangle \quad , \tag{1.63}$$

where

$$\bar{\varepsilon} = \frac{(2\varepsilon_\perp + \varepsilon_\|)}{3} \tag{1.64}$$

and ρ is the number of molecules per unit volume. The anisotropy of the dielectric constant

$$\Delta\varepsilon = \varepsilon_\| - \varepsilon_\perp \tag{1.65}$$

is, thus, given as

$$\Delta\varepsilon = (\bar{\varepsilon} + 2)\frac{1}{3\varepsilon_0}\rho(\langle\beta_{zz}\rangle - \langle\beta_{xx}\rangle) \tag{1.66}$$

$$= (\bar{\varepsilon} + 2)\frac{1}{3\varepsilon_0}\rho\Delta\beta S_2 \quad .$$

The birefringence follows from

$$\Delta\varepsilon = \Delta(n^2) \approx 2\bar{n}\Delta n \quad , \tag{1.67}$$

and is thus given by

$$\Delta n \approx \frac{\bar{n}^2 + 2}{\bar{n}}\frac{1}{6\varepsilon_0}\rho\Delta\beta S_2 \tag{1.68}$$

$$= \Delta n_{\text{max}} \cdot S_2 \quad . \tag{1.69}$$

It has, therefore, been proved that Δn is, as stated above, proportional to S_2.

For the range over which a nematic phase exists, the degree of birefringence changes with temperature. Figure 1.19 shows the temperature dependence

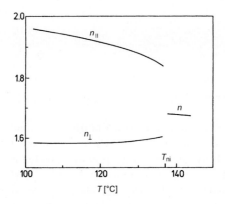

Fig. 1.19. The temperature dependence of the refraction indices parallel and perpendicular to the director, as measured for PAA at a wavelength of 509 nm (from Chatelain and Germain [4]).

of the two refraction indexes of PAA which exhibits a particularly strong birefringence: Δn and, thus, S_2 decrease with increasing temperature.

S_2 also determines the anisotropy of the diamagnetic susceptibility tensor $\Delta\kappa = \kappa_\parallel - \kappa_\perp$. An analogous derivation to that just given leads to

$$\Delta\kappa = \Delta\kappa_{max}S_2 \quad . \tag{1.70}$$

Figure 1.20 shows an example: A loss in the degree of orientation upon increasing temperature up to the clearing point is reflected in $\kappa_\parallel$ and S_2. The S_2 plot shows typical values for a nematic phase, namely $S_2 = 0.7 - 0.8$ and $S_2 = 0.4 - 0.5$ at the lower end of the nematic phase and at the clearing point, respectively.

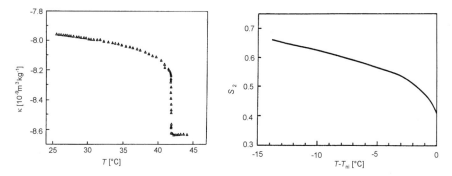

Fig. 1.20. The temperature dependence of the diamagnetic susceptibility in the direction of the director, as measured in the nematic and isotropic-liquid phases of the nematogen 7CB (*left*). These results are used to determine the temperature dependence of the nematic order parameter S_2 (*right*) (from Vertogen and de Jeu [5]).

1.4 Polymers

1.4.1 Chemical Structure and Chain Conformation

Polymers, or as is often said **macromolecules**, are made up of a very large number of molecular building blocks, which are linked together by covalent bonds into a chain. Polymers are, thus, predominantly organic molecules with carbon atoms as the principal component, together with hydrogen, oxygen, nitrogen, and halogens, etc. Figure 1.21 shows an example with a particularly simple structure, namely the chemical structure of polyethylene, where the structural unit is the CH_2 (methylene) group. The number of structural units, also known as **monomers**, determines the **degree of polymerisation**. Polymers are obtained chemically by a polymerisation process, which starts

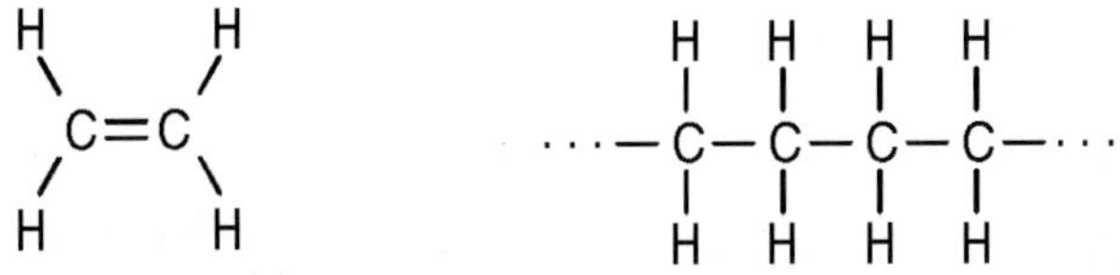

Fig. 1.21. The chemical structure of ethylene and polyethylene.

with reactive molecules. The name polyethylene hints at the fact that this polymer normally arises from ethylene by means of a gas phase reaction; the figure also gives the chemical structure of the starting material. As a second example, Fig. 1.22 shows the chemical structure of polystyrene. As in the case of polyethylene, the backbone consists of a chain of carbon atoms to which benzene rings are added as side groups. Table 1.2 gives the names and usual abbreviations as well as the structures of the basic repeating units for some other commonly used polymers.

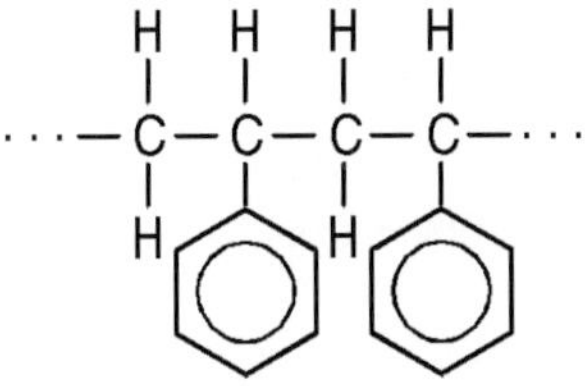

Fig. 1.22. The chemical structure of polystyrene.

Polymerisation reactions always lead to mixtures of macromolecules with different molecular weights, with the width of the molecular weight distribution depending on the chosen process. Stated molecular weights are always average values, and they are usually given with a parameter describing the **non-uniformity** of the distribution.

In the case of polyethylene, the monomers couple together as a chain in one unique way; for polystyrene, this is not the case. In principle, either the part with the benzene ring or the CH_2 group can be attached to the growing chain. Moreover, there are two different ways in which the benzene ring can be positioned relative to the carbon backbone. It is, thus, by no means certain that a chemical reaction will lead to a chain with a sterically regular structure. When there is no regularity in the chemical structure, the polymer is referred to as **atactic**, while **isotactic** and **syndiotactic** refers to completely regular and alternating attachment patterns, respectively.

Polyethylene and polystyrene, our two examples above, are both built up from a single type of building block. Polymerisation reactions can also

Table 1.2. The chemical structure of some other polymers.

$\left[\text{CH}_2-\overset{\overset{\displaystyle\text{CH}_3}{\textstyle\mid}}{\text{CH}}\right]_n$	polypropylene 'PP'
$\left[\text{CH}_2-\overset{\overset{\displaystyle\text{Cl}}{\textstyle\mid}}{\underset{\underset{\displaystyle\text{H}}{\textstyle\mid}}{\text{C}}}\right]_n$	poly(vinyl chloride) 'PVC'
$\left[\text{CF}_2-\text{CF}_2\right]_n$	poly(tetrafluoroethylene) 'PTFE'
$\left[\text{CH}_2-\overset{\overset{\displaystyle\overset{\displaystyle O}{\parallel}}{\text{C}-\text{O}-\text{CH}_3}}{\underset{\underset{\displaystyle\text{CH}_3}{\textstyle\mid}}{\text{C}}}\right]_n$	poly(methylmethacrylate) 'PMMA'
$\left[\overset{\overset{\displaystyle O}{\parallel}}{\text{C}}-\bigcirc-\overset{\overset{\displaystyle O}{\parallel}}{\text{C}}-\text{O}-\text{CH}_2-\text{CH}_2-\text{O}\right]_n$	poly(ethylene terephthalate) 'PET'
$\left[\text{O}-\bigcirc-\overset{\overset{\displaystyle\text{CH}_3}{\textstyle\mid}}{\underset{\underset{\displaystyle\text{CH}_3}{\textstyle\mid}}{\text{C}}}-\bigcirc-\text{O}-\overset{\overset{\displaystyle O}{\parallel}}{\text{C}}\right]_n$	'polycarbonate' 'PC'
$\left[\text{CH}_2-\overset{\overset{\displaystyle\text{O}-\overset{\overset{\displaystyle O}{\parallel}}{\text{C}}-\text{CH}_3}{\textstyle\mid}}{\underset{\underset{\displaystyle\text{H}}{\textstyle\mid}}{\text{C}}}\right]_n$	poly(vinylacetate) 'PVAc'
$\left[\text{CH}_2-\overset{\overset{\displaystyle\text{CH}_3}{\textstyle\mid}}{\underset{\underset{\displaystyle\text{CH}_3}{\textstyle\mid}}{\text{C}}}\right]_n$	polyisobutylene 'PIB'
$\left[\overset{\overset{\displaystyle\text{H}}{\textstyle\mid}}{\text{N}}-(\text{CH}_2)_6-\overset{\overset{\displaystyle\text{H}}{\textstyle\mid}}{\text{N}}-\overset{\overset{\displaystyle O}{\parallel}}{\text{C}}-(\text{CH}_2)_4-\overset{\overset{\displaystyle O}{\parallel}}{\text{C}}\right]_n$	poly(hexamethylene adipamide) nylon 6,6
$\left[\overset{}{\underset{\underset{\displaystyle\text{CH}_3}{\textstyle\mid}}{\text{C}}}=\text{CH}-\text{CH}_2-\text{CH}_2\right]_n$	polyisoprene 'PI'

be performed where there is a mixture of reactive building blocks – this is referred to as a **copolymerisation**. A commercially important example is the polymerisation of ethylene and propylene together. This system is addressed in Fig. 1.23. The **statistical copolymer** results when the two monomers in the mixture together undergo a polymerisation reaction. There is, however, also a completely different way to carry out a reaction: Ethylene and propylene can first be polymerised separately, and the two products can then be joined together by a coupling reaction to give a **block-copolymer**. Such a product naturally possesses completely different properties.

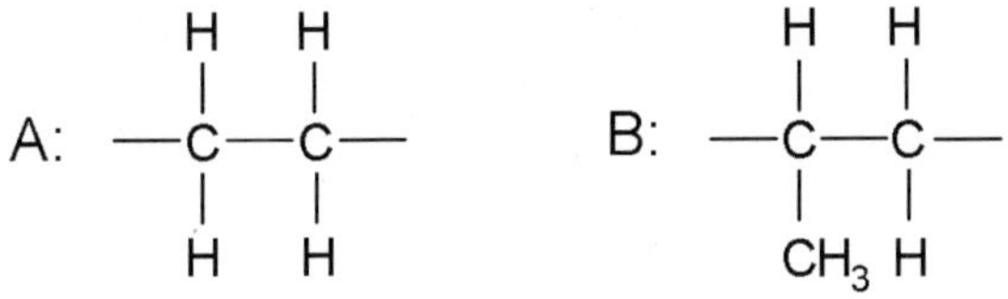

$$
\text{A: } -\overset{\overset{\displaystyle H}{|}}{\underset{\underset{\displaystyle H}{|}}{C}}-\overset{\overset{\displaystyle H}{|}}{\underset{\underset{\displaystyle H}{|}}{C}}- \qquad\qquad \text{B: } -\overset{\overset{\displaystyle H}{|}}{\underset{\underset{\displaystyle CH_3}{|}}{C}}-\overset{\overset{\displaystyle H}{|}}{\underset{\underset{\displaystyle H}{|}}{C}}-
$$

......ABAABABBAABA......   statistical

......AAAAABBBBBB......   block structure

Fig. 1.23. The monomers from which ethylene-propylene copolymers are constructed. The structure of a statistical copolymer and of a block copolymer.

So far we have only considered linear chains. A larger group of polymers, which includes commercially used materials, contains branches, for example side chains of different lengths. All factors together, the monomers, the regularity in the chemical and steric structure, as well as the linear or branched architecture determine the properties of polymer materials. The great variation provided by chemical synthesis thus offers many possibilities for producing specialised polymer materials, suited for a wide range of different applications.

Strong intermolecular interactions are of much importance for atoms and molecules in the condensed state. This is, of course, also the case for polymer systems, but an additional significant factor comes here into play. Since there is a large number of internal degrees of freedom associated with each macromolecule because of the very large number of monomer units, intramolecular properties contribute significantly to the observed behaviour.

Figure 1.24 again shows polyethylene; this time, the spatial arrangement of the atoms is illustrated. A segment of the chain, which is in the conformation with the lowest energy, is shown: All carbon atoms lie in the plane, and the C-C bonds adopt a zig-zag arrangement. A macromolecule such as polyethylene can completely change its conformation through rotations per-

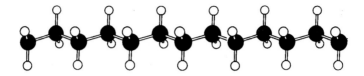

Fig. 1.24. The equilibrium conformation of polyethylene.

formed about the C-C bonds. These rotations occur in a simple way because of the energetics of the **rotation potential**. Figure 1.25 shows the expected change in the potential energy associated with a rotation about one particular C-C bond starting from the zig-zag equilibrium conformation. The potential energy shows two local minima at rotations of 120° and 240°. The starting point which has the lowest energy is referred to as **trans** – the zig-zag chain is 'all trans' – while the local minima which are separated by potential barriers are termed **gauche**$^+$ and **gauche**$^-$. By considering the given energies, it is possible to determine the population distribution. The conclusion is that a C-C bond either adopts a trans conformation or is in one of the two gauche conformations, the latter having a lower probability. All other angles, especially those corresponding to the energy barriers, are not found. Nevertheless, the height of the barriers is important, since it determines the rate of transitions between the trans and both gauche conformations. The population probabilities of the **rotational isomeric states**, trans, gauche$^+$ and gauche$^-$, for the C-C bonds as a function of temperature is given by the

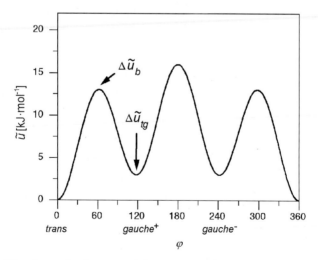

Fig. 1.25. The change in the potential energy in polyethylene for a rotation about a C-C bond.

Boltzmann distribution. For an energy difference $\Delta\tilde{u}_{\text{tg}} \simeq 2 - 3 \text{ kJmol}^{-1}$, the population probability for the two local gauche minima at room temperature is given by

$$w_{\text{g}} = \frac{2\exp(-\Delta\tilde{u}_{\text{tg}}/\tilde{R}T)}{1 + 2\exp(-\Delta\tilde{u}_{\text{tg}}/\tilde{R}T)} \simeq 0.5 \qquad (1.71)$$

($\tilde{R}$ is the universal gas constant). This equation assumes that each C-C bond can arrange itself independent of the others. Although this is not really true, Eq. (1.71) does provide a reasonable estimate for w_{g}.

In the liquid state, i.e., in a polyethylene melt or the melt of some other polymer, all the different rotational isomeric states can be occupied. Virtually all the chains exist in forms which can be referred to as coil-like. As they are densely packed they must interpenetrate each other. A polymer melt thus represents a dense packing of entangled coiled chains.

1.4.2 Polymer Melts

In a polymer melt, the chains take up all possible conformations in a statistical distribution. The exact description of this distribution seems at first a very difficult problem. It could be thought that it is necessary, for each polymer, to consider its particular chemical structure, which then determines the distribution over the rotational isomeric states for the bonds comprising the polymer backbone. All details such as the position of the local minima as well as bond lengths and bond angles which are important for structural properties would then have to be considered. This would mean that each polymer would amount to a new problem to be solved. The situation is actually much simpler. For many polymer properties, the behaviour in the Å range is not important; it is the structure and dynamics over length-scales greater than 10 nm which are decisive. As is easily appreciated, differences between different polymer chains disappear over such mesoscopic length scales. Figure 1.26 shows how a typical polymer coil would then appear: All details of the chemical structure have disappeared and only a worm-like object is observed.

The question arises as to the extent to which a polymer molecule extends itself in a melt, i.e., we ask what is the diameter of the sphere which is just big enough to encompass a coiled polymer molecule? The latter corresponds, for a large number of polymer conformations, to the case where both chain ends, i.e., the two most distant structural building blocks along the chain, lay within the sphere. The average distance between the chain ends can be estimated by calculating the square root of the mean squared chain-end distance $\langle R^2 \rangle$. The distance vector between the chain ends, $\boldsymbol{R}$, is shown in Fig. 1.26. In order to calculate $\langle R^2 \rangle$, we break the chain up into N_s segments, which have, as shown, end-to-end distance vectors $\boldsymbol{a}_1, \boldsymbol{a}_2, \ldots, \boldsymbol{a}_{N_s}$. The vectors $\boldsymbol{a}_j$ of different segments are completely uncorrelated in their orientation provided that the chain segments are sufficiently long. It is this property which allows $\langle R^2 \rangle$ as well as the distribution function of the vector $\boldsymbol{R}$ to be determined.

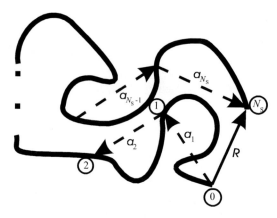

Fig. 1.26. A polymer molecule at low resolution and an associated model chain made up of freely jointed segments.

The problem is equivalent to that of the motion of a Brownian colloidal particle in a liquid. The motion of a Brownian particle away from its starting point takes place by means of a large number of uncorrelated single steps, hence, in the same way as the end-to-end distance vector of a polymer chain is set-up. Representing $\boldsymbol{R}$ as the sum of uncorrelated steps $\boldsymbol{a}_j$,

$$\boldsymbol{R} = \sum_{j=1}^{N_s} \boldsymbol{a}_j \quad , \tag{1.72}$$

the probability distribution for the motion of a Brownian particle away from its starting point can be directly used for the polymer chain. As will be explained in Sect. 5.2 (Eqs. (5.157) and (5.158)), the Brownian motion is described by a Gaussian distribution function. The probability

$$w(\boldsymbol{R})\mathrm{d}^3\boldsymbol{R} \quad ,$$

that the second end of a polymer chain finds itself in a volume element $\mathrm{d}^3\boldsymbol{R}$ at a distance $\boldsymbol{R}$ from the first end is, thus, also given by a Gaussian distribution, of the form

$$w(\boldsymbol{R}) = \left(\frac{3}{2\pi \langle R^2 \rangle} \right)^{3/2} \exp - \frac{3R^2}{2 \langle R^2 \rangle} \qquad (R = |\boldsymbol{R}|) \quad . \tag{1.73}$$

The pre-factor comes from the normalisation condition

$$\int w(\boldsymbol{R})\mathrm{d}^3\boldsymbol{R} = 1 \quad . \tag{1.74}$$

The only free parameter in the expression for the distribution is $\langle R^2 \rangle$, which is defined as

$$\langle R^2 \rangle = \int\limits_{R=0}^{\infty} w(\boldsymbol{R}) R^2 4\pi R^2 \mathrm{d}R \quad . \tag{1.75}$$

Considering the breaking up of the chain into segments as described by Eq. (1.72), it follows that

$$\langle R^2 \rangle = \left\langle \left| \sum_{j=1}^{N_s} \boldsymbol{a}_j \right|^2 \right\rangle = \left\langle \sum_{j,j'=1}^{N_s} \boldsymbol{a}_j \cdot \boldsymbol{a}_{j'} \right\rangle \quad . \tag{1.76}$$

The absence of any correlation in the orientation of the different segments means that

$$\langle \boldsymbol{a}_j \cdot \boldsymbol{a}_{j'} \rangle = \langle | \boldsymbol{a}_j |^2 \rangle \delta_{jj'} \quad , \tag{1.77}$$

giving the result

$$\langle R^2 \rangle = N_s \langle | \boldsymbol{a}_j |^2 \rangle \quad . \tag{1.78}$$

It was stated above that the square root of $\langle R^2 \rangle$ can serve as a measure of the extension of a polymer molecule in the melt, and we choose for this quantity the symbol R_0:

$$R_0 = \langle R^2 \rangle^{1/2} \quad . \tag{1.79}$$

The number of segments N_s is proportional to the degree of polymerisation N. We therefore obtain for the size of a polymer chain in a melt the characteristic power law

$$R_0 \propto N^{1/2} \quad .$$

Upon introducing the constant of proportionality a_0, we write

$$R_0 = a_0 N^{1/2} \quad . \tag{1.80}$$

a_0 corresponds to an effective length per structural unit and depends on all the microscopic properties of the chain, such as the bond lengths, bond angles, and the stiffness of the chain – the last property is determined by the occupation probabilities of the rotational isomeric states of the main-chain bonds.

Our attention above was on the distribution function for the chain end distance. It is clear that the distance distribution for any two points in the chain is also Gaussian-like provided that the points are far enough apart and are thus separated by many individual segments $\boldsymbol{a}_j$; this is true for all length-scales for which the chemical structure is no longer of importance. This has an important consequence, namely, that no change in the general appearance is seen when observing the interior parts of a chain at different resolutions. The structures always look the same, i.e., they always have the appearance of a Gaussian coil and hence, are similar to each other. Polymer chains in a melt are **self-similar objects**.

For self-similar objects, the question about their **fractal dimension** comes up. The answer is found in the power law of Eq. (1.80). If applied to a volume which encloses only a part of the polymer chain – say, a sphere of diameter r – we obtain for the number of structural units n which are found on average

$$n \propto r^2 \quad . \tag{1.81}$$

The power to which the distance is raised corresponds, by definition, to the fractal dimension. Thus, for polymer chains in a melt, the fractal dimension takes the value two. Polymer chains are objects which exist in three dimensions; however, since they only partially fill the available space, a lower dimension is found.

The self-similarity property of the objects applies only over a limited range: The upper limit is set by the size of the whole molecule, i.e., R_0, while the lower limit is reached for microscopic regions where the chemical structure becomes visible.

1.4.3 Solid Polymers

Upon cooling to sufficiently low temperatures, it is found, also for polymer systems, that the melt solidifies. There are two completely different processes which can cause a solidification. Polymers which are sufficiently regular in their chemical and steric structures can crystallise, although this process follows rules which differ fundamentally from those for the case of the crystallization of atomic and molecular systems. Alternatively, if the chain structure is irregular, it is impossible that a crystalline structure forms; a solidification process occurs, nevertheless, but this time by a transition to a rigid glassy state. In the following, both processes are considered.

The Semi-Crystalline State. A little thought allows us to envisage how a three-dimensional crystal with a periodic structure can be constructed from polymer chains. The polymer chains are fully stretched out and transferred into the state with the lowest conformational energy which is always periodic in structure. A three-dimensional structure, i.e., a polymer crystal, forms if such chains are packed together parallelly in a regular side-by-side fashion. Such a crystal is characterised by a marked internal anisotropy with strong covalent bonds in one direction and weak van der Waals forces in the other two directions. In principle, it would be possible to form single crystals out of polymers if all the chains had the same molecular weight and, thus, the same length. The end groups could then be arranged together in the upper and lower surfaces of a crystal formed from all the extended chains. In fact, this doesn't happen, and not only because, as stated above, polymer systems always exhibit a molecular weight distribution. The actual factor which prevents the transition into such an ideal crystalline state is present in the melt. Here the coiled molecules interpenetrate each other and, thus, form a large number of entanglements which cannot easily be resolved. The complete disentangling with a separation of all molecules from each other as would be required to attain the ideal crystalline state needs a much too long time. It would involve a passage over a large number of chain conformations with a low probability, and would therefore meet an extremely high entropic barrier. What happens instead? If a polymer is cooled to a temperature which lies below the equilibrium melting point, i.e., the temperature at which a large ideal polymer single crystal would melt, a state is attained which is only semi-crystalline. Layer-like crystallites form, which are separated by regions which remained liquid-like, with the layered crystals being, on the whole, packed parallel to each other. Such stacks of alternating crystalline and liquid layers are distributed isotropically in space and completely fill the sample volume.

This is illustrated for the example of polyethylene in Figs. 1.27 and 1.28, which show two images obtained using different electron microscopy techniques. Figure 1.27 shows an image, obtained using a carbon-film replica technique, of a polyethylene surface after a fracture of the material: The stacked layered crystals are clearly visible. Figure 1.28 shows in a clear fashion that the layered crystals are separated by liquid-like regions. This image was recorded for a thin film – contrast was achieved by means of a contrast agent which

Fig. 1.27. An electron microscope image of a fractured polyethylene surface obtained using a carbon-film replica technique (from Eppe and Fischer [6]).

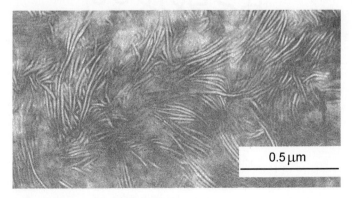

Fig. 1.28. An electron microscope image of a thin layer of polyethylene treated with a contrast agent (from Kanig [7]).

was taken up by the liquid region only. The liquid and crystalline regions, therefore, appear dark and light, respectively.

The reason why this kind of a semi-crystalline state forms is easily understood. As mentioned, the chain entanglements which exist in the melt can neither be untangled nor integrated into the crystal. So they become accumulated in liquid-like regions above and below a laterally expanding crystallite. Their presence permanently prevents any further crystal growth in chain direction and, thus, stabilises a layered semi-crystalline state. The formation of the two-phase crystalline-amorphous structure resembles a demixing process. Those parts of the chain which can extend themselves and can, thus, be incorporated into a growing crystal are separated from the other parts which form entanglements and must, therefore, remain liquid. Also chain segments which contain defects like the presence of co-monomers or steric defects which arose during the polymerisation process must remain in the liquid-like regions, i.e., all parts which are unable to undergo crystallisation concentrate there. Figure 1.29 shows a schematic view of a layered crystal with its liquid surroundings. Its surface exhibits some **chain foldings**, which appear when unperturbed chain sequences return immediately into the crystal. In addition, there are larger loops, whose hooked-up nature prevents their crystallisation.

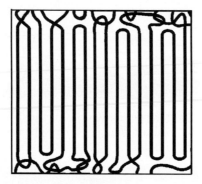

Fig. 1.29. A layer-like polymer crystallite with its amorphous surroundings.

Figure 1.30 shows a view into the interior of a crystalline region, namely the unit cell of polyethylene, where there is a regular arrangement of the zig-zag chains. Two chains, which are rotated by $90°$ with respect to each other, run through the unit cell. The lattice constants are $a_1 = 7.4\,\text{Å}$, $a_2 = 4.9\,\text{Å}$ and $a_3 = 2.5\,\text{Å}$.

The structure of the crystalline phase for low molecular mass, i.e., atomic or molecular, systems is fixed by the laws of equilibrium thermodynamics and corresponds to the state with the lowest Gibbs free energy. By contrast, the semi-crystalline state of polymers is thermodynamically only metastable and is selected in a completely different way: It represents the end result of

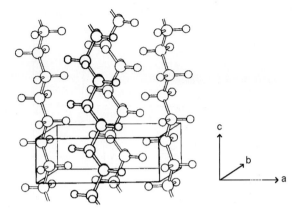

Fig. 1.30. The unit cell of polyethylene crystals (orthorhombic with two C_2H_4 groups per cell).

a kinetically favoured process with the structure which develops being that which forms quickest under the given conditions, i.e., the chosen temperature. Indeed, the adopted structure changes with the chosen crystallisation temperature. Moreover, since it is only metastable, it is not really fixed and changes can always occur over long times.

Typical values for the thickness of polymer layered crystals are in the nm range. The melting point for this type of thin crystal is significantly lower than the equilibrium value for an ideal macroscopic single crystal; the thinner the crystal is, the lower the observed melting point. Upon cooling from the melt, crystals of different thickness normally arise. Figure 1.31 shows the typical consequences which are seen upon observing the crystallisation and a subsequent melting process in a calorimeter. In contrast to low molecular weight systems, for which the transition from a crystal into the melt occurs at a well-defined temperature, it is found that polymers exhibit a wide melting range. It is also apparent that the crystallisation process upon cooling sets in at a markedly lower temperature relative to the equilibrium melting point of polyethylene, the latter being 143 °C. Figure 1.32 shows schematically the manner in which crystallisation occurs upon cooling from a polymer melt. Initially, here at a temperature T_0, crystals with a particular thickness form, which fill the sample volume to a certain extent. The crystallisation process only continues upon a further cooling, with the crystals becoming ever thinner with decreasing temperature. Upon reheating, the crystals melt in the reverse order, beginning with those that had formed last.

A range of structural parameters are required to describe the semi-crystalline state of a polymer. The most important is the degree of crystallisation, i.e., the amount of crystallised material in the sample. A second important parameter is the crystallite thickness, and it is generally found that

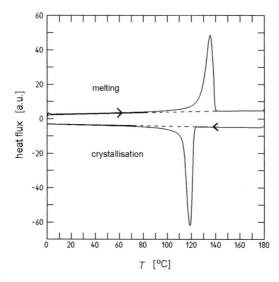

Fig. 1.31. The crystallisation and melting curve of polyethylene.

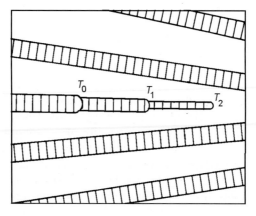

Fig. 1.32. The mechanism of secondary crystallisation which occurs upon cooling a polymer sample.

this increases with increasing crystallisation temperature. In contrast to low molar mass substances where crystallisation always sets in immediately below the equilibrium melting point, for polymers, the temperature of crystallisation can be chosen over a broad range. This choice selects the kinetics and, thus, determines the resulting structure.

The Glassy State. As is the case for all liquids, the internal mobility in a polymer melt is reduced upon decreasing the temperature. A first simple reason for this is, as shown in Fig. 1.25, that conformational changes require jumps over barriers and these must be thermally activated. There is, however, a second important factor: For a conformational change to occur, there must be a certain minimum space in the surroundings which must be made available at least during a short time as **free volume**. This free volume is a part of the total volume and is reduced upon cooling. It, thus, becomes ever more difficult to satisfy the space requirement for a conformational change. The phenomenon can be followed by means of a parameter which is determined by the internal mobility, namely the viscosity η. Its temperature dependence is described by the equation

$$\eta(T) \propto \exp \frac{T_A}{T - T_V} \quad , \tag{1.82}$$

which is referred to as the Vogel–Fulcher equation. It is the **Vogel temperature**, T_V, which makes the Vogel–Fulcher equation different to the Arrhenius law $\exp(T_A/T)$ – where $k_B T_A$ is the activation energy – the latter normally applying to activated processes. The increase in the viscosity as the temperature decreases is faster than that which is to be expected for an activated process, this being because of the decreasing free volume. The free volume is not an externally determined parameter, but rather it arises as a consequence of conformational chain dynamics. Free volume and chain dynamics are intimately linked and affect each other in a thermodynamic equilibrium.

What is expected to occur upon cooling a liquid? The free volume and the reorientation dynamics become slower. After each cooling step, a finite time is required before thermodynamic equilibrium is reached. This time becomes ever longer, the lower the temperature is – since the adjustment time depends on the viscosity, which is described by the Vogel–Fulcher equation. It is clear that as the adjustment time increases, eventually a point is reached where the equilibrium conformation can no longer be achieved in the course of normal observation times. The melt then remains in the last adopted state and hardens into a glass. Experience shows that this occurs when the viscosity has reached the value of

$$\eta(T_g) = 10^{12} \text{ Nm}^{-2}\text{s} \quad . \tag{1.83}$$

The associated temperature is termed the **glass transition temperature**, T_g. The Vogel–Fulcher equation is no longer valid beyond T_g.

The solidification process is very clearly observable if the specific volume or specific heat is measured during the cooling of a melt – such experimental data for poly(vinyl acetate) are shown in Figs. 1.33 and 1.34. Upon cooling at a rate of 20 K min^{-1}, the heat capacity changes in a 10 K wide temperature range around a temperature of $33 \,^\circ\text{C}$ – it is over this range that poly(vinylacetate) undergoes a transition into the glassy state. The reverse process of a softening is observed during a subsequent heating and occurs over the same temperature range. It is apparent that the position of the **glass transition** depends on the heating rate: Softening occurs earlier for a slower heating rate. Figure 1.34 shows that the glassy solidification is also associated with a change

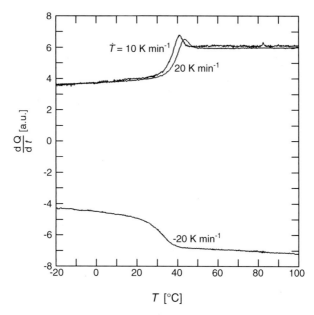

Fig. 1.33. The glass transition in poly(vinyl acetate) as observed by measuring with a calorimeter the heat capacity upon cooling and during heating with two different heating rates.

in the thermal expansion coefficient. The results are again for poly(vinyl acetate), where the glass temperature is in the region of 30 °C. The reason for the characteristic decrease in the heat capacity and in the thermal expansion coefficient is easy to appreciate: Above T_g, the molecular conformations together with the free volume change with increasing temperature, while below T_g, they are frozen in. In a glass, heat only flows into the vibrational degrees of freedom, and the thermal expansion is based solely on the anharmonicity of the chain vibrations. From the change at T_g, it is straightforward to determine the proportion of, first, the take-up of heat and, second, the thermal expansion of the liquid which is due to the conformational degrees of freedom.

Although glasses behave like crystals as far as their specific heat and thermal expansion coefficients are concerned, they are very different when viewed from a thermodynamic perspective. The freezing process which occurs at the glass transition temperature is a purely kinetic phenomenon and is therefore not to be compared with the phase transition between two equilibrium states. For a system at thermodynamic equilibrium, e.g., in a liquid, all possible conformations are explored over the course of the experimental time, i.e., the system is 'ergodic'. The glassy solidification transforms it into a conformational state which is chosen by chance, and its ergodicity is lost. The energy take-up or loss is now limited and concerns only the vibrational degrees of

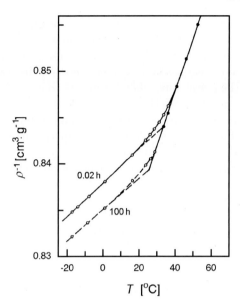

Fig. 1.34. The glass transition in poly(vinyl acetate) as observed by temperature dependent measurements of the specific volume after 0.02 and 100 hours storage at $-20\,°C$ (from Kovacs [8]).

freedom. A complete thermodynamic equilibrium, which would encompass all degrees of freedom, no longer exists.

In fact, it is incorrect to state that no conformational changes at all occur in the glassy state. Local reorientations of side groups or also short sections of the main chain remain possible. Through the residual motion, the system is able, if enough time is available, to come closer to thermal equilibrium. This is the reason for the observed time effect in Fig. 1.34: Storing at $-20\,°C$ for 100 hours led to a contraction of the sample. An interesting observation is that melting occurs earlier upon reheating for the densified glass, exactly at the point where the two lines for the glass and the melt intersect.

Considering the finite width of the solidification region as well as the dependence on the cooling or heating rate, it is evident that the glass transition temperature T_g is not an exactly defined parameter. It is therefore always necessary to state the conditions by which it was measured. Calorimetry and dilatometry are the most commonly used methods, and here it is necessary to specify the employed heating or cooling rate.

Is solidification into the glassy state a special property of polymers? The answer is no. In principle, for every liquid, there is a temperature at which solidification into the glassy state would occur; however, in nearly all cases, it is not observed, since crystallisation already sets in at higher temperatures. A glass is only obtained if the crystallisation process can be prevented. It is well known that this occurs normally for silicate melts, from which all standard glasses arise. There are also some molecular glasses, e.g. Salol, where crystallisation occurs so slowly that it is completely suppressed by a rapid cooling. It is also possible at some extremely fast cooling rates to force some metal alloys into the glassy amorphous state. As soon as the system is in the glassy state, there is no longer the motion present which is required to form a crystal. Therefore, also systems which are capable of undergoing crystallisation remain, for temperatures below T_g, in an amorphous state. The fact that the glassy state is so often encountered with polymers and is indeed considered here to be a normal state is a consequence of the especially high viscosity of these systems under normal conditions. This itself is a consequence of the linking of the monomers into a chain, which forces a co-operativity onto the molecular dynamics, hence very much slowing down the dynamics as compared to the case of a low molar mass system.

1.5 Structural Investigations Using Scattering Experiments

Among a large number of methods with which the structure of condensed matter can be investigated, there is a method of particular importance, namely scattering experiments, which is extensively applicable and for which no special preparation procedures are required. Such experiments are usually performed using electromagnetic radiation, though neutron or electron beams can also be used. The same principal set-up is employed in all cases, and it is schematically represented in Fig. 1.35: Primary radiation, which is generated in a radiation source and which is characterised by a frequency ω_0, a wavevector $\boldsymbol{k}$, and an intensity I_0, hits a sample, and, thus, generates spherical scattering waves. The resulting total scattered intensity I depends generally on the observation direction and is detected using a suitable detector at a distance R, which is large as compared to the sample diameter. For reasons which become clear later, the change in angle upon scattering is expressed as 2θ, where θ is the **Bragg scattering angle**.

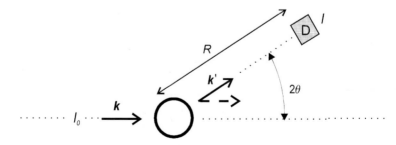

Fig. 1.35. The general set-up for a scattering experiment.

In order to obtain information about the microscopic structure of a sample, it is necessary to choose the wavelength of the employed radiation to be of the same or nearly the same order of magnitude as the length-scale of the sample under investigation. The microscopic structure of condensed matter is usually on a length-scale from Å to nm, and X-ray radiation is suitable here. Neutrons with energies in the range of 10 meV, which are known as **thermal neutrons**, also have suitable wavelengths. For colloidal systems, much larger sizes and distances up to the μm range are encountered, and in this case light radiation can be used. For very high resolution, electron beams are suitable, since the wavelengths are less than 1 Å.

The sample dimensions must be adapted to the radiation, since the radiation should be only weakly absorbed. Neutrons with their strong penetration can be used with large samples. For X-rays, the sample dimensions must be under 1 mm, while, for electrons, which are strongly absorbed, the sample thickness cannot exceed a few tens of nanometers.

The information about the sample structure is contained in the change of the scattered intensity with the observation direction. The changes result from the interference of the scattered waves arising from the different particles in the sample. In the following, we will discuss, for the example of the X-ray scattering experiment, how interference occurs and how it can be treated.

1.5.1 Interference

In an X-ray scattering experiment, monochromatic radiation is usually employed, for example, the characteristic K_α radiation of a Cu anode. The primary radiation can be described as a plane wave with an amplitude E_0:

$$E_0 \exp(-i\omega_0 t + i\boldsymbol{k}\boldsymbol{r}) \ . \tag{1.84}$$

The wave hits the particles, atoms or molecules, and excites dipolar vibrations. Consider first a vibrating dipole at the origin of the coordinate system. If the dipole vibrates together with the electric field in the z direction, the scattered wave emitted by it has a field strength at the position of the detector given by:

$$E'(R, 2\theta, t) = \cos(2\theta) \frac{1}{4\pi\varepsilon_0 c_1^2} \frac{1}{R} \ddot{p}\left(t - \frac{R}{c_1}\right) \ . \tag{1.85}$$

The field strength depends on the second derivative with respect to time of the induced dipole moment at an earlier time corresponding to the time needed for the wave to propagate from the dipole to the detector point (c_1 is the speed of light). The $\cos(2\theta)$ term describes the directional dependence of the dipolar scattering. The X-ray radiation frequencies are much larger than the frequencies of the outer electrons of the atoms, and thus these behave as free electrons upon an excitation. The second derivative of the generated dipole moment for a single free electron which vibrates in the electromagnetic field of the primary radiation is given as:

$$\ddot{p}\left(t - \frac{R}{c_1}\right) = \frac{e^2}{m_e}E_0 \exp\left(-i\omega_0\left(t - \frac{R}{c_1}\right)\right) = \frac{e^2}{m_e}E_0 \exp(-i\omega_0 t + ik'R) \quad.$$

(1.86)

The emitted scattered wave is, thus, given by

$$E'(R, 2\theta, t) = \cos(2\theta) \cdot r_e \frac{E_0}{R} \exp(-i\omega_0 t + ik'R)$$

(1.87)

with

$$r_e = \frac{e^2}{4\pi\varepsilon_0 m_e c_1^2} \quad,$$

(1.88)

where m_e and r_e denote the mass and the classical radius of an electron, respectively. For scattered waves due to an atom or molecule j, there is an additional factor, as compared to the expression for a free electron, of f_j; this is referred to as the atomic or molecular form factor. Scattering experiments are normally carried out with unpolarised X-ray radiation, in which case the $\cos(2\theta)$ term is no longer valid, and instead an angle dependent **polarisation factor**, denoted by F_p, is introduced. The scattered wave due to a single atom or molecule located at the origin then has the form:

$$E'(R, 2\theta, t) = F_p r_e f_j \frac{E_0}{R} \exp(-i\omega_0 t + ik'R) \quad.$$

(1.89)

Consider now the interference of scattered waves due to two particles at different locations. Such a situation is sketched in Fig. 1.36. Provided that the distance to the detector is large compared to the sample diameter, the

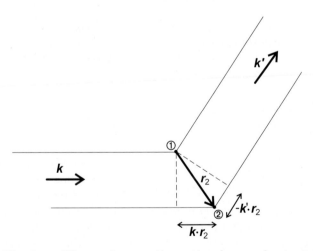

Fig. 1.36. The phase difference between the scattered waves due to atoms 1 and 2.

wavevectors k' of the two scattered waves have the same direction. As is shown in the figure, there is a phase difference between the scattered waves due to particles 1 and 2, located at the origin and the position r_2, respectively, which is easy to calculate. The phase delay of the scattered wave due to particle 2 corresponds to the sum of the two specified components in Fig. 1.36 and is given by

$$\mathrm{i}\Delta\varphi_2 = \mathrm{i}r_2(k - k') \quad . \tag{1.90}$$

The combined field strength generated by the two scattered waves is thus

$$E'(R, 2\theta, t) = F_\mathrm{p}\frac{r_\mathrm{e}E_0}{R}\exp(-\mathrm{i}\omega_0 t + \mathrm{i}k'R)f_j(1 + \exp[-\mathrm{i}r_2(k' - k)]) \quad . \tag{1.91}$$

It is apparent how the field strength of the total scattered radiation due to the interaction of the primary radiation with all particles in the sample can be calculated: It is the sum over all scattered waves due to all particles j, and the parameter of interest, the amplitude, E_0', is given by

$$E_0' = F_\mathrm{p}\frac{r_\mathrm{e}E_0}{R}\sum_j f_j \exp[-\mathrm{i}r_j(k' - k)] \quad . \tag{1.92}$$

It can be seen that the scattered amplitude depends on the difference between the wavevector k', which all individual scattered waves have in common, and the wavevector of the primary radiation k. This variable which is decisive as regards the result of a scattering experiment,

$$q = k' - k \quad , \tag{1.93}$$

is termed the **scattering vector**, q. Using the thus defined scattering vector leads to the following simple equation:

$$E_0'(q) \propto \sum_j f_j \exp(-\mathrm{i}qr_j) \quad . \tag{1.94}$$

For the scattering of X-rays by condensed matter, the wavelength and frequency of the detected radiation is the same, at least within the experimental precision, as that of the primary radiation, i.e.,

$$|k'| \approx |k| = \frac{2\pi}{\lambda} \quad . \tag{1.95}$$

Figure 1.37 illustrates the geometric relationship which, then, exists between k', k, and q. The magnitude of the scattering vector is given by

$$|q| = \frac{4\pi}{\lambda}\sin\theta \quad . \tag{1.96}$$

Therefore, $|q|$ depends on the half of the angle between k and k', i.e., the Bragg scattering angle θ, as well as the wavelength λ.

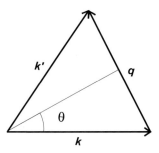

Fig. 1.37. Definition of the scattering vector q.

It is the intensity of the scattered radiation which is measured by the detector, and this is given by

$$I(\boldsymbol{q}) \propto \left\langle |E_0'(\boldsymbol{q})|^2 \right\rangle \propto \sum_{j,k} f_j f_k \left\langle \exp[-\mathrm{i}\boldsymbol{q}(\boldsymbol{r}_j - \boldsymbol{r}_k)] \right\rangle \quad . \tag{1.97}$$

It is important to note the presence of the triangular brackets in the above expression. Remember, first, that experimental measurements always require a definite time, and, second, that, in general, the particles in a sample are moving, which means that the scattered intensity also fluctuates over time. This fluctuation is not detected in an X-ray scattering experiment, rather it is the time-averaged value which is measured, and it is this that the triangular brackets express.

Diffraction experiments are very often carried out on samples where there is only one type of particle. This means that the atomic or molecular form factors f_j are all the same and do not need to be considered further. In this case, only the sum over all phase terms is important

$$C(\boldsymbol{q}) = \sum_j \exp(-\mathrm{i}\boldsymbol{q}\boldsymbol{r}_j) \quad , \tag{1.98}$$

and we are led to the following expression for the scattered intensity:

$$I(\boldsymbol{q}) \propto \left\langle |C(\boldsymbol{q})|^2 \right\rangle \quad . \tag{1.99}$$

Since the total scattered intensity must be proportional to the number of particles, denoted $\mathcal{N}$, which cause a scattering, it is useful to introduce the following reduced function:

$$S(\boldsymbol{q}) = \frac{1}{\mathcal{N}} \left\langle |C(\boldsymbol{q})|^2 \right\rangle \quad . \tag{1.100}$$

Substituting Eq. (1.98) into Eq. (1.100) gives

$$S(\boldsymbol{q}) = \frac{1}{\mathcal{N}} \sum_{j,k=1}^{\mathcal{N}} \left\langle \exp[-\mathrm{i}\boldsymbol{q}(\boldsymbol{r}_j - \boldsymbol{r}_k)] \right\rangle \quad . \tag{1.101}$$

$S(\boldsymbol{q})$ is a well-defined function. It appears in the literature under different names such as the **interference function**, the **scattering function**, or, since it characterises the structure, the **structure function**.

Since the atomic and molecular form factors are no longer included, it follows that the type of radiation, be it electromagnetic radiation or neutron or electrons, is no longer of relevance – it only needs to be considered in the evaluation of experimental data, i.e., the calculation of the interference function from measured scattered intensities. Equation (1.101) is generally applicable and can be applied to all types of condensed matter, be it liquid or solid – this indeed will be done in the following sections.

1.5.2 Scattering by Liquids

Atomic Liquids. Consider first the scattering by an atomic liquid, for the characterisation of which the pair distribution function $g_2(\boldsymbol{r})$ was introduced above. This pair distribution function can now be used to calculate the interference function. Introducing g_2, Eq. (1.101) becomes

$$S(\boldsymbol{q}) = \frac{1}{\mathcal{N}} \left[\mathcal{N} + \mathcal{N} \int_{\mathcal{V}} \exp(-i\boldsymbol{q}\boldsymbol{r}) g_2(\boldsymbol{r}) \mathrm{d}^3\boldsymbol{r} \right] \quad . \tag{1.102}$$

The integration volume $\mathcal{V}$ is chosen such that all distances in the sample are included. The first term in the square brackets, $\mathcal{N}$, is the sum over all terms with $j = k$ in Eq. (1.101), while the second term is the sum over all pairs with $j \neq k$. The pair distribution function g_2 has the property that all pairs are weighted according to the statistical probability of the given distance, and it is exactly this which is required to calculate the average value in Eq. (1.101).

The integral has the form of a Fourier transformation with respect to the pair distribution function g_2. In evaluating the integral, it is to be realised that g_2 does not tend to zero as $r \to \infty$, but rather the asymptotic value is given by the particle density ρ. Introducing this limiting value gives

$$S(\boldsymbol{q}) - 1 = \int_{\mathcal{V}} \exp(-i\boldsymbol{q}\boldsymbol{r}) \cdot (g_2(\boldsymbol{r}) - \rho) \mathrm{d}^3\boldsymbol{r} + \rho \int_{\mathcal{V}} \exp(-i\boldsymbol{q}\boldsymbol{r}) \, \mathrm{d}^3\boldsymbol{r} \quad . \tag{1.103}$$

The second term with the prefactor ρ corresponds to the Fourier transformation over a macroscopic volume. Contributions to this term are only found in the forward direction near $\boldsymbol{q} = 0$. In a scattering experiment, the detector must not be exposed to the primary radiation, and, thus, this part of the forward scattering is not detected. The second term in Eq. (1.103) can therefore be omitted.

Equation (1.103) is the fundamental equation by which scattering experiments in liquids are analysed, and it shows that the interference function corresponds to the Fourier transformation of the pair distribution function.

Reversely, the pair distribution function is determined by the back Fourier transformation of the $S(\boldsymbol{q})$ function measured in a scattering experiment:

$$g_2(\boldsymbol{r}) - \rho = \left(\frac{1}{2\pi}\right)^3 \int \exp(i\boldsymbol{qr})(S(\boldsymbol{q}) - 1)\mathrm{d}^3\boldsymbol{q} \ . \tag{1.104}$$

The space of the scattering vectors $\boldsymbol{q}$ is referred to as **reciprocal space**. Equations (1.103) and (1.104) express mathematically the Fourier relation between the pair distribution function and the interference function, which are defined in the direct space and in reciprocal space, respectively.

Liquids and, thus, the corresponding pair distribution functions are isotropic, i.e.,

$$g_2(\boldsymbol{r}) = g_2(|\boldsymbol{r}| = r) \ . \tag{1.105}$$

Therefore, it follows that the interference function in reciprocal space is also isotropic, i.e.,

$$S(\boldsymbol{q}) = S(|\boldsymbol{q}| = q) \ . \tag{1.106}$$

For the case of an isotropic function, the Fourier transformation in Eq. (1.103) is given by

$$S(q) - 1 = \int\limits_{r=0}^{\infty} (g_2(r) - \rho)4\pi r^2 \frac{\sin(qr)}{qr}\mathrm{d}r \tag{1.107}$$

while the back Fourier transformation becomes

$$g_2(r) - \rho = \left(\frac{1}{2\pi}\right)^3 \int\limits_{r=0}^{\infty} (S(q) - 1)4\pi r^2 \frac{\sin(qr)}{qr}\mathrm{d}r \ . \tag{1.108}$$

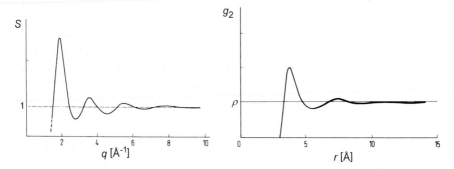

Fig. 1.38. The scattering function obtained for liquid argon at a density of $\rho_{\mathrm{m}} = 0.982$ g cm^{-3} (*left*) and the calculated pair distribution function (*right*).

Figure 1.38 shows as an example the interference function obtained for liquid argon and the pair distribution function derived by a Fourier transformation. The pair distribution function exhibits a maximum in the immediate vicinity of the excluded region and then other maxima which are due to shells of higher order. Such a pair distribution function leads to an interference function with a series of maxima. In comparison to the sharp reflections observed for crystals to be discussed later on, the maxima are broad and only appear at low q values.

Macromolecules. Equation (1.107) can be used to evaluate the scattering function for polymer chains in the coiled state found in a melt. This state is characterised by the Gaussian distribution of the distances, the end-to-end distances as well as the distances between two arbitrary points j, k in the chain. The distribution function for each such distance r_{jk} is given by

$$w(\boldsymbol{r}_{jk}) = \left(\frac{3}{2\pi \left\langle r_{jk}^2 \right\rangle} \right)^{3/2} \exp - \frac{3 r_{jk}^2}{2 \left\langle r_{jk}^2 \right\rangle} \qquad (r_{jk} = |\boldsymbol{r}_{jk}|) \ , \qquad (1.109)$$

where $\left\langle r_{jk}^2 \right\rangle$ denotes the mean squared distance. As was described in Sect. 1.4.2, a polymer chain can be divided into N_s segments – these segments are chosen here as the individual particles which lead to scattering. The pair distribution function for such particles, g_2, is the sum over the distance distribution functions for all the pairs within the chain, divided by the total number of particles:

$$g_2(|\boldsymbol{r}| = r) = \frac{2}{N_s} \sum_{l=1}^{N_s - 1} (N_s - l) \left(\frac{3}{2\pi l a_s^2} \right)^{3/2} \exp \left(- \frac{3 r^2}{2 l a_s^2} \right) \ . \qquad (1.110)$$

In deriving Eq. (1.110), it was considered that there are $2(N_s - l)$ pairs of particles separated by l segments along the chain and that the corresponding distance fluctuation $\left\langle r_{jk}^2 \right\rangle$ has, in analogy to Eq. (1.78), the value

$$\left\langle r_{jk}^2 \right\rangle = |j - k| a_s^2 = l a_s^2 \ , \qquad (1.111)$$

where

$$a_s^2 = \left\langle | \boldsymbol{a}_j |^2 \right\rangle \ . \qquad (1.112)$$

Equation (1.107) can now be used. Since, for a single polymer chain, the pair distribution function disappears at sufficiently large distances, i.e., $\rho = 0$, the following is obtained

$$S(q) - 1 = \int_{r=0}^{\infty} g_2(r) 4\pi r^2 \frac{\sin(qr)}{qr} \mathrm{d}r$$

$$= \frac{2}{N_s} \sum_{l=1}^{N_s - 1} (N_s - l) \exp \left(- \frac{l a_s^2 q^2}{6} \right)$$

and thus

$$S(q) = \frac{1}{N_s} \sum_{l=-(N_s-1)}^{N_s-1} (N_s - |l|) \exp\left(-\frac{|l| a_s^2 q^2}{6}\right)$$

$$\approx \frac{2}{N_s} \int_{l=0}^{N_s} (N_s - l) \exp\left(-\frac{l a_s^2 q^2}{6}\right) dl \quad . \tag{1.113}$$

Introducing the dimensionless variables

$$v' = \frac{l a_s^2 q^2}{6} \qquad v = \frac{N_s a_s^2 q^2}{6} = \frac{R_0^2 q^2}{6} \tag{1.114}$$

gives the following

$$S(q) = N_s \frac{2}{v} \cdot \int_{v'=0}^{v} \left(1 - \frac{v'}{v}\right) \exp(-v') dv' \quad . \tag{1.115}$$

The scattered intensity I is proportional to the number of segments in the chain, and is, thus, given by

$$I(q) \propto N_s S(q) = N_s^2 S_D(q) \quad , \tag{1.116}$$

where

$$S_D\left(v = \frac{R_0^2 q^2}{6}\right) = \frac{2}{v^2}[\exp(-v) + v - 1] \quad . \tag{1.117}$$

The function S_D is normalised, i.e., $S_D(0) = 1$. It was first introduced by Debye and is hence referred to as the **Debye scattering function**. In the limit of small scattering vectors, it can be shown that

$$S_D(v \to 0) = 1 - \frac{1}{v} \quad . \tag{1.118}$$

Using this equation, it is straightforward to determine the size R_0 of a macromolecule in a melt: R_0^2 can be determined from the slope of a plot of S_D^{-1} against q^2.

How is it possible to measure the scattering factor of an individual polymer chain in a melt? In a scattering experiment using X-rays on a polymer melt, there is no differentiation between monomers within a chain and those on different chains. The scattering curve is, therefore, the same as that of a molecular liquid and shows no chain character. Neutron scattering experiments provide a way to make the individual chains visible provided that the experiment is performed on a sample in which a small proportion of the chains have been deuterated. Since neutrons are scattered by the atomic nuclei, contrast in terms of the scattering behaviour is achieved between the deuterated part and the rest. There are now two contributions to the measured scattering intensity, and one of them is the structure factor of the individual chains, i.e., the Debye scattering factor, S_D. Figure 1.39 shows the results of the first measurements of this kind on poly(methylmethacrylate) with S_D being extracted.

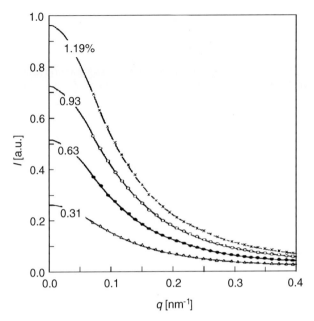

Fig. 1.39. The results of neutron scattering experiments on mixtures of PMMA and deuterated PMMA (from Kirste [9]).

1.5.3 Crystal Diffraction

The Ideal Crystal – The Reciprocal Lattice. By far the largest number of scattering experiments are carried out on crystals with the aim being the elucidation of the structure, i.e., to determine, first, the lattice constants, i.e., the Bravais lattice, and, second, the arrangement of the atoms in the unit cell. Usually, it is the latter which is of most relevance, this always being the case where it is the structure adopted by a given molecule which is under investigation. The crystalline state in which all molecules are arranged in a parallel fashion is then the best possible situation for a structural investigation.

The interference function $S(\boldsymbol{q})$ was introduced for the condition that the individual particles which are the scatterers in the investigated sample are all of the same type. In the case of a crystal, the unit cells play the role of these individual particles. A starting expression for the scattering intensity distribution, $I(\boldsymbol{q})$, in the reciprocal space for a crystal is then given, using Eqs. (1.97) and (1.101), as

$$I(\boldsymbol{q}) \propto \mathcal{N}_1\mathcal{N}_2\mathcal{N}_3 |f_{\mathrm{c}}(\boldsymbol{q})|^2 S(\boldsymbol{q}) \quad . \tag{1.119}$$

$\mathcal{N}_1, \mathcal{N}_2$ and $\mathcal{N}_3$ specify the numbers of unit cells along the three axes of the Bravais lattice. The scattering properties of a single unit cell are described by the cell structure factor $f_{\mathrm{c}}(\boldsymbol{q})$. It arises from a superposition of the scattering waves emanating from all n atoms in the cell, resulting in

$$f_c(\boldsymbol{q}) = \sum_{j=1}^{n} f_j \exp(-i\boldsymbol{q}\boldsymbol{r}_j) \quad . \tag{1.120}$$

The interference function $S(\boldsymbol{q})$ describes the interference between all unit cells of the crystal. As the starting point in its calculation, we use Eq. (1.103)

$$S(\boldsymbol{q}) - 1 = \int_{\mathcal{V}} \exp(-i\boldsymbol{q}\boldsymbol{r})(g_2(\boldsymbol{r}) - \rho_c)\mathrm{d}^3\boldsymbol{r} \quad , \tag{1.121}$$

where ρ_c denotes the number of unit cells per unit volume. ρ_c is equal to the inverse of the cell volume, V_c:

$$\rho_c = \frac{1}{V_c} \quad . \tag{1.122}$$

The corners of the unit cells are located at the points of the Bravais lattice

$$\boldsymbol{R}_{uvw} = u\boldsymbol{a}_1 + v\boldsymbol{a}_2 + w\boldsymbol{a}_3 \quad . \tag{1.123}$$

The pair distribution function g_2 is, thus, given as

$$g_2(\boldsymbol{r}) = \sum_{uvw \neq 000} \delta(\boldsymbol{r} - \boldsymbol{R}_{uvw}) \quad . \tag{1.124}$$

Making use of

$$1 = \int_{\mathcal{V}} \exp(-i\boldsymbol{q}\boldsymbol{r})\delta(\boldsymbol{r})\mathrm{d}^3\boldsymbol{r} \tag{1.125}$$

leads to the following expression for the interference function:

$$S(\boldsymbol{q}) = \int_{\mathcal{V}} \exp(-i\boldsymbol{q}\boldsymbol{r}) \sum_{uvw} \delta(\boldsymbol{r} - \boldsymbol{R}_{uvw})\mathrm{d}^3\boldsymbol{r} - \rho_c \int_{\mathcal{V}} \exp(-i\boldsymbol{q}\boldsymbol{r}) \, \mathrm{d}^3\boldsymbol{r} \quad . \tag{1.126}$$

As was described above, the second term in the above expression corresponds to a contribution in the forward direction only, which is not measured in the scattering experiment. The measurable part of the interference function is, thus, given by

$$S(\boldsymbol{q}) = \sum_{uvw} \exp(-i\boldsymbol{q} \cdot \boldsymbol{R}_{uvw}) = \sum_{u=-\mathcal{N}_1/2}^{\mathcal{N}_1/2} \exp(-iu\boldsymbol{q} \cdot \boldsymbol{a}_1) \tag{1.127}$$

$$\times \sum_{v=-\mathcal{N}_2/2}^{\mathcal{N}_2/2} \exp(-iv\boldsymbol{q} \cdot \boldsymbol{a}_2) \sum_{w=-\mathcal{N}_3/2}^{\mathcal{N}_3/2} \exp(-iw\boldsymbol{q} \cdot \boldsymbol{a}_3)$$

The sums can be determined exactly by expressing the exponentials as geometric series. The result is, however, obvious. A consideration of Eq. (1.127) reveals that a non-zero value is only obtained when the following conditions are fulfilled:

$$\boldsymbol{q} \cdot \boldsymbol{a}_1 = 2\pi h$$
$$\boldsymbol{q} \cdot \boldsymbol{a}_2 = 2\pi k \qquad (1.128)$$
$$\boldsymbol{q} \cdot \boldsymbol{a}_3 = 2\pi l \; .$$

Physically, this means that the scattered waves from all cells must be in phase. Even in the case where there is only a tiny phase difference between neighbouring cells, there is a cancelling out when the sum is over the enormous number of cells.

The values of $\boldsymbol{q}$ which fulfil these conditions can be directly identified. Use is made of a concept which is of central importance in solid-state physics and is known as the **reciprocal lattice**. The reciprocal lattice is spanned by three basis vectors $\hat{\boldsymbol{a}}_1, \hat{\boldsymbol{a}}_2$, and $\hat{\boldsymbol{a}}_3$, which are related in a well-defined fashion to the basis vectors of the Bravais lattice $\boldsymbol{a}_1, \boldsymbol{a}_2$, and $\boldsymbol{a}_3$. The equations defining the relation are

$$\boldsymbol{a}_i \cdot \hat{\boldsymbol{a}}_j = 2\pi \delta_{ij} \; . \qquad (1.129)$$

Each basis vector of the reciprocal lattice is thus perpendicular to two basis vectors of the Bravais lattice. Points in the reciprocal lattice are described by the vectors

$$\boldsymbol{G}_{hkl} = h\hat{\boldsymbol{a}}_1 + k\hat{\boldsymbol{a}}_2 + l\hat{\boldsymbol{a}}_3 \; , \qquad (1.130)$$

where h, k, l are integers.

It is immediately apparent that the reciprocal lattice vectors $\boldsymbol{G}_{hkl}$ have the property

$$\boldsymbol{G}_{hkl} \cdot \boldsymbol{a}_1 = 2\pi h$$
$$\boldsymbol{G}_{hkl} \cdot \boldsymbol{a}_2 = 2\pi k \qquad (1.131)$$
$$\boldsymbol{G}_{hkl} \cdot \boldsymbol{a}_3 = 2\pi l \; .$$

These vectors are, thus, exactly those which satisfy Eq. (1.128). Hence, constructive interference between all unit cells is achieved when the scattering vector $\boldsymbol{q}$ equals one of the vectors of the reciprocal lattice, $\boldsymbol{G}_{hkl}$:

$$\boldsymbol{q} = \boldsymbol{G}_{hkl} \; . \qquad (1.132)$$

The above equation is referred to as the **Laue reflection condition**.

The Laue reflection condition represents a strong limitation in the scattering power of a crystal and constitutes a selection rule. It means that normally no scattered radiation can be expected to be observed if the orientation of the crystal in the primary beam and the direction of the detector are arbitrarily chosen. Generally, the scattering vector will not correspond to a vector of the reciprocal lattice. Only for certain orientations of the crystal and, at the same time, correct locations of the detector, will scattered radiation be detected, with the radiation being very intense in this case. Using the Laue condition, the interference function S_{L} of the lattice can be written as

$$S_{\mathrm{L}}(\boldsymbol{q}) \propto \sum_{hkl \neq 000} \delta(\boldsymbol{q} - \boldsymbol{G}_{hkl}) \ . \tag{1.133}$$

A full treatment leads to the complete form

$$S_{\mathrm{L}}(\boldsymbol{q}) = \rho_{\mathrm{c}} \sum_{hkl \neq 000} \delta(\boldsymbol{q} - \boldsymbol{G}_{hkl}) \ . \tag{1.134}$$

The scattered intensity distribution in the reciprocal space is then given by

$$I(\boldsymbol{q}) \propto \mathcal{N}_1 \mathcal{N}_2 \mathcal{N}_3 |f_{\mathrm{c}}(\boldsymbol{q})|^2 \rho_{\mathrm{c}} \sum_{hkl \neq 000} \delta(\boldsymbol{q} - \boldsymbol{G}_{hkl}) \ . \tag{1.135}$$

The **Ewald construction** can be used to determine whether a crystal is oriented such that it can reflect the primary beam, and it also shows the direction of the reflected beam. Figure 1.40 illustrates the construction in the two-dimensional case. The points are the points of the reciprocal lattice. The wavevector of the primary beam $\boldsymbol{k}$ points to the lattice origin. Since $|\boldsymbol{k}'| = |\boldsymbol{k}|$, the wavevectors of all possible scattering rays must end on the drawn circle. The Laue condition is fulfilled when a reciprocal lattice point lies on the circle. The scattered vector can then be chosen such that, as shown, it corresponds to a vector of the reciprocal lattice. It is clear how the representation can be expanded into a third dimension: It is necessary to consider a sphere, and determine whether the reciprocal lattice points lie on the surface of the sphere. If that is the case, there is reflected radiation in the direction of the associated wavevector $\boldsymbol{k}'$.

By means of a controlled rotation of the crystal together with a simultaneous independent rotation of the detector, it is possible to determine experimentally when a reflection is observed. Each reflection yields, by means

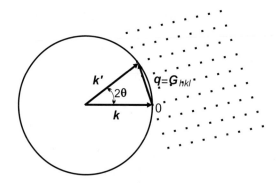

Fig. 1.40. The Ewald construction, by means of which the fulfilment of the Laue reflection conditions can be investigated.

of its associated scattering vector, a reciprocal lattice vector. When many reflections have been located, it is possible to establish the reciprocal lattice, from which the Bravais lattice can be directly determined, using Eq. (1.129) or equivalently

$$\widehat{\boldsymbol{a}}_1 = 2\pi \frac{\boldsymbol{a}_2 \times \boldsymbol{a}_3}{(\boldsymbol{a}_1 \times \boldsymbol{a}_2) \cdot \boldsymbol{a}_3} \qquad \text{etc.} \tag{1.136}$$

The term 'reciprocal' expresses also the fact that the volume of the unit cell of the Bravais lattice is inversely proportional to that of the reciprocal lattice, specifically,

$$V_c = \frac{(2\pi)^3}{\widehat{V}_c} \quad . \tag{1.137}$$

The above equation is proved as follows: Expressing $\boldsymbol{a}_i$ and $\widehat{\boldsymbol{a}}_i$ in the same orthogonal coordinate system defined by the unit vectors $\boldsymbol{i}, \boldsymbol{j}$ and $\boldsymbol{k}$ as

$$\boldsymbol{a}_i = a_{ix}\boldsymbol{i} + a_{iy}\boldsymbol{j} + a_{iz}\boldsymbol{k} \tag{1.138}$$

$$\widehat{\boldsymbol{a}}_i = \widehat{a}_{ix}\boldsymbol{i} + \widehat{a}_{iy}\boldsymbol{j} + \widehat{a}_{iz}\boldsymbol{k} \quad , \tag{1.139}$$

it follows that

$$V_c = \boldsymbol{a}_1 \cdot (\boldsymbol{a}_2 \times \boldsymbol{a}_3) = \begin{vmatrix} a_{1x} & a_{1y} & a_{1z} \\ a_{2x} & a_{2y} & a_{2z} \\ a_{3x} & a_{3y} & a_{3z} \end{vmatrix} \tag{1.140}$$

and

$$\widehat{V}_c = \begin{vmatrix} \widehat{a}_{1x} & \widehat{a}_{1y} & \widehat{a}_{1z} \\ \widehat{a}_{2x} & \widehat{a}_{2y} & \widehat{a}_{2z} \\ \widehat{a}_{3x} & \widehat{a}_{3y} & \widehat{a}_{3z} \end{vmatrix} = \begin{vmatrix} \widehat{a}_{1x} & \widehat{a}_{2x} & \widehat{a}_{3x} \\ \widehat{a}_{1y} & \widehat{a}_{2y} & \widehat{a}_{3y} \\ \widehat{a}_{1z} & \widehat{a}_{2z} & \widehat{a}_{3z} \end{vmatrix} \tag{1.141}$$

and thus

$$V_c \cdot \widehat{V}_c = \begin{vmatrix} a_{1x} & a_{1y} & a_{1z} \\ a_{2x} & a_{2y} & a_{2z} \\ a_{3x} & a_{3y} & a_{3z} \end{vmatrix} \cdot \begin{vmatrix} \widehat{a}_{1x} & \widehat{a}_{2x} & \widehat{a}_{3x} \\ \widehat{a}_{1y} & \widehat{a}_{2y} & \widehat{a}_{3y} \\ \widehat{a}_{1z} & \widehat{a}_{2z} & \widehat{a}_{3z} \end{vmatrix} = \begin{vmatrix} 2\pi & 0 & 0 \\ 0 & 2\pi & 0 \\ 0 & 0 & 2\pi \end{vmatrix} = (2\pi)^3 \quad .$$

The reciprocal lattice vectors possess a special property: Each one is perpendicular to a certain set of parallel lattice planes in the Bravais lattice. A **set of lattice planes** is a stack of planes with a constant separation, which has the property that all points of the Bravais lattice are encompassed. Sets of parallel lattice planes can be created using the 'Miller construction', as shown in Fig. 1.41. The basis vectors $\boldsymbol{a}_1$, $\boldsymbol{a}_2$, and $\boldsymbol{a}_3$ are divided into h, k, and l segments, respectively. The three points picked out in this way define a plane, whose repetition throughout the whole lattice yields the set. The numbers h, k, and l are known as the **Miller indices** of the set of parallel lattice planes. It is easy to see that the vector normal to the set of parallel lattice planes with the Miller indices hkl is parallel to the vector $\boldsymbol{G}_{hkl}$ of the

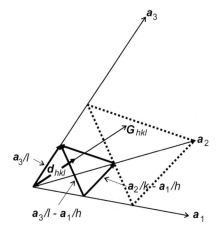

Fig. 1.41. The relationship between the hkl-set of parallel lattice planes and the reciprocal lattice vector $\boldsymbol{G}_{hkl}$.

reciprocal lattice. The perpendicularity is expressed mathematically by the vanishing scalar products

$$(h\widehat{\boldsymbol{a}}_1 + k\widehat{\boldsymbol{a}}_2 + l\widehat{\boldsymbol{a}}_3) \cdot \left(\frac{\boldsymbol{a}_2}{k} - \frac{\boldsymbol{a}_1}{h}\right) = 0 \tag{1.142}$$

and

$$(h\widehat{\boldsymbol{a}}_1 + k\widehat{\boldsymbol{a}}_2 + l\widehat{\boldsymbol{a}}_3) \cdot \left(\frac{\boldsymbol{a}_3}{l} - \frac{\boldsymbol{a}_1}{h}\right) = 0 \ . \tag{1.143}$$

Moreover, the following is valid

$$d_{hkl}|\boldsymbol{G}_{hkl}| = \frac{\boldsymbol{a}_3}{l} \cdot \boldsymbol{G}_{hkl} = \frac{\boldsymbol{a}_3}{l} \cdot (h\widehat{\boldsymbol{a}}_1 + k\widehat{\boldsymbol{a}}_2 + l\widehat{\boldsymbol{a}}_3) = 2\pi \ , \tag{1.144}$$

i.e., the magnitude of a reciprocal lattice vector is inversely proportional to the separation of the planes in the corresponding set of lattice planes. Equation (1.144) leads directly to **Bragg's law**

$$2d_{hkl}\sin\theta = \lambda \ . \tag{1.145}$$

To derive Eq. (1.145), it is simply necessary to consider that the Laue condition leads to

$$|\boldsymbol{G}_{hkl}| = |\boldsymbol{q}| \ , \tag{1.146}$$

where $|\boldsymbol{q}|$ is given by Eq. (1.96).

Bragg's law is well suited for the evaluation of scattering experiments performed on crystalline powders. In a powder, all crystal orientations are present, and, therefore, the Laue reflection condition for a particular reflection hkl will always be fulfilled. All scattered rays which belong to a reflection

hkl leave the powdered sample along the surface of a cone which makes a constant angle 2θ with respect to the primary beam. Bragg's law relates θ to the distance d_{hkl} of the associated set of lattice planes.

As was noted above, the main aim in performing an X-ray scattering experiment is usually the determination of the atomic arrangement in the unit cell. Information about this is contained in the cell structure factor f_c. Each reflection with the indices hkl yields, by means of its associated integrated reflection intensity I_{hkl}, a value for the cell structure factor. The following applies

$$
\begin{aligned}
I_{hkl} &\propto \left| f_c \left(\boldsymbol{q} = \boldsymbol{G}_{hkl} \right) \right|^2 \\
&= \left| \sum_j f_j \exp \left[-i \left(h\hat{\boldsymbol{a}}_1 + k\hat{\boldsymbol{a}}_2 + l\hat{\boldsymbol{a}}_3 \right) \cdot \left(x_{1j}\boldsymbol{a}_1 + x_{2j}\boldsymbol{a}_2 + x_{3j}\boldsymbol{a}_3 \right) \right] \right|^2 \\
&= \left| \sum_j f_j \exp \left[-i2\pi \left(hx_{1j} + kx_{2j} + lx_{3j} \right) \right] \right|^2 .
\end{aligned}
\tag{1.147}
$$

In order to determine the positions of the atoms in the unit cell, it is necessary to measure the intensities for a sufficiently large number of reflections. Equation (1.147) provides the basis by which the determination of the atomic arrangements can thus be attempted – to achieve this, there exists a range of effective and well-established methods.

Real Crystals – The Debye–Waller Factor. In the above considerations, ideal conditions have been assumed, i.e., the corners of the unit cells correspond exactly to points in the Bravais lattice. In fact, because of thermal motion in the crystal, this is not exactly correct, and the following needs to be considered. Fortunately, the essential features of the scattering by ideal crystals remain, and the changes due to the real conditions are also straightforward to predict. Lattice vibrations lead, generally, to a deviation of all atoms from their equilibrium positions. As will be discussed in Sect. 5.1.1, it is possible to distinguish displacements of atoms relative to each other within each unit cell from the case where the whole unit cell is displaced. The discussion here is limited to the latter types of displacement, which is the most common and is caused by acoustic lattice vibrations.

The displacements of the unit cells from the points of the Bravais lattice lead to a modification of the pair distribution function. Equation (1.124) becomes

$$
g_2(\boldsymbol{r}) = \sum_{uvw \neq 000} w(\boldsymbol{r} - \boldsymbol{R}_{uvw}) ,
\tag{1.148}
$$

whereby the δ-function has been replaced by a probability distribution $w(\boldsymbol{r} - \boldsymbol{R}_{uvw})$ for the displacement from the equilibrium position. It is helpful to express the pair distribution function in the form of a convolution. Convolutions relate two functions by means of the integration operation

$$\Phi_1(\boldsymbol{r}) \otimes \Phi_2(\boldsymbol{r}) = \int \Phi_1(\boldsymbol{r} - \boldsymbol{r}')\Phi_2(\boldsymbol{r}')\mathrm{d}^3\boldsymbol{r}' \quad . \tag{1.149}$$

Equation (1.148) can then be expressed as

$$g_2(\boldsymbol{r}) = \int w(\boldsymbol{r} - \boldsymbol{r}') \sum_{uvw \neq 000} \delta(\boldsymbol{r}' - \boldsymbol{R}_{uvw})\mathrm{d}^3\boldsymbol{r}' = w(\boldsymbol{r}) \otimes \sum_{uvw \neq 000} \delta(\boldsymbol{r} - \boldsymbol{R}_{uvw}) \quad . \tag{1.150}$$

There is a simple rule for evaluating the Fourier transformation of a convolution:

$$\mathrm{Ftr}\ \Phi_1 \otimes \Phi_2 = \mathrm{Ftr}\ \Phi_1 \cdot \mathrm{Ftr}\ \Phi_2 \quad , \tag{1.151}$$

where the following short-hand notation is employed:

$$\mathrm{Ftr}\ \Phi(\boldsymbol{r}) = \int \exp(-\mathrm{i}\boldsymbol{q}\boldsymbol{r})\Phi(\boldsymbol{r})\mathrm{d}^3\boldsymbol{r} \quad . \tag{1.152}$$

The interference function is then given by

$$S(\boldsymbol{q}) = 1 + \mathrm{Ftr}\ \left(w \otimes \sum_{uvw \neq 000} \delta(\boldsymbol{r} - \boldsymbol{R}_{uvw}) \right)$$

$$= \mathrm{Ftr}\ w \cdot \mathrm{Ftr} \sum_{uvw} \delta(\boldsymbol{r} - \boldsymbol{R}_{uvw}) - \mathrm{Ftr}\ w \cdot \mathrm{Ftr}\ \delta(\boldsymbol{r}) + 1$$

$$= \mathrm{Ftr}\ w \cdot S_{\mathrm{L}}(\boldsymbol{q}) + (1 - \mathrm{Ftr}\ w) \quad . \tag{1.153}$$

The presence of the interference function S_{L} of an ideal crystal in the above expression is to be noted. This means that the same reflections are observed for a real crystal as would be expected for an ideal crystal – the intensities are weakened, though, by a factor $\mathrm{Ftr}\ w$. In addition, there is a continuous scattering component, which is described by the second term.

In many cases, it is possible to approximate the probability distribution of the displacements of the lattice points, $\Delta\boldsymbol{r}$, with a Gaussian function

$$w(\Delta\boldsymbol{r}) = \left(\frac{3}{2\pi \langle \Delta r^2 \rangle} \right)^{3/2} \exp\left(-\frac{3|\Delta\boldsymbol{r}|^2}{2 \langle \Delta r^2 \rangle} \right) \tag{1.154}$$

whose Fourier transformation is

$$\mathrm{Ftr}\ w = \exp\left(-\langle \Delta r^2 \rangle \frac{q^2}{6} \right) \quad . \tag{1.155}$$

The interference function is then given by

$$S(\boldsymbol{q}) = \exp\left(-\frac{\langle \Delta r^2 \rangle q^2}{6} \right) \cdot S_{\mathrm{L}}(\boldsymbol{q}) + \left[1 - \exp\left(-\frac{\langle \Delta r^2 \rangle q^2}{6} \right) \right] \quad . \tag{1.156}$$

This expression implies that the reflection intensities become increasingly weaker for larger scattering angles. This is accompanied by an increase in the continuous 'diffuse' scattering component. The exponential term which describes the weakening in the reflection intensity is termed the **Debye–Waller factor**.

It can be shown that the total scattering in the reciprocal space, which is given by the integral of the interference function, is not affected by the lattice vibrations. There is solely a redistribution of the intensity from the reflections into the q regions lying in between. Actually, this can be shown to be generally valid for all scattering experiments: Changes under constant density in the arrangement of the particles which give rise to scattering always only lead to a redistribution of the scattered intensity in the reciprocal space.

1.6 Exercises

1. Consider crystals made up of touching hard spheres with a radius b. Calculate, first, the fraction ϕ of the volume which is filled, and, second, the coordination number z (the number of next-nearest neighbours) for the following cases:

   (a) A primitive cubic lattice (one sphere per cell),
   (b) A body-centred cubic lattice (two spheres per cell),
   (c) A face-centred cubic lattice (four spheres per cell).

2. Determine the group of symmetry elements for the molecules

   (a) Water H_2O (bent),
   (b) Carbon Dioxide CO_2 (linear),
   (c) Ammonia NH_3 (pyramidal),
   (d) Methane CH_4 (tetrahedral),
   (e) Benzene C_6H_6 (hexagonal),
   (f) Cyclohexane C_6H_{12} (the chair- and boat-forms), as well as the
   (g) monoclinic and the
   (h) tetragonal Bravais lattices.

3. A Bravais lattice is closed, i.e., the sum of two lattice vectors is also an element of the Bravais lattice. Show for the case of a two-dimensional lattice that this property is not met if there is five-fold rotation symmetry.

4. Calculate the nematic order parameter for a liquid crystal, whose molecules are, with respect to their long axes,

   (a) all oriented parallel to the reference axis,
   (b) all oriented perpendicular to the reference axis,
   (c) isotropically distributed.

5. The molecular weight distribution or the distribution of the degree of polymerisation N of a polymer depends on the polymerisation reaction mechanism. Narrow distributions are achieved for reactions by which individual monomers are attached to the chain ends. A Poisson distribution is to be expected in this case:

$$w(N) = \exp\left(-\overline{N}\right) \frac{\overline{N}^N}{N!} \ .$$

Determine the variance $\sigma^2 = \left\langle (N - \langle N \rangle)^2 \right\rangle$ of the distribution and the relationship between σ^2 and the normally employed non-uniformity parameter U. What non-uniformity is expected for a polymer with a degree of polymerisation of 100?

$$U = \frac{M_w}{M_n} - 1 \ \text{with} \ \begin{array}{l} M_n = \displaystyle\int w(N)M(N)dN \\[2ex] M_w = \displaystyle\int w(N)M(N)\cdot M(N)dN \Big/ \int w(N)M(N)dN \end{array}$$

M_n denotes the number average, while M_w is the weight average of the distribution.

6. The molecular weight and size of polymers can be determined by means of scattering experiments on dilute polymer solutions. Here, it is sufficient to determine the scattering behaviour at small scattering vectors. Calculate the scattered intensity $I(q)$ of a polymer chain for small scattering vectors q in the quadratic approximation, such that the following general expression for $I(q)$ for small q is obtained

$$I(q) = \sum_{j,k=1}^{N} f^2 \left\langle \exp\left[-i\boldsymbol{q}\left(\boldsymbol{r}_j - \boldsymbol{r}_k\right)\right]\right\rangle \ .$$

$\boldsymbol{r}_j$ are the position vectors of the monomers, which all have the same molecular form factor f. Show that $I(q)$ can be written in the following form:

$$I(q) = f^2 N^2 \left(1 - \frac{R_g^2 q^2}{3}\right) \ .$$

The 'gyration radius' R_g is defined as follows:

$$R_g^2 = \frac{1}{2N^2} \sum_{j,k=1}^{N} \left\langle (\boldsymbol{r}_j - \boldsymbol{r}_k)^2 \right\rangle \ .$$

Use the following in the above derivation:

$$\left\langle [\boldsymbol{q}\left(\boldsymbol{r}_j - \boldsymbol{r}_k\right)]^2 \right\rangle = \frac{1}{3}q^2 \left\langle (\boldsymbol{r}_j - \boldsymbol{r}_k)^2 \right\rangle \ .$$

Show that R_g can be expressed in the following way:

$$R_g^2 = \frac{1}{2N^2} \sum_{j,k=1}^{N} \left\langle (\boldsymbol{r}_j - \boldsymbol{r}_k)^2 \right\rangle = \frac{1}{N} \sum_{j=1}^{N} \left\langle (\boldsymbol{r}_j - \boldsymbol{r}_c)^2 \right\rangle \quad,$$

where $\boldsymbol{r}_c = \sum_{j=1}^{N} \boldsymbol{r}_j / N$ denotes the centre of gravity of the polymer molecule.

7. A unit cell has the following parameters:

$$a_1 = 0.5\,\text{nm}, \quad a_2 = 1\,\text{nm}, \quad a_3 = 1.5\,\text{nm}; \quad \alpha_1 = \alpha_2 = 90°, \quad \alpha_3 = 120° \quad.$$

The distances of the (321) planes are to be determined. At which angle would X-rays be scattered if Cu-K$_\alpha$ radiation with a wavelength of $\lambda = 0.154$ nm was used?

8. Calculate the structure factor $f_c(\boldsymbol{q}_{hkl})$ for the face-centred cubic structure. For which indices (hkl) are systematic absences found in the scattered radiation?

2

Moduli, Viscosities and Susceptibilities

In this chapter, we will consider how condensed matter reacts macroscopically when an external field is applied. Included in the discussion will be mechanical fields, which cause deformations in solids and flow phenomena in liquids, as well as electric fields which polarise materials, and magnetic fields which lead to a magnetisation. Provided that the fields are not too strong, the reactions follow a common pattern: There is a linear relationship between the strength of the field and the generated response. The discussion will first consider in turn how mechanical, electric, and magnetic fields act. In the final section, a summarising overview of **linear responses** is presented , and some generally valid laws are derived.

2.1 Mechanical Fields

2.1.1 Hookian Elasticity and Newtonian Viscosity

If a mechanical force is applied to a crystal or a glass, the resulting deformation is given by **Hooke's law**. A tensile stress

$$\sigma_{zz} = f/A \ ,\tag{2.1}$$

where f and A are the force in the z direction and the sample cross section on which the force acts, respectively, leads to a strain

$$e_{zz} = \frac{\Delta L_z}{L_z} \ ,\tag{2.2}$$

where ΔL_z and L_z denote the extension and the original length, respectively. Hooke's law expresses a linear relationship between the tensile stress and the strain:

$$\sigma_{zz} = E_t e_{zz} \ .\tag{2.3}$$

The constant of proportionality E_t is referred to as **Young's modulus**. Alternatively, if the strain is expressed as a function of the tensile stress, the resulting expression

$$e_{zz} = D_t \sigma_{zz} \qquad (2.4)$$

contains the **tensile compliance** D_t as the constant of proportionality.

If a force is applied along the upper face of a cubic object whose bottom face is fixed in place, a shearing results. For Hookian solids, the following applies

$$\sigma_{zx} = G \tan \gamma$$
$$\approx G\gamma \ . \qquad (2.5)$$

σ_{zx} describes the shear stress (the force in the x direction, which works on a surface with its normal in the z direction), γ is the shear angle, and G denotes the **shear modulus**. Figure 2.1 shows such a shear deformation.

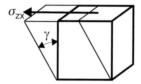

Fig. 2.1. Shear stress σ_{zx} and shear angle γ.

If a shear stress σ_{zx} is applied to a liquid layer, it begins to flow, with the deformation velocity following **Newton's law**

$$\sigma_{zx} = \eta \frac{\mathrm{d}}{\mathrm{d}t} \tan \gamma \ . \qquad (2.6)$$

The material-dependent parameter η is referred to as the **viscosity**. The force causes a laminar flow, whereby the flow velocity v_x changes linearly in the z direction. Equation (2.6) can, thus, be re-expressed as

$$\sigma_{zx} = \eta \frac{\mathrm{d}v_x}{\mathrm{d}z} \ . \qquad (2.7)$$

2.1.2 Liquid Crystals: Frank Moduli and Miesowicz Viscosities

Figure 2.2 shows a possible experiment. Consider a macroscopic rod-like object, e.g., a small piece of a glass fibre, in a liquid crystal with a homogeneous director field, the latter forming a thin layer upon a glass plate with lengthways grooves. Initially, as is shown on the left, the glass fibre is parallel to the director. When the fibre itself has lengthways grooves, the liquid-crystalline

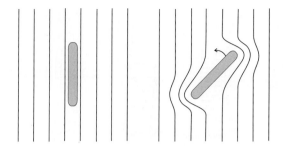

Fig. 2.2. A rod in a homogeneous director field (*left*). A local distortion of the director field caused by a small rotation of the rod leads to a restoring torque (*right*).

molecules at its surface will likewise align themselves in this direction, and the director field is only disturbed in the vicinity of the ends of the fibre. If the fibre is now rotated using tweezers, the director field is deformed in a manner which is qualitatively represented by the sketch on the right-hand side of the figure. If the glass fibre is then released, it is observed that a torque acts to return the system to its initial position. Such a behaviour means that elastic forces can be active in liquid crystals and these arise when the director field is deformed, i.e., when it loses its homogeneity.

How can this elasticity be described? The general approach is the same as that in continuum mechanics: The change in the Helmholtz free energy of the system caused by the deformation of the director field $\boldsymbol{n}(\boldsymbol{r})$ must be considered. The terms to lowest order, to which we can at first limit ourselves, depend on the simple derivatives $\partial n_i/\partial r_j$. If the homogeneous equilibrium state is the ground state, then the change in the Helmholtz free energy upon a deformation can only contain these terms squared. In addition, there are other conditions which the expression to be constructed must satisfy, e.g.,

- it must be invariant under $\boldsymbol{n} \to -\boldsymbol{n}$,
- since liquid crystals have a local uniaxial symmetry, it must be invariant under a rotation around the director direction,
- $\boldsymbol{n}$ is a unit vector and, therefore, must not change its length.

Frank carried out a theoretical analysis considering these conditions and found that the modes of distortion of nematic liquid crystals can always be broken up into three basic types, which give three independent from each other contributions to the Helmholtz free energy. Figure 2.3 shows these three basic modes of distortion, namely a splaying, twisting, or bending of the director field. The associated expression for the increase of the Helmholtz free energy density, denoted f_{el}, is given by

$$f_{\mathrm{el}} = \frac{1}{2} \left[K_1 (\mathrm{div}\ \boldsymbol{n})^2 + K_2 (\boldsymbol{n} \cdot \mathrm{curl}\ \boldsymbol{n})^2 + K_3 |\boldsymbol{n} \times \mathrm{curl}\ \boldsymbol{n}|^2 \right] \quad . \tag{2.8}$$

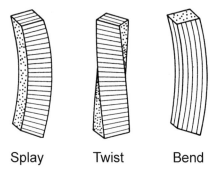

Splay   Twist   Bend

Fig. 2.3. The three fundamental modes of distortion of the director field: splay, twist, and bend.

K_1, K_2 and K_3 are referred to as the **Frank moduli** and describe the stiffness of the system with respect to splaying, twisting, and bending.

How can the Frank moduli be determined experimentally? Figure 2.4 shows an experimental set-up by which this can in principle be achieved. The initial state is again a liquid-crystalline film which is homogeneously planar oriented because of the interaction with a substrate-surface. A magnetic field is then applied parallel to the surface normal, such that the director field changes as illustrated in the figure. We assume here that the interaction of the liquid crystal with the magnetic field leads to the favouring of a parallel arrangement. At a sufficiently large distance from the surface, the magnetic field becomes the dominant factor; at the surface, it loses out to the stronger surface forces. Let the direction of the director be described by the angle Ψ:

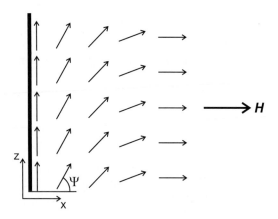

Fig. 2.4. The director field resulting from the application of a magnetic field to an initially homogeneously planar nematic sample. The *arrows* show the local director orientations.

it changes in a continuous fashion from $\Psi = 90°$ at the surface to a final value of $\Psi = 0°$ at a sufficiently large distance from the surface.

The size of the region over which the transition occurs depends on the relative magnitudes of the elastic energy and the interaction energy with the magnetic field. The situation can be analysed in an exact fashion, we only require for the analysis an expression for the interaction energy. Liquid crystals are nearly always diamagnetic and, in addition, anisotropic, and therefore possess different components of the magnetic susceptibility parallel, $\kappa_\parallel$, and perpendicular, $\kappa_\perp$, to the director. If the magnetic field is also broken up into two components parallel and perpendicular to the director, $H_\parallel$ and $H_\perp$, the following expressions for the two components of the generated magnetisation can be given:

$$M_\parallel = \kappa_\parallel H_\parallel \quad , \tag{2.9}$$

$$M_\perp = \kappa_\perp H_\perp \quad . \tag{2.10}$$

For the equilibrium state which is adopted for a given fixed field H, it is necessary to go from the Helmholtz free energy to the Gibbs free energy, and the reduced form $\hat{g}$ will be used (see Appendix A, Eq. (A.12)). The deformation part remains unchanged:

$$\hat{g}_{\rm el} = f_{\rm el} \quad . \tag{2.11}$$

The second part, $\hat{g}_{\rm m}$, which arises upon switching on the magnetic field, is given by

$$\hat{g}_{\rm m} = -\mu_0 \int M_\parallel {\rm d}H_\parallel - \mu_0 \int M_\perp {\rm d}H_\perp \tag{2.12}$$

$$= -\mu_0 \int_0^H (\kappa_\parallel \cos^2 \Psi + \kappa_\perp \sin^2 \Psi) H' {\rm d}H' \tag{2.13}$$

$$= -\frac{\mu_0}{2}(\kappa_\parallel - \kappa_\perp) H^2 \cos^2 \Psi - \frac{\mu_0}{2}\kappa_\perp H^2 \quad . \tag{2.14}$$

In the following, only the change $\Delta\hat{g}_{\rm m}$ of the energy in a magnetic field upon rotation of the director is important. Neglecting the final term in the previous equation, this is given by

$$\Delta\hat{g}_{\rm m} = -\frac{\mu_0}{2}\Delta\kappa(\boldsymbol{n} \cdot \boldsymbol{H})^2 \quad . \tag{2.15}$$

$\Delta\kappa$ denotes the anisotropy of the diamagnetic susceptibility

$$\Delta\kappa = \kappa_\parallel - \kappa_\perp \quad . \tag{2.16}$$

It is evident that the deformation considered here corresponds to a combination of splaying and bending, while a twisting is not present. For the bending part, the following applies:

$$\boldsymbol{n} \times \operatorname{curl} \boldsymbol{n} = \operatorname{curl} \boldsymbol{n} \quad . \tag{2.17}$$

Let us make the simplifying assumption that

$$K_1 \approx K_3 = K \quad , \tag{2.18}$$

i.e., differences between splay and bend stiffnesses are neglected. The following expression is then immediately obtained for the change $\Delta \hat{g}$ in the local density of the Gibbs free energy associated with the new equilibrium state generated upon applying a magnetic field:

$$\Delta \hat{g} = \hat{g}_{\mathrm{el}} + \Delta \hat{g}_{\mathrm{m}} = \frac{K}{2} \left[(\operatorname{div} \boldsymbol{n})^2 + |\operatorname{curl} \boldsymbol{n}|^2 - \frac{\mu_0 \Delta \kappa}{K} (\boldsymbol{H} \cdot \boldsymbol{n})^2 \right] \quad . \tag{2.19}$$

The director field can be described by

$$n_x = \cos \Psi(x) \tag{2.20}$$
$$n_y = 0 \tag{2.21}$$
$$n_z = \sin \Psi(x) \quad . \tag{2.22}$$

For this, the following is true

$$\operatorname{div} \boldsymbol{n} = \frac{\partial}{\partial x} \cos \Psi(x) = -\sin \Psi \frac{\mathrm{d}\Psi}{\mathrm{d}x} \tag{2.23}$$

while we can write for the only non-zero component of the second term in Eq. (2.19)

$$(\operatorname{curl} \boldsymbol{n})_y = -\frac{\partial}{\partial x} n_z = -\cos \Psi \frac{\mathrm{d}\Psi}{\mathrm{d}x} \quad . \tag{2.24}$$

The change in the Gibbs free energy for the whole of the film of surface area A, considered per surface unit, is given by the following integration:

$$\frac{\Delta \widehat{\mathcal{G}}}{A} = \frac{K}{2} \int_0^\infty \left[\left(\frac{\mathrm{d}\Psi}{\mathrm{d}x} \right)^2 - \frac{\mu_0 \Delta \kappa}{K} H^2 \cos^2 \Psi \right] \mathrm{d}x \quad . \tag{2.25}$$

The director field takes up exactly that form for which the Gibbs free energy $\Delta \widehat{\mathcal{G}}$ assumes its minimum value, i.e., we are looking for the director field for which the following is valid for any arbitrary variation $\delta \Psi(x)$:

$$\frac{\delta(\Delta \widehat{\mathcal{G}})}{\delta \Psi(x)} = 0 \quad . \tag{2.26}$$

The solution of this basic problem in variation calculus is given by the Euler–Lagrange equation. The expression in Eq. (2.25) is of the form of the functional

$$\frac{\Delta \widehat{\mathcal{G}}}{A} = \int_0^\infty \Phi \left(\Psi(x), \frac{\mathrm{d}\Psi}{\mathrm{d}x} \right) \mathrm{d}x \tag{2.27}$$

with integrand Φ. Applying the Euler–Lagrange equation

$$\frac{\mathrm{d}}{\mathrm{d}x}\frac{\partial\Phi}{\partial(\frac{\mathrm{d}\Psi}{\mathrm{d}x})} - \frac{\partial\Phi}{\partial\Psi} = 0 \tag{2.28}$$

to this integrand yields

$$\frac{\mathrm{d}}{\mathrm{d}x}\frac{\mathrm{d}\Psi}{\mathrm{d}x} - \frac{\mu_0\Delta\kappa}{K}H^2\cos\Psi\sin\Psi = 0 \ . \tag{2.29}$$

In order to solve this differential equation, multiply both sides by $\mathrm{d}\Psi/\mathrm{d}x$:

$$\frac{\mathrm{d}\Psi}{\mathrm{d}x}\frac{\mathrm{d}}{\mathrm{d}x}\frac{\mathrm{d}\Psi}{\mathrm{d}x} = \frac{\mu_0\Delta\kappa H^2}{K}\cos\Psi\sin\Psi\frac{\mathrm{d}\Psi}{\mathrm{d}x} \ , \tag{2.30}$$

which then gives

$$\frac{1}{2}\frac{\mathrm{d}}{\mathrm{d}x}\left(\frac{\mathrm{d}\Psi}{\mathrm{d}x}\right)^2 = \frac{1}{2\xi_{\mathrm{H}}^2}\frac{\mathrm{d}}{\mathrm{d}x}\sin^2\Psi \ . \tag{2.31}$$

In doing so, a length ξ_{H} has been introduced by means of the definition

$$\xi_{\mathrm{H}}^2 = \frac{K}{\mu_0\Delta\kappa H^2} \ . \tag{2.32}$$

Upon integration, a first constant of integration C_1 appears:

$$\left(\frac{\mathrm{d}\Psi}{\mathrm{d}x}\right)^2 = \frac{1}{\xi_{\mathrm{H}}^2}\sin^2\Psi + C_1 \ . \tag{2.33}$$

Because of the boundary condition

$$\lim_{x\to\infty}\frac{\mathrm{d}\Psi}{\mathrm{d}x} = \lim_{x\to\infty}\Psi = 0 \tag{2.34}$$

it is found that C_1 equals zero. The so-obtained equation

$$\frac{\mathrm{d}\Psi}{\sin\Psi} = \pm\frac{\mathrm{d}x}{\xi_{\mathrm{H}}} \tag{2.35}$$

can be integrated to give, upon choosing the negative prefactor,

$$\ln\tan\frac{\Psi}{2} = -\frac{x}{\xi_{\mathrm{H}}} + C_2 \ . \tag{2.36}$$

A further constant of integration, C_2, is thereby introduced, into Eq. (2.36) and also in a modified form, as C_3, in the resulting equation

$$\tan\frac{\Psi}{2} = C_3\exp-\frac{x}{\xi_{\mathrm{H}}} \ . \tag{2.37}$$

The boundary condition

$$\Psi(0) = \frac{\pi}{2} \qquad (2.38)$$

fixes the value of C_3 to be one, and the final result is thus

$$\Psi = 2 \arctan\left(\exp -\frac{x}{\xi_H}\right) \quad . \qquad (2.39)$$

If the positive prefactor in Eq. (2.35) is chosen, a second solution is obtained:

$$\Psi = 2 \arctan\left(\exp \frac{x}{\xi_H}\right) \quad . \qquad (2.40)$$

The director field in Fig. 2.4 corresponds to the first solution. The transition region has a thickness on the order of ξ_H. Therefore, the determination of this thickness yields the Frank modulus K, provided that $\Delta\kappa$ is known.

The set-up in Fig. 2.4, thus, allows the determination of the average value of the Frank moduli for splaying and bending. The Frank modulus for twisting, K_2, can be determined by means of a similar experiment: It is simply necessary to apply the magnetic field in the y direction, i.e., perpendicular to the director in the surface plane. A transition region with a characteristic width ξ_H again arises, from which K_2 can then be determined.

The functioning of a display element depends strongly on its switching time. Upon switching off the external field, which is always electrical in nature for switching elements, elastic forces lead to the recovery of the director. The necessary time for this depends on the Frank moduli as well as, very importantly, the viscosity. The viscosity of a liquid crystal is direction dependent because of the structural anisotropy. Figure 2.5 shows this for three different situations. The flow is in the x direction and the gradient of the flow velocity is perpendicular to it in the z direction for all three cases; they differ with respect to the orientation of the director. It is obvious that different viscosities must result for the three different arrangements. They are referred to here as η_a, η_b, η_c and are termed the **Miesowicz viscosity coefficients**. In order to determine them, first, a flow velocity field as shown must be realised, e.g., by

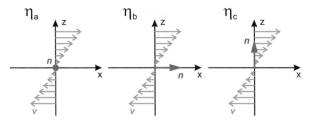

Fig. 2.5. Flow pattern and director orientations associated with the three Miesowicz viscosity η_a, η_b and η_c coefficients of a liquid crystal.

means of two plates which move against one another, and, second, the director must at the same time be held in place by means of a sufficiently strong magnetic field.

The figure shows only three important special cases. In the general case, the following applies

$$\eta(\vartheta, \varphi) = \eta_c \sin^2 \vartheta \cos^2 \varphi + \eta_b \cos^2 \vartheta + \eta_a \sin^2 \vartheta \sin^2 \varphi + \eta_d \sin^2 \vartheta \cos^2 \vartheta \cos^2 \varphi \ . \tag{2.41}$$

The angles ϑ and φ fix, in terms of spherical coordinates, the orientation of the director relative to the flow direction. η_b, η_c, η_a correspond to $\vartheta = 0°$, $\vartheta = 90°$, $\varphi = 0°$ (director in the z direction), and $\vartheta = 90°$, $\varphi = 90°$ (director in the y direction), respectively. The three Miesowicz coefficients alone are insufficient to determine every viscosity. An additional term, denoted η_d in Eq. (2.41), is important in the case of oblique director orientations. In addition, there is a fifth coefficient, γ_1, which is of particular importance. It exists because of the fact that viscous forces can arise in liquid crystals even if there is no flow of matter. This occurs when a field does no more than cause the director to rotate. Viscous forces appear for such a rotation of all molecules, and a torque $\boldsymbol{T}$ arises which maintains the equilibrium with respect to the external torque. Its magnitude is proportional to both the angular velocity of the director and the sample volume $\mathcal{V}$

$$\boldsymbol{T} = \gamma_1 \omega \mathcal{V} \ . \tag{2.42}$$

The constant of proportionality is the **rotational viscosity** γ_1. Figure 2.6 shows how γ_1 can be measured. A cylinder-shaped container, which is filled with the liquid crystal, is hung by means of a twistable thread. The sample is then exposed to a magnetic field which rotates with the frequency ω. This causes a homogeneous orientation of the liquid crystal with a director which

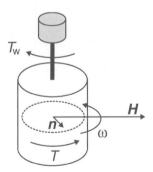

Fig. 2.6. The determination of the rotational viscosity γ_1 using a rotating magnetic field. Equilibrium is maintained with respect to the torque $\boldsymbol{T}$ of the field $\boldsymbol{H}$ through the restoring torque $\boldsymbol{T}_w$ of a twisted wire. Both are equal to the torque associated with the viscous forces.

rotates together with the magnetic field. $\boldsymbol{H}$ and $\boldsymbol{n}$ are not parallel to each other, as is shown in the figure. A torque

$$\boldsymbol{T} = \mathcal{V}\mu_o\boldsymbol{M} \times \boldsymbol{H} \propto \boldsymbol{n} \times \boldsymbol{H} \tag{2.43}$$

is thus exerted on the liquid crystal by the magnetic field. This equals the torque due to friction of Eq. (2.42) which acts in the opposite direction. An equilibrium between the torques is likewise established for the whole system, i.e., the liquid crystal in the container whose rotation energy does not change any further after a short adjustment process. Therefore, the torque due to the magnetic field equals the torque $\boldsymbol{T}_{\mathrm{w}}$ due to the twist of the thread. The desired viscosity coefficient γ_1 can thus be obtained from the latter without requiring any knowledge of the magnetic properties of the liquid crystal.

Figure 2.7 shows two typical sets of experimental results, namely the Miesowicz viscosities of MBBA and the rotational viscosity of PCH5 (pentyl-cyclohexyl-benzonitrile). It is always found for the nematic phase that

$$\eta_{\mathrm{c}} > \eta_{\mathrm{a}} > \eta_{\mathrm{b}} \quad , \tag{2.44}$$

with the differences decreasing with increasing temperature. In this way, $\eta_{\mathrm{a}}, \eta_{\mathrm{b}}$ and η_{c} approach the viscosity η of an isotropic liquid. The rotational viscosity exhibits a different behaviour. Since γ_1 does not exist any more in the isotropic phase, γ_1 cannot simply become equal to η as is the case for $\eta_{\mathrm{a}}, \eta_{\mathrm{b}}$ and η_{c}, but, rather, it becomes ever smaller as the order parameter is reduced and finally vanishes. The experimental data show this.

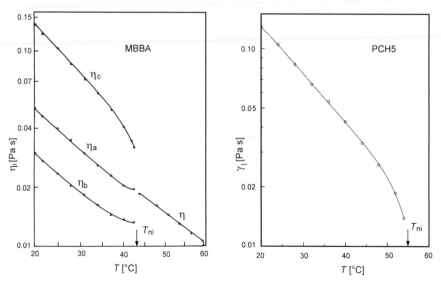

Fig. 2.7. The temperature dependence of the Miesowicz viscosity coefficients of MBBA (from Gähwiller ([10] *left*) and the rotational viscosity of PCH5 ([11] *right*)).

2.1.3 Polymers: Viscoelasticity

Polymers are widely used materials, with their behaviour under mechanical loads being of particular interest. As far as their overall behaviour is concerned, they differ clearly from other materials such as metallic or ceramic solids or, on the other hand, simple liquids. Polymers bring together elastic and viscous properties over a wide range, and are therefore characterised as **viscoelastic**. Viscoelasticity does not only mean a simple overlap of viscous and elastic behaviours. Typical is the appearance of a behaviour referred to as **anelasticity**, in which elastic and viscous forces exist in close interaction. This leads to the fact that a finite time is required for the system to adjust for part of a deformation, even though it is reversible and disappears when the force is removed. It is typical for polymers that the mechanical behaviour varies not only among different polymer materials, but that there are also marked changes as a function of temperature. The temperature region within which a given polymer material can be used is therefore always limited.

Creep Experiments, Stress Relaxation Experiments and Dynamic-Mechanical Experiments. How can the deformation behaviour of a viscoelastic object – polymers are the main representatives of this class of materials – be experimentally characterised? Different approaches are used when mechanically characterising polymer materials, corresponding to the various conditions imposed on them. As samples are often exposed to a constant stress it is of interest to investigate the resulting time-dependent deformation. Such a test is referred to as a **creep experiment**, and it can be performed for different types of loading, e.g., a uniaxial stress or a shear. The force is applied quickly at time $t = 0$ and the resulting deformation is followed as a function of time. In contrast to a Hookian solid, where a definite final extension would immediately result, a complex time-dependent extension curve is usually observed for a polymer sample. Three different contributions must always be considered, as is shown schematically in Fig. 2.8 for the case of a tensile force. As well as the immediate elastic reaction, an anelastic component acts, i.e., there is an adjustment process, which only becomes effective after a certain time delay. Finally, a viscous **plastic** flow is observed. The first two parts of the deformation are reversible, i.e., they completely disappear upon removing the force, whereas the viscous flow leads to an irreversible change. Figure 2.8 also shows the changes resulting from an unloading. They constitute a second part of the test, usually referred to as the **strain recovery**.

The result of a tensile creep test is a material specific time-dependent extension curve $e_{zz}(t)$. The following important observation is to be noted: The measurements show, for a certain range of not too high loading forces, that the change in $e(t)$ is proportional to the applied stress σ_{zz}^0. It is, thus, possible, in this **linear viscoelastic region**, to fully describe the results of a tensile creep experiment by means of a function, which is independent of the applied stress, namely with the form

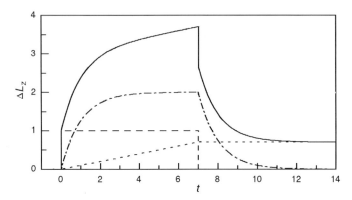

Fig. 2.8. The general time-dependent change in the extension of a polymer upon applying a force (*the creep curve*) and then removing it (*recovery curve*). The total extension is the result of an overlap of Hookian elasticity (*dashed*), delayed anelastic adjustment processes (*dot-dashed*) and irreversible viscous flow (*dotted*).

$$e_{zz}(t) = D_t(t)\sigma_{zz}^0 \quad . \tag{2.45}$$

The function $D_t(t)$ is material dependent and is termed the tensile **creep compliance**.

Sometimes materials are not placed under a constant stress, but rather a constant deformation is maintained by fixing the sample. The behaviour of samples under such conditions is studied in tensile **stress relaxation experiments**. The sample is held under a constant well-defined extension and the experiment follows the resulting, usually time-dependent stress. Figure 2.9 shows the typical form of this time dependence. A high value of the stress is observed immediately after the deformation. There is then a fall off caused, first, by the time-delayed reaction mechanisms and, second, by the plastic flow. It is also found in stress relaxation experiments that the resulting time-dependent tension is proportional over a certain range to the deformation, e_{zz}^0. The constant of proportionality in these experiments is the **time-dependent Young's modulus** $E_t(t)$:

$$\sigma_{zz}(t) = E_t(t)e_{zz}^0 \quad . \tag{2.46}$$

There are many practical situations where a periodically time-dependent load as opposed to a constant stress or strain is applied. This is the case with **dynamic-mechanical experiments**. For example, consider that a sample is exposed to a periodically changing tensile stress

$$\sigma_{zz}(t) = \sigma_{zz}^0 \exp(-\mathrm{i}\omega t) \quad . \tag{2.47}$$

It is clear that a periodically changing strain results

$$e_{zz}(t) = e_{zz}^0 \exp(-\mathrm{i}\omega t) \quad . \tag{2.48}$$

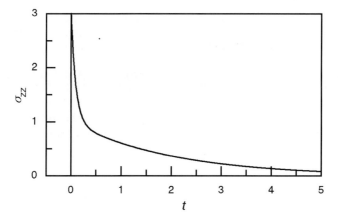

Fig. 2.9. The time dependence of the stress in a polymer sample which was exposed to a sudden extension.

Figure 2.10 shows that generally there is a time delay associated with the strain $e_{zz}(t)$ because of the sample viscosity, i.e., there is a phase shift relative to the stress $\sigma_{zz}(t)$. We deliberately choose here the complex notation for $\sigma_{zz}(t)$ and $e_{zz}(t)$. The relationship between the two functions can then simply be described with the help of the complex expression

$$D_{t}(\omega) = \frac{e_{zz}(t)}{\sigma_{zz}(t)} = \frac{e_{zz}^{0}}{\sigma_{zz}^{0}} = D_{t}'(\omega) + iD_{t}''(\omega) \quad . \tag{2.49}$$

$D_{t}(\omega)$ is referred to as the **dynamic tensile compliance**. Dynamic-mechanical experiments are usually carried out for a series of different frequencies,

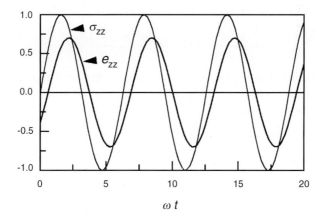

Fig. 2.10. The time dependencies of the stress and the strain for a dynamic-mechanical experiment carried out on a viscoelastic sample.

which cover many orders of magnitude. Such experiments yield the frequency dependence of the real and imaginary parts of the complex compliance, $D'_t(\omega)$ and $D''_t(\omega)$, respectively. The phase shift, δ, is given by

$$\tan \delta(\omega) = \frac{D''_t(\omega)}{D'_t(\omega)} \quad . \tag{2.50}$$

As an alternative to the dynamic tensile compliance, the **dynamic Young's modulus** can also be used to describe the experimental results. It is defined by

$$E_t(\omega) = \frac{\sigma^0_{zz}}{e^0_{zz}} = E'_t(\omega) - iE''_t(\omega) \tag{2.51}$$

and corresponds to the inverse of the compliance:

$$E_t(\omega) = \frac{1}{D_t(\omega)} \quad . \tag{2.52}$$

By convention, a negative sign is chosen for the imaginary part.

In the previous examples, we have only discussed reactions to tensile stresses. There are, of course, corresponding expressions for a shear, i.e., a time-dependent shear compliance $J(t)$, a time-dependent shear modulus $G(t)$ and the two corresponding frequency-dependent and in general complex functions $J(\omega)$ and $G(\omega)$.

Dynamic-mechanical experiments have the advantage, as compared to time-dependent experiments, that they provide a particular physical insight. During the deformation, work is done on the sample by the external force. On the one hand, this increases the potential energy, on the other hand, it leads to the generation of heat. The real and imaginary parts of the dynamic compliance correspond exactly to this separation. That this is so can be easily demonstrated. It is only necessary to analyse the power provided by the external force. Taken per unit volume the power is given by

$$\frac{dw}{dt} = \Re(\sigma_{zz}(t)) \frac{d\Re(e_{zz}(t))}{dt} \tag{2.53}$$

(for the calculation of a product a transition from the complex notation to the actual physical quantity given by the real part is necessary). The time-dependent extension is given by

$$e_{zz}(t) = D_t(\omega)\sigma^0_{zz} \exp(-i\omega t) = (D'_t + iD''_t)\sigma^0_{zz}[\cos(\omega t) - i\sin(\omega t)] \quad . \tag{2.54}$$

The real part is then given by

$$\Re(e_{zz}(t)) = D'_t \sigma^0_{zz} \cos(\omega t) + D''_t \sigma^0_{zz} \sin(\omega t) \quad . \tag{2.55}$$

Therefore, the following expression is obtained for the power

$$\frac{dw}{dt} = \sigma^0_{zz} \cos(\omega t)[-\omega\sigma^0_{zz}D'_t \sin(\omega t) + \omega\sigma^0_{zz}D''_t \cos(\omega t)] \tag{2.56}$$

or, upon applying a well-known trigonometric relation

$$\frac{\mathrm{d}w}{\mathrm{d}t} = -\frac{(\sigma_{zz}^0)^2}{2}\omega D_t' \sin(2\omega t) + (\sigma_{zz}^0)^2 \omega D_t'' \cos^2(\omega t) \quad . \tag{2.57}$$

It is apparent that there are two contributions to the power. The first term oscillates between positive and negative values with double the frequency of the stress. This expresses an exchange: Work, which is stored up in the sample during a quarter period, goes back out in the next quarter period. This part clearly describes the storing and release of elastic potential energy. Its magnitude is given only by the real part of the dynamic compliance. The second part, which is proportional to the imaginary part, behaves completely differently. It describes a take up of power which is always positive, with a time-averaged magnitude

$$\overline{\frac{\mathrm{d}w}{\mathrm{d}t}} = \frac{1}{2}(\sigma_{zz}^0)^2 \omega D_t'' \quad . \tag{2.58}$$

What does this mean? Generally, the internal energy of the sample $\mathcal{U}$ changes as work is done and heat is exchanged according to

$$\mathrm{d}\mathcal{U} = \mathcal{V}\mathrm{d}w + \mathrm{d}\mathcal{Q} \quad . \tag{2.59}$$

If the experiment is, as usual, performed under isothermal conditions, there is no change in the internal energy of the sample. The supplied work must therefore be completely released as heat:

$$\mathcal{V}\overline{\frac{\mathrm{d}w}{\mathrm{d}t}} = -\overline{\frac{\mathrm{d}\mathcal{Q}}{\mathrm{d}t}} \quad . \tag{2.60}$$

The Relationships Between the Response Functions. With the time-dependent compliance $D_t(t)$, the time-dependent Young's modulus $E_t(t)$, the frequency-dependent dynamic compliance $D_t(\omega)$, the dynamic Young's modulus $E_t(\omega)$, and the corresponding functions for shearing, a range of material functions has been introduced which describe the mechanical reaction of samples in the linear viscoelastic region. It is naturally to be expected that relationships exist between these different **response functions**, and they will now be derived. In order to make the expressions independent of the type of load, we denote the force as ξ and the deformation as X.

The above discussion has dealt with static and periodically changing conditions. These are definitely the most important cases, but nevertheless the question arises as to whether it is possible to treat the general case of a completely arbitrary time dependence. In fact, this is possible, the condition being solely the knowledge of a further response function of particular importance: It must be known how a sample reacts upon applying a force as a short pulse

$$\xi(t) = \xi_t \delta(t) \quad . \tag{2.61}$$

ξ_t characterises here the strength of the pulse. In the linear-viscoelastic region, the resulting time-dependent deformation $X(t)$ is given by

$$X(t) = \xi_t \alpha_{\mathrm{p}}(t) \ . \tag{2.62}$$

$\alpha_{\mathrm{p}}(t)$ is referred to as the **pulse response function**. As a further choice in addition to the other response functions, $\alpha_{\mathrm{p}}(t)$ can also be used to describe the viscoelastic properties of a sample. Figure 2.11 shows this for a range of examples. The curves plotted for the different materials have obvious forms: A Hookian elastic body will only be deformed during the length of the pulse (a), a plastic body will suffer a certain permanent deformation (b), while an anelastically reacting body will subsequently slowly remove again the initially built-up deformation (c). The particular importance of the pulse response functions lies in the fact that they allow the deformation of a probe under the action of an arbitrary time-dependent force $\xi(t)$ to be specified. The following equation applies for the resulting time-dependent deformation in this general case:

$$X(t) = \int_{-\infty}^{t} \alpha_{\mathrm{p}}(t - t')\xi(t')\mathrm{d}t' \ . \tag{2.63}$$

The physical basis of the relationship can be immediately recognised. In order to formulate this equation, use was made of the fact that the superposition principle applies for linear systems, i.e., that a superposition of different solutions of the equation of motion always gives a new solution. When writing Eq. (2.63), the arbitrary time-dependent force is expressed as a sum of pulses, and the complete reaction $X(t)$ then corresponds to the sum of their individual responses. Equation (2.63) is generally referred to as the **Boltzmann superposition principle**.

This general equation now allows the relationships between the different response functions to be determined. First, the following expression can be written for the creep of a sample after the application of a constant force ξ_0 at time $t = 0$

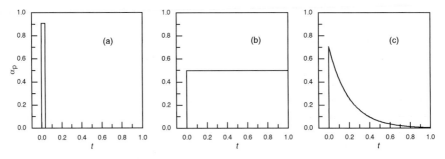

Fig. 2.11. Pulse response functions after excitation for a Hookian elastic body (a), a plastic body (b), and an anelastic body exhibiting simple relaxation (c).

$$X(t) = \int_0^t \alpha_{\mathrm{p}}(t - t')\xi_0 \mathrm{d}t' \quad . \tag{2.64}$$

It, thus, follows that the creep compliance, which is denoted in this general case by $\alpha_{\mathrm{c}}(t)$, is given by

$$\alpha_{\mathrm{c}}(t) = \int_0^t \alpha_{\mathrm{p}}(t - t')\mathrm{d}t' = \int_0^t \alpha_{\mathrm{p}}(t'')\mathrm{d}t'' \quad . \tag{2.65}$$

Differentiating both sides yields

$$\alpha_{\mathrm{p}}(t) = \frac{\mathrm{d}\alpha_{\mathrm{c}}}{\mathrm{d}t}(t) \quad . \tag{2.66}$$

The stress relaxation experiment is described by

$$X_0 = \int_0^t \alpha_{\mathrm{p}}(t - t')\xi(t')\mathrm{d}t' \quad , \tag{2.67}$$

where X_0 is the strain imposed on the sample and $\xi(t')$ is the resulting time-dependence of the force. On introducing the time-dependent modulus – it is denoted by $a(t)$ in this general notation – we obtain

$$1 = \int_0^t \alpha_{\mathrm{p}}(t - t')a(t')\mathrm{d}t' \quad . \tag{2.68}$$

The combination of Eqs. (2.66) and (2.68) yields the relationship between $\alpha_{\mathrm{c}}(t)$ and $a(t)$.

In a dynamic-mechanical experiment, there is a periodically changing force

$$\xi(t) = \xi_0 \exp(-\mathrm{i}\omega t) \tag{2.69}$$

which leads to a corresponding periodic strain

$$X(t) = X_0 \exp(-\mathrm{i}\omega t) \quad . \tag{2.70}$$

Inserting these two expressions into Eq. (2.63) gives

$$X_0 \exp(-\mathrm{i}\omega t) = \int_{-\infty}^t \alpha_{\mathrm{p}}(t - t')\xi_0 \exp(-\mathrm{i}\omega t')\mathrm{d}t' \quad . \tag{2.71}$$

It, thus, follows that the dynamic compliance, denoted in the general case by $\alpha(\omega)$, is given by

$$\alpha(\omega) = \int\limits_{-\infty}^{t} \alpha_{\mathrm{p}}(t - t') \exp \mathrm{i}\omega(t - t') \, \mathrm{d}t' \quad . \tag{2.72}$$

Making the substitution

$$t - t' = t''$$

yields

$$\alpha(\omega) = \int\limits_{0}^{\infty} \alpha_{\mathrm{p}}(t'') \exp(\mathrm{i}\omega t'') \mathrm{d}t'' \quad . \tag{2.73}$$

It is thus apparent that the compliance equals the Fourier transform of the pulse response function.

As is to be expected, the response functions are, thus, found to be related to each other by one-to-one correspondencies. It is therefore, in principle, possible to transform a frequency- into a time-dependent experiment and vice versa. There is, however, a condition to be fulfilled which is only rarely met in practice: The time- and frequency-dependence over all relevant times and frequencies must be known, and this is difficult to achieve.

From a Glass to a Liquid Via a Rubbery Elastic State. We will now consider some typical experimental results, which also illustrate the different experimental techniques. Figure 2.12 shows the results of a comprehensive series of measurements of the time-dependent shear compliance of polystyrene. The measurements extend over a temperature range of almost $600\,\mathrm{K}$ and a time range of eight orders of magnitude and thus cover all the significant aspects of mechanical behaviour. The simplest behaviour is found in the limits of the lowest and highest temperatures. At the lowest temperature, polystyrene is in a rigid glassy state and the mechanical properties are those of an elastic Hookian body. There is a small solid-like compliance which does not change at all over time. In the other limit, namely at the highest temperature, polystyrene exists as a melt. Here, a deformation with a constant shear rate is observed, like that in a Newtonian liquid. A behaviour which is specific to polymers is found in between these two limits. Marked time-dependent effects are apparent in the temperature region around $100\,^{\circ}\mathrm{C}$. It is evident that in addition to the immediate elastic deformation of about $10^{-9}\,\mathrm{N}^{-1}\,\mathrm{m}^2$, which is typical for the solid state, there is another time-delayed component. The final value of the compliance is around $10^{-6}\,\mathrm{N}^{-1}\,\mathrm{m}^2$ – this is no longer that of an amorphous, rigid glassy solid, but rather that of a rubber. In this temperature range, which corresponds to the glass transition, the sample behaves like a rubber with a high internal friction. What is the origin of the rubber-like behaviour of polystyrene, in which, in contrast to a rubber, there are no chemical cross-links? It is the **entanglements** which unavoidably exist in a polymer melt. They prevent, in the first instant, a free movement of the macromolecule in all directions and lead to the temporary formation of knots,

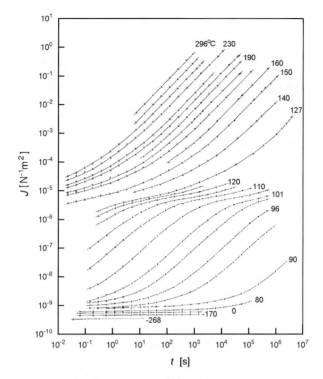

Fig. 2.12. The time-dependent shear compliance of polystyrene, measured at different temperatures between $-268\,^{\circ}$C and $296\,^{\circ}$C (from Schwarzl [12]).

thus giving rise to a quasi-rubber-elastic behaviour. At high temperatures, a sliding and disentangling of the chains begins and a viscous flow sets in. Thus, according to the particular temperature, polystyrene behaves as either a Hookian solid, a rubber, or a Newtonian liquid.

The curves in Fig. 2.12 give the impression that there is a systematic change in the mechanical behaviour as a function of temperature. Expressed more precisely, it seems that a change in the temperature alone leads to a shifting of the curve along the logarithmic time axis. This equivalence in the temperature and time variation opens up an interesting way of handling the data: By means of a corresponding shift of the individual experimental curves along the time axis until they overlap, it is possible to generate a curve which represents the whole reaction of the sample. Such a complete plot is presented in schematic form in Fig. 2.13. This result would be obtained for an experiment carried out at a temperature of $100\,^{\circ}$C, which begins at very short times and ends with extremely long times. The complete plot now contains all three basic parts in one, namely the elastic deformation component at very short times, the overlapping with the anelastic component in the middle region, with this component causing the transition from a glass to a rubber, and finally at very

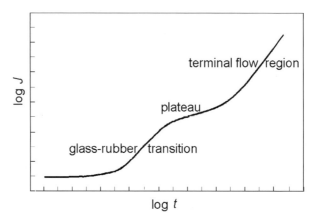

Fig. 2.13. The putting together of the single measurements from Fig. 2.12 to give the complete creep curve.

long times in the terminal region, the remaining now clearly apparent flow component.

As a second example, Fig. 2.14 shows the result of a stress relaxation experiment carried out on polyisobutylene. At room temperature, polyisobuty-

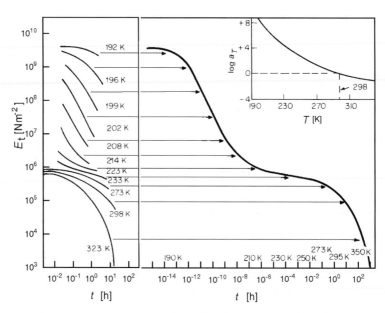

Fig. 2.14. The variation with temperature of the Young's modulus of polyisobutylene: Individual measurements at different temperatures and the complete stress relaxation curve $E_t(t)$ (from Castiff and Tobolsky [13]).

lene is in a cross-linked state a typical rubber. In spite of the fact that the experimental results shown here were obtained for a non-cross-linked melt, the presence of entanglements leads like in the case of polystyrene to the observation of a temporary rubber-elastic behaviour, in the region around T_g ($\simeq 200\,\text{K}$). Putting the individual curves together in an empirical fashion again gives the total representation (shown on the right) of the time-dependence of Young's modulus. The result of this stress relaxation experiment shows the same main features as the creep experiment discussed above: At short times, polyisobutylene behaves like a glass with Young's moduli values on the order of $10^9\,\text{Nm}^{-2}$, there then follows a decay until the Young's modulus takes the value of a rubber, $10^6\,\text{Nm}^{-2}$, and at still longer times, this stress reduces further because of the lasting viscous flow.

The plot in the top right of the figure shows the temperature-dependent shifts applied to the individual experimental curves in order to generate the complete stress relaxation function. The shown curve corresponds to a temperature of $T = 298\,\text{K}$. The **shift factor** a_T has a characteristic temperature dependence. We have actually encountered this same dependence in an earlier section. The Vogel–Fulcher equation which describes the temperature dependence of the viscosity was introduced in the discussion of the glass transition (Eq. (1.82)), and it is this function which also describes the shift factor:

$$a_T = \exp\left(\frac{T_A}{T - T_V} - \frac{T_A}{T_0 - T_V}\right) \quad . \tag{2.74}$$

T_0 denotes the reference temperature, which is chosen here to be $298\,\text{K}$. The temperature dependence of the viscosity, thus, also determines the temperature dependence of the time which is required for the rubber-elastic component in the relaxation function to appear. At first, this may be surprising, but it is qualitatively completely understandable. Fundamentally, both processes, namely the flow motion of macromolecules in a shear field and the extension of the chains upon stretching a rubber sample, involve the movement of chain segments in the melt, with the segments being exposed to the same frictional forces resulting from their environment. The parameters T_A and T_V from the Vogel–Fulcher equation are related to exactly these local frictional forces and determine their temperature dependence.

Simple Relaxation Processes. As a third example, Fig. 2.15 shows the results of a dynamic-mechanical experiment carried out on poly(cyclohexyl-methacrylate) (PCMA) under shear stress in the rigid glassy state. In contrast to the investigated polymers in the other examples, polystyrene and polyisobutylene, for PCMA, it is evident that a residual and well-defined molecular motion remains in the glassy state. As can be seen from a consideration of the size of the imaginary part of the shear compliance $J''(\omega)$, it leads to a mechanical loss. For each temperature, the loss goes through a maximum as a function of the frequency, with the position of the maximum loss moving

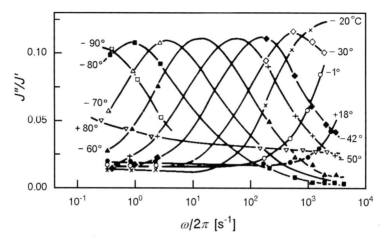

Fig. 2.15. The frequency dependence of the imaginary part of the dynamic shear compliance ($J' \approx$ const), as measured on poly(cyclohexylmethacrylate) at different temperatures in the glassy state (from Heijboer [14]).

to higher frequency upon increasing temperature. How can this apparently so simple result be understood? Actually, we have here a fine example of a **simple relaxation process**. The cyclohexyl side-groups in PCMA can, exactly like in the molecule cyclohexane, change back and forth between two different conformations known as the 'chair' and 'boat' forms; there is only a slight energy difference between the two conformations. The application of a mechanical field changes the distribution: One of the forms is preferred according to the local situation. The redistribution upon applying a stress requires a definite time, is accompanied by a growing deformation and thus leads to the observed delayed, i.e., anelastic, reaction.

In fact, it can be shown that the adjustment time can be extracted from the experiment: The frequency of the maximum in $J''(\omega)$ exactly reflects the rate with which the redistribution over the conformations occurs. This can be seen by considering a simple model. Rather than immediately looking at the dynamic-mechanical experiment of interest, consider a thought creep experiment on the same sample. If the redistribution over the conformations requires a time τ and results at the end, i.e., at equilibrium, in an additional shearing

$$\Delta\gamma(t \to \infty) = \Delta J \, \sigma^0_{zx} \quad,$$

it is possible to describe the adjustment process in the following way:

$$\frac{\mathrm{d}\Delta\gamma}{\mathrm{d}t} = -\frac{1}{\tau}(\Delta\gamma(t) - \Delta J \, \sigma^0_{zx}) \quad. \tag{2.75}$$

This relaxation equation implies that a system, which is forced from equilibrium (here, this is achieved by applying a stress), returns to equilibrium at a rate which increases proportionally with the perturbation. The solution of the differential equation is

$$\Delta\gamma(t) = \Delta J \, \sigma_{zx}^0 \left(1 - \exp -\frac{t}{\tau}\right) \tag{2.76}$$

and it describes an adjustment process with a time constant τ.

Consider now the dynamic-mechanical experiment. The relaxation equation can also be applied in this case. Replacing the time-independent stress σ_{zx}^0 by

$$\sigma_{zx}(t) = \sigma_{zx}^0 \exp(-\mathrm{i}\omega t) \tag{2.77}$$

gives

$$\frac{\mathrm{d}\Delta\gamma}{\mathrm{d}t} = -\frac{1}{\tau}[\Delta\gamma(t) - \Delta J\sigma_{zx}^0 \exp(-\mathrm{i}\omega t)] \ . \tag{2.78}$$

A solution with a periodically changing extension $\Delta\gamma(t)$ is required, and we therefore write

$$\Delta\gamma(t) = \sigma_{zx}^0[\Delta J'(\omega) + \mathrm{i}\Delta J''(\omega)]\exp(-\mathrm{i}\omega t) \ . \tag{2.79}$$

This leads to the following expression for the complex compliance $\Delta J(\omega)$:

$$\Delta J'(\omega) + \mathrm{i}\Delta J''(\omega) = \frac{\Delta J(0)}{1 - \mathrm{i}\omega\tau} \tag{2.80}$$

$$= \frac{\Delta J}{1 + \omega^2\tau^2} + \mathrm{i}\frac{\Delta J \, \omega\tau}{1 + \omega^2\tau^2} \ . \tag{2.81}$$

Figure 2.16 shows the frequency dependence of $\Delta J'(\omega)$ and $\Delta J''(\omega)$. The

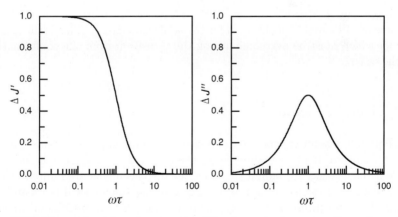

Fig. 2.16. The real and imaginary parts of the dynamic shear compliance for a simple relaxation process.

observation that the imaginary part, for a logarithmic frequency scale, corresponds to a symmetric 'Lorentzian' curve is apparent if it is re-expressed as

$$\Delta J''(\omega) = \frac{\Delta J}{10^{-\log(\omega\tau)} + 10^{\log(\omega\tau)}} \quad . \tag{2.82}$$

An important conclusion can be drawn from this result, namely that the maximum loss occurs when

$$\omega\tau = 1 \tag{2.83}$$

i.e., exactly when the frequency of the external force which causes the excitation corresponds to the relaxation rate τ^{-1}. Around the same condition, $\Delta J'(\omega)$ decreases in a step-like fashion from one to zero. In our example of the cyclohexyl side groups, the relaxation rate is identical to the jump rate with which the side groups change between the two possible conformations. It is to be emphasised that the applied force does not alter this jump rate. At thermodynamic equilibrium, jumps between the two conformations take place at the same rate, and this determines the time required for the system to react to external changes. Thus, dynamic-mechanical experiments allow the investigation of the dynamics of conformational redistributions, i.e., the determination of the time associated with a particular process. The method is, therefore, sometimes referred to as **relaxation spectroscopy**.

The temperature-dependent change in the rate of the transition between the chair and boat conformations can be determined from the shift in the location of the loss maxima in Fig. 2.15. An analysis shows that the temperature dependence of this process is consistent with the Arrhenius law

$$\tau^{-1} \propto \exp -\frac{\Delta\tilde{u}_{\mathrm{b}}}{\tilde{R}T} \quad . \tag{2.84}$$

A value of 47 kJ mol^{-1} is obtained for the activation energy. This is the energy barrier which must be overcome during the transition from one conformation to the other.

2.2 Electric Fields

If a liquid or a solid is exposed to an electric field, charges will be displaced. When there are freely mobile charges present, a flow of current is induced, while if the charges are bound together, the displacements are limited. These local charge displacements in the Å range give rise to electric dipoles and the sample becomes **polarised**. Freely mobile charges and the associated currents will be discussed in Chap. 4. Here, we consider the subject of polarisation, which means the reaction behaviour of non-conducting **dielectrics**.

2.2.1 Dielectric Susceptibility

The Maxwell equations in the form appropriate to the treatment of the interaction of electromagnetic fields with matter use two equivalent terms in order to describe polarisation effects, the polarisation $\boldsymbol{P}$, which gives the field-induced electric dipole moment per unit volume, and the electric displacement $\boldsymbol{D}$. $\boldsymbol{D}$, $\boldsymbol{P}$ and the electric field $\boldsymbol{E}$ are related to each other through the equation

$$\boldsymbol{D} = \varepsilon_0 \boldsymbol{E} + \boldsymbol{P} \tag{2.85}$$

(ε_0 is the electric constant). Two different parameters can be used to describe the polarisation properties of a particular material. The **dielectric constant** ε relates $\boldsymbol{D}$ to the electric field, by means of the expression

$$\boldsymbol{D} = \varepsilon_0 \varepsilon \boldsymbol{E} \;, \tag{2.86}$$

while the dielectric susceptibility χ relates $\boldsymbol{P}$ to the electric field, by

$$\boldsymbol{P} = \varepsilon_0 \chi \boldsymbol{E} \;. \tag{2.87}$$

The two equations imply the following link between χ and ε:

$$\varepsilon = \chi + 1 \;. \tag{2.88}$$

Equations (2.86) and (2.87) express a linear dependence, and this is valid in most cases of practical relevance.

In addition to static conditions, the equations can also be used in the case of electric fields which change with time, and this applies to all frequencies, up to those of X-rays. ε and χ are then complex functions with a marked frequency dependence. Generally a fluctuating electric field

$$E(t) = E_0 \exp(-\mathrm{i}\omega t) \tag{2.89}$$

induces a polarisation with the same oscillating frequency

$$P(t) = P_0 \exp(-\mathrm{i}\omega t) \;. \tag{2.90}$$

The dielectric susceptibility determines, now as a complex variable, the relation between the amplitudes E_0 and P_0:

$$P_0 = \varepsilon_0 \chi(\omega) E_0 \;. \tag{2.91}$$

Being in general complex, $\chi(\omega)$ takes into account the possibility that a phase shift can appear between the polarisation and the field:

$$P_0 = \varepsilon_0 [\chi'(\omega) + \mathrm{i}\chi''(\omega)] E_0 \;. \tag{2.92}$$

2.2.2 Orientational and Distortional Polarisation

The polarisation of liquid or solid matter in an electric field arises from several contributions, which differ very markedly in their frequency dependence. This is illustrated by the schematic plot of the frequency dependence of the real part of the dielectric susceptibility in Fig. 2.17. A first contribution, which can be very strong, is found for liquids made up of polar molecules. In the absence of an external field, the permanent dipole moments carried by all the molecules are distributed over all directions. In a field, there is a preferred orientation and this results in an **orientational polarisation** $P_{or} \propto \varepsilon_0 \chi_{or}$. It is clear that at too high frequencies the molecules will no longer be able to follow the field, because of their moment of inertia. Empirically, this is found to occur when the frequency significantly exceeds $10^{12} \, s^{-1}$. After this, as is indicated in the figure, no further orientational polarisation is observed.

A second mechanism which contributes to the polarisation can be observed in ionic crystals, molecular crystals and molecular liquids. In molecules, there are always charge centres. An electric field can push these centres of positive and negative charge in opposite directions and, in this way, induce dipole moments. Particularly strong polarisation effects are to be expected in the infra-red region, since these frequencies correspond to the eigenfrequencies of

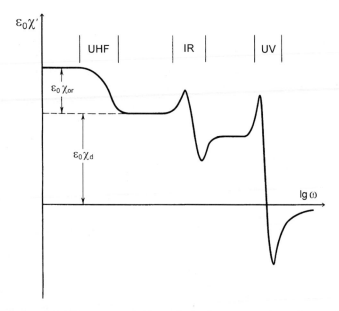

Fig. 2.17. A schematic representation of the frequency dependence of the real part of the dielectric susceptibility: The contributions due to dipolar orientation (χ_{or}, up to the UHF region) as well as a deformation of the nuclear skeleton and displacements of the electrons (χ_d, in the IR and UV region).

the vibrations of the molecular skeleton. As is indicated in the figure, this polarisation mechanism is effectively switched off when the molecular vibration frequencies are exceeded.

There remains a final polarisation mechanism which is always present in the case of atomic systems, namely the displacement of the electron shells relative to the nuclei. The eigenfrequencies of the displaceable outermost electrons lie in the visible light and ultra-violet region. Polarisation effects are thus particularly large in this frequency range. In the visible range, the refractive index n is usually used as the material-specific parameter in place of ε or χ. n is related to ε and χ by means of the Maxwell relation

$$n^2 = \varepsilon = 1 + \chi \ . \tag{2.93}$$

Like ε and χ, n is in the general case a complex variable.

The frequency dependence of the orientational polarisation and that of the **distortional polarisation** $P_{\mathrm{d}} \propto \varepsilon_0 \chi_{\mathrm{d}}$ – this encompasses the other two contributions – differ in a characteristic manner. While only a simple decay is observed for the orientational polarisation, resonance effects are observed for the contributions to the distortional polarisation. Using classical equations, it is possible to describe in a straightforward manner the frequency dependence of both processes. The orientational polarisation P_{or} is correctly described using the following relaxation equation:

$$\frac{\mathrm{d}P_{\mathrm{or}}}{\mathrm{d}t} = -\frac{1}{\tau}(P_{\mathrm{or}} - \varepsilon_0 \chi_{\mathrm{or}} E_0) \ . \tag{2.94}$$

The starting point is identical with that which was introduced for the treatment of anelastic deformations in Eq. (2.75). The time constant τ now has the meaning of the reorientation time of the polar molecule. The solution of the relaxation equation has been given above and can be directly used again: Upon switching on a static field E_0, an orientational polarisation appears according to

$$P_{\mathrm{or}} = \varepsilon_0 \chi_{\mathrm{or}} E_0 \left(1 - \exp -\frac{t}{\tau}\right) \ . \tag{2.95}$$

For a field oscillating at a frequency ω, the (complex) amplitude P_0 is given by

$$P_0 = \frac{\varepsilon_0 \chi_{\mathrm{or}}}{1 - \mathrm{i}\omega\tau} E_0 \ . \tag{2.96}$$

This leads for the real and imaginary part of the orientational dielectric susceptibility to

$$\frac{P_0}{E_0} = \varepsilon_0 \chi'_{\mathrm{or}}(\omega) + \mathrm{i}\varepsilon_0 \chi''_{\mathrm{or}}(\omega)$$

$$= \frac{\varepsilon_0 \chi_{\mathrm{or}}}{1 + \omega^2\tau^2} + \mathrm{i}\frac{\varepsilon_0 \chi_{\mathrm{or}}\omega\tau}{1 + \omega^2\tau^2} \ . \tag{2.97}$$

Plots of these functions have been shown previously in Fig. 2.16. Evidently the situation here again corresponds to a simple relaxation process. In fact, simple relaxation processes were first theoretically treated for the case of this dipolar relaxation, by Debye, and are therefore also generally addressed as **Debye processes**.

Corresponding experimental data obtained for poly(vinylacetate) are shown in Fig. 2.18. This material contains polar sidegroups, which can be reoriented. As was discussed in the previous section, the position of the step in the real part and the maximum in the imaginary part directly gives the relaxation rate, which is in this case the reorientation rate of the side groups. In polymers – due to the high inner viscosity – these rates are lowered by several orders of magnitude as compared to those in simple liquids.

In order to describe the characteristic shape of $\chi(\omega)$ in the ranges of the distortional polarisations, the following differential equation is suitable:

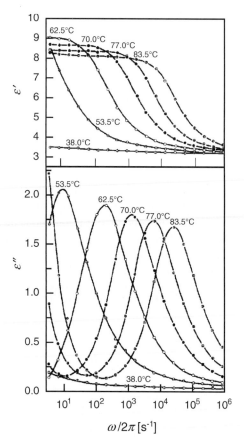

Fig. 2.18. The frequency dependence of the real and imaginary parts of the dielectric function, as measured for a sample of poly(vinylacetate) at various temperatures above T_{g} (from Ishida et al. [15]).

$$\tau'^2 \frac{d^2 P_d}{dt^2} = -\tau \frac{dP_d}{dt} - P_d + \varepsilon_0 \chi_d E \quad . \tag{2.98}$$

It differs from the relaxation equation through the term on the left-hand side which takes the inertia effects into account. Equation (2.98) is the equation of motion for a driven damped harmonic oscillator. For a varying electric field

$$E(t) = E_0 \exp(-i\omega t) \tag{2.99}$$

there exists a stationary solution for P_d,

$$P_d(t) = P_0 \exp(-i\omega t) \quad . \tag{2.100}$$

Inserting it into Eq. (2.98) leads to

$$\left(-\omega^2 \tau'^2 - i\omega\tau + 1\right) P_0 = \varepsilon_0 \chi_d E_0 \tag{2.101}$$

and, thus, to the following expression for the frequency-dependent distortional part of the dielectric susceptibility:

$$\frac{P_0}{E_0} = \varepsilon_0 \chi_d'(\omega) + i\varepsilon_0 \chi_d''(\omega) \tag{2.102}$$

$$= \frac{\varepsilon_0 \chi_d \left(1 - \omega^2 \tau'^2\right)}{\left(1 - \omega^2 \tau'^2\right)^2 + \omega^2 \tau^2} + i \frac{\varepsilon_0 \chi_d \omega \tau}{\left(1 - \omega^2 \tau'^2\right)^2 + \omega^2 \tau^2} \quad . \tag{2.103}$$

In formulating this equation, the resonance curve associated with a single oscillation process of the molecular skeleton or of the electrons relative to the nucleus is considered; the total curve for all distortional polarisation processes corresponds to the sum of a large number of such contributions. The real part, for each individual contribution, has the form shown in Fig. 2.17, while the imaginary part is a bell-shaped curve with its maximum near the eigenfrequency and a width being given by τ'.

In Sect. 2.1, it was shown for the dynamic compliance that the imaginary part describes the loss process, i.e., the part of the work which is dissipated. The same is true for the polarisation induced by an electric field, and this regardless of whether the orientational or the distortional part is being concerned. The imaginary part of the dielectric susceptibility specifies always to what extent the energy of the electromagnetic field is dissipated when interacting with the material. Thus, $\chi''(\omega)$ represents the absorption spectrum of a sample.

The Local Field – The Clausius–Mosotti Equation. For liquids made up of polar molecules, i.e., molecules with a permanent dipole moment, we find all the different contributions to the polarisation. We consider now the question: What value for the dielectric constant ε would be measured in a capacitor under static conditions? Both the orientational and distortional part are proportional to the electric field strength E_{loc} at the location of the molecule – that which acts there to polarise the molecule – and we express this as

$$P_{\mathrm{or}} = \rho \beta_{\mathrm{or}} E_{\mathrm{loc}} \quad , \qquad (2.104)$$

$$P_{\mathrm{d}} = \rho \beta_{\mathrm{d}} E_{\mathrm{loc}} \quad . \qquad (2.105)$$

In addition to the particle density ρ, the equations contain the coefficients β_{or} and β_{d}. They are termed **polarisabilities**, and describe in an empirical way the reaction of an individual molecule to the electric field. The total polarisation is given by

$$P = \rho \beta E_{\mathrm{loc}} \quad , \qquad (2.106)$$

where the polarisability β contains now the contributions from all mechanisms.

At first, it could be thought that the electric field in the capacitor, which is given by the voltage and the plate separation, acts directly on the molecule and is identical to E_{loc}. This is not, however, the case. Figure 2.19 shows the way one must proceed in order to determine the local electric field. Think about the selection of a particular 'on-molecule', placed in the centre of a sphere of mesoscopic size with a radius a. This opens up the possibility of explicitly considering the fields arising from the neighbouring molecules, which can be separated from the continuous charge distribution in the dielectric and on the capacitor plates. The figure illustrates that the local field can then be split up into four contributions:

$$E_{\mathrm{loc}} = E_0(\sigma_{\mathrm{P}}) + E_1(\sigma_{\mathrm{M}} = P) + E_{2z}(\sigma_{\mathrm{L}}) + E_{3z} \quad . \qquad (2.107)$$

All fields are oriented in the same direction, which is here chosen to be z. E_0 is the homogeneous field which is generated by the charges on the capacitor plates, with the surface density being σ_{P}. A second homogeneous field, E_1, is due to the surface charges of the dielectric (charge density σ_{M}) directly at the plates. The two fields together, E_0 and E_1, which act against each other,

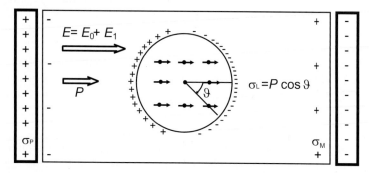

Fig. 2.19. The sources of the local electric field at the location of a molecule inside a block of material in a capacitor.

determine the average field strength E within the capacitor. As will become apparent from the following discussion, it is the contribution E_{2z} from the charges at the surface of the sphere which is decisive for the local field. This surface charge density, σ_L, is proportional to the polarisation and depends on the angle ϑ according to

$$\sigma_L = P \cos \vartheta \ . \tag{2.108}$$

The fourth and final contribution to consider comes from the fields arising from the molecular dipoles inside the sphere; these add up to give E_{3z}.

E_{2z} is referred to as the **Lorentz field** and can be calculated in the following way. An amount of charge $\mathrm{d}Q$ on the surface of the sphere gives rise to a contribution to the Lorentz field of magnitude

$$\mathrm{d}E_2 = \frac{1}{4\pi\varepsilon_0} \frac{\mathrm{d}Q}{a^2} \ . \tag{2.109}$$

Only the component in the z direction has an effect, and its magnitude is given by

$$\mathrm{d}E_{2z} = \cos \vartheta \frac{1}{4\pi\varepsilon_0} \frac{\mathrm{d}Q}{a^2} \ . \tag{2.110}$$

Using the spherical coordinates ϑ and φ, the integration over the charge distribution on the surface of the sphere can be directly carried out, and gives

$$E_2 = \frac{1}{4\pi\varepsilon_0} \int_\varphi \int_\vartheta \cos \vartheta \frac{\mathrm{d}Q}{a^2} = \frac{1}{4\pi\varepsilon_0} \int_\varphi \int_\vartheta \frac{\cos \vartheta}{a^2} P \cos \vartheta \ a^2 \sin \vartheta \mathrm{d}\vartheta \mathrm{d}\varphi$$

$$= \frac{2\pi P}{4\pi\varepsilon_0} \int_\vartheta \cos^2 \vartheta \sin \vartheta \mathrm{d}\vartheta = \frac{1}{3\varepsilon_0} P \ . \tag{2.111}$$

This result shows that the Lorentz field E_2 is, first, oriented in the same direction as the external field E and, second, proportional to the polarisation.

It can easily be shown that the field E_{3z} due to the molecular dipoles within the sphere disappears in the case of isotropic liquids and also symmetric crystals, and this applies to the figure. If the molecule is located at the position $\boldsymbol{r}_j$ inside the region cut out by the sphere and carries a dipole moment $\boldsymbol{p}$, the strength of the field $\boldsymbol{E}_3$ at the centre of the sphere is given by

$$\boldsymbol{E}_3 = \frac{1}{4\pi\varepsilon_0} \left\langle \sum_j \frac{3(\boldsymbol{p} \cdot \boldsymbol{r}_j)\boldsymbol{r}_j - r_j^2 \boldsymbol{p}}{r_j^5} \right\rangle \ . \tag{2.112}$$

For a liquid, note that it is necessary to carry out the shown averaging. On the basis of symmetry, it follows that the two components perpendicular to $\boldsymbol{E}$ are zero

$$E_{3x} = E_{3y} = 0 \tag{2.113}$$

and, thus, the z-component is given by

$$E_{3z} \propto \left\langle \sum_j \frac{3z_j^2 - r_j^2}{r_j^5} \right\rangle \quad . \tag{2.114}$$

Upon expanding the numerator, it follows that

$$E_{3z} \propto \left\langle \sum_j \frac{3z_j^2 - x_j^2 - y_j^2 - z_j^2}{r_j^5} \right\rangle = 0 \quad , \tag{2.115}$$

since the following is true from symmetry arguments:

$$\left\langle \sum_j \frac{x_j^2}{r_j^5} \right\rangle = \left\langle \sum_j \frac{y_j^2}{r_j^5} \right\rangle = \left\langle \sum_j \frac{z_j^2}{r_j^5} \right\rangle \quad . \tag{2.116}$$

Putting all the different contributions together yields the sought-after result: The local field which acts upon an individual molecule is a sum of the capacitor field E and the Lorentz field:

$$E_{\text{loc}} = E + \frac{1}{3\varepsilon_0} P \quad . \tag{2.117}$$

The polarisation can now be calculated by substituting Eq. (2.117) into Eq. (2.106):

$$P = \rho\beta E_{\text{loc}} = \rho\beta \left(E + \frac{1}{3\varepsilon_0} P \right) \quad . \tag{2.118}$$

P appears on both sides of this equation, which indicates that a feedback mechanism is acting. The following expression is obtained for the dielectric constant

$$\frac{P}{E} = \frac{\rho\beta}{1 - \rho\beta/3\varepsilon_0} = \varepsilon_0(\varepsilon - 1) \quad . \tag{2.119}$$

Solving for the polarisability gives

$$\frac{1}{3\varepsilon_0} \rho\beta = \frac{\varepsilon - 1}{\varepsilon + 2} \quad . \tag{2.120}$$

The above discussion has been restricted to monomolecular liquids; in the general case, different types of molecules with densities ρ_j and polarisabilities β_j can exist, and Eq. (2.120) then becomes

$$\frac{1}{3\varepsilon_0} \sum_j \rho_j \beta_j = \frac{\varepsilon - 1}{\varepsilon + 2} \quad . \tag{2.121}$$

This is referred to as the **Clausius–Mosotti equation**.

Orientational Polarisability. The orientation of the dipoles in a polar liquid in a static electric field is incomplete because of thermal rotational motion. The higher the temperature, the lower the observed degree of orientation. The form of this dependence can be directly calculated. First, it is known that the potential energy u of a molecule with a permanent dipole p_0 in an external field of strength E_{loc} depends on the included angle ϑ according to

$$u = -p_0 E_{\mathrm{loc}} \cos \vartheta \quad . \tag{2.122}$$

The orientational distribution function follows from Boltzmann statistics, as

$$w \sin \vartheta \mathrm{d}\varphi \mathrm{d}\vartheta \propto \exp -\frac{u}{k_{\mathrm{B}} T} \sin \vartheta \mathrm{d}\varphi \mathrm{d}\vartheta = \exp \frac{p_0 E_{\mathrm{loc}} \cos \vartheta}{k_{\mathrm{B}} T} \sin \vartheta \mathrm{d}\varphi \mathrm{d}\vartheta \tag{2.123}$$

($w \sin \vartheta \mathrm{d}\varphi \mathrm{d}\vartheta$ is the fraction of dipoles in the angular interval $\mathrm{d}\varphi \mathrm{d}\vartheta$). From this, we obtain the orientational part of the polarisation P_{or} as

$$P_{\mathrm{or}} = \rho p_0 \langle \cos \vartheta \rangle = \rho p_0 \frac{1}{\mathcal{Z}} \int_0^\pi \cos \vartheta \exp \frac{p_0 E_{\mathrm{loc}} \cos \vartheta}{k_{\mathrm{B}} T} 2\pi \sin \vartheta \mathrm{d}\vartheta \tag{2.124}$$

(as always, ρ denotes the particle density). The partition function $\mathcal{Z}$ is introduced for normalisation purposes; it is a function of the variable

$$x = \frac{p_0 E_{\mathrm{loc}}}{k_{\mathrm{B}} T} \quad , \tag{2.125}$$

and can be calculated:

$$\mathcal{Z}(x) = \int_0^\pi \exp(x \cos \vartheta) 2\pi \sin \vartheta \mathrm{d}\vartheta$$

$$= \frac{2\pi}{x} [\exp x - \exp(-x)] \quad . \tag{2.126}$$

It can be recognised that P_{or} is given by

$$P_{\mathrm{or}} = \rho p_0 \frac{1}{\mathcal{Z}} \frac{\mathrm{d}\mathcal{Z}}{\mathrm{d}x} \quad . \tag{2.127}$$

To obtain the orientational part of the polarisability, β_{or}, expand $\mathcal{Z}$ as a power series and keep only the lowest order terms

$$\mathcal{Z} = 4\pi + \frac{2\pi x^2}{3} + \dots \quad . \tag{2.128}$$

This leads to

$$P_{\mathrm{or}} \approx \rho p_0 \frac{x}{3} = \rho \frac{p_0^2}{3 k_{\mathrm{B}} T} E_{\mathrm{loc}} \tag{2.129}$$

and, thus, to the following expression for β_{or}

$$\beta_{\text{or}} = \frac{p_0^2}{3k_{\text{B}}T} \ .$$

(2.130)

The total polarisability of a polar liquid is a sum of the distortional part β_{d} and β_{or}:

$$\beta = \beta_{\text{d}} + \frac{p_0^2}{3k_{\text{B}}T} \ .$$

(2.131)

This result shows that temperature dependent measurements using the Clausius–Mosotti equation (2.121) provide a means of determining β_{d} and p_0.

2.2.3 The Piezo Effect

The subject of the first section of this chapter was the deformation which results from the application of a mechanical field. The second section then discussed the polarisation which arises due to an electric field. There are some crystals, where cross correlations are observed in the sense that, first, mechanical stresses cause not only deformations but also generate polarisation, and, second, electric fields do not simply polarise the sample but also cause deformations at the same time. This is referred to as the **Piezo effect**. The Piezo effect is of technological importance and has been broadly applied. The most important material in this respect is quartz. Small single crystals of quartz serve as ultra-sound sources, stabilisers in oscillating circuits – in this application, they are found in almost all watches, and transmitting and receiving devices – or as positioning elements in situations where a precision in the nm or Å range is required. The construction of an atomic force microscope, in which quartz crystals are used to shift the cantilever and to make it oscillate, would be unthinkable without the piezoelectric properties of quartz. In addition to quartz, poly(vinylidenefluoride) is becoming increasingly more important. If a foil made from this polymer is stretched at high temperatures, where the chains are sufficiently mobile, in an electric field, and then cooled rapidly to room temperature, the electric dipoles carried by the CF_2 groups retain a permanent preferred direction. The so-achieved permanently polarised foil shows a piezoelectric effect which surpasses that of quartz. In order to use it, thin-layer electrodes are placed on both sides of the foil. Upon applying a voltage, changes in the thickness and also the length and width are achieved. As an example, this allows the construction of spherical loud speakers.

In order to describe how piezoelectric materials function, it is necessary to extend Eqs. (2.87) and (2.4). Equation (2.87) is replaced by

$$P = \varepsilon_0 \chi E + d_1 \sigma_{zz}$$

(2.132)

while the following applies instead of Eq. (2.4):

$$e_{zz} = d_2 E + D\sigma_{zz} \ .$$

(2.133)

The strength of the piezo effect depends on the coefficients d_1 and d_2.

Which crystals exhibit the piezo effect? The answer is that it is simply a question of crystal symmetry. Stated more exactly, there are 20 out of 32 crystal classes for which the piezo effect is observable. They have in common that all the associated point groups do not contain the 'inversion' symmetry element. The reason why this is a pre-requisite for piezo behaviour can be immediately appreciated: If a crystal has inversion symmetry, then it is of course true that this symmetry remains under the application of pressure. In this case, no polarisation can arise since this is characterised by a unique direction.

Figure 2.20 shows an example which provides immediate insight. It involves a crystal, which is made up of planar groups having a three-legged appearance. In the middle of each group, there is an atom A, which is symmetrically surrounded by three B atoms. A and B are meant to represent ions with opposite charges. The crystal has a three-fold rotation-reflection axis and a two-fold rotation axis, but no centre of inversion. On the right-hand side, it is shown how such a crystal reacts to a uniaxial pressure. There is a charge displacement for each of the AB_3 ion groups and a polarisation, thus, arises. Alternately, the deformation of the ion groups can be achieved by a field, and this leads to a macroscopic change in the length. It is clear that piezo effects are only observed in this system when a load is applied to the crystal in the shown direction. Placing a load perpendicular to the plane of the ion groups would have no effect.

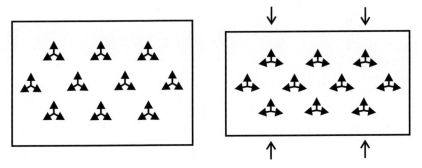

Fig. 2.20. An example of a crystal with piezoelectric properties. The crystal consists of groups of ions, whereby in each ion a triply charged atom is symmetrically surrounded by three singly charged atoms. Under uniaxial pressure, the symmetry is lost and a polarisation arises.

2.3 Magnetic Fields

Generally, magnetic fields arise from electric currents. This can involve both the macroscopic freely flowing currents in conductors as well as currents which flow inside atoms or molecules and thus remain limited to this microscopic region. Normally, these atomic currents do not lead to any external effect, since they compensate each other inside the sample. This situation changes upon applying a magnetic field, which can be achieved for example by placing a cylinder of material in a coil through which current flows. The atomic currents react to this and the sample is **magnetised**.

2.3.1 Magnetic Susceptibility

Electrodynamics distinguishes between the external magnetic field, $\boldsymbol{H}$, the magnetisation $\boldsymbol{M}$, which describes the field-induced magnetic moment per unit volume, and the resulting total magnetic field, given by the vector $\boldsymbol{B}$. $\boldsymbol{B}$ is the sum of $\boldsymbol{H}$ and $\boldsymbol{M}$:

$$\boldsymbol{B} = \mu_0(\boldsymbol{H} + \boldsymbol{M}) \tag{2.134}$$

(μ_0 denotes the magnetic constant). In order to describe the reaction of liquid or solid matter, either the relationship between $\boldsymbol{B}$ and $\boldsymbol{H}$

$$\boldsymbol{B} = \mu_0\mu\boldsymbol{H} \quad, \tag{2.135}$$

which is fixed by the magnetic permeability μ, or that between $\boldsymbol{M}$ and $\boldsymbol{H}$ which alternatively employs the magnetic susceptibility κ

$$\boldsymbol{M} = (\mu - 1)\boldsymbol{H} = \kappa\boldsymbol{H} \tag{2.136}$$

can be used. In many cases, it is true that

$$M \ll B \quad.$$

It, thus, follows that

$$\kappa = \frac{\mathrm{d}M}{\mathrm{d}H} \tag{2.137}$$

can usually be replaced by

$$\kappa \approx \mu_0 \frac{\mathrm{d}M}{\mathrm{d}B} \quad. \tag{2.138}$$

2.3.2 Dia- and Paramagnetism

In Sect. 2.2, it was shown that electric fields always generate a polarisation in the same direction. The reaction of matter to magnetic fields is variable: There are some materials which amplify the external field ($\mu > 1$, $\kappa > 0$) and others which reduce it ($\mu < 1$, $\kappa < 0$). The usual case is the second behaviour, which is referred to as **diamagnetism**. As will be shown in the following, this underlying mechanism is always present. A field amplification is referred to as **paramagnetism** and is only observed for particular materials, namely those for which the atoms have a permanent magnetic dipole. If this is the case, the effect of the presence of a permanent dipole is dominant and it surpasses the diamagnetism which is nevertheless still present.

Diamagnetism. There is a generally applicable law, referred to as the **Larmor theorem**, which states that the introduction of atoms into a magnetic field B sets the electron shell of the atom into a rotational motion. This motion involves the shell as a whole and occurs with a well-defined rotational frequency, namely the Larmor frequency

$$\omega_{\mathrm{L}} = \frac{eB}{2m_{\mathrm{e}}} \quad . \tag{2.139}$$

When an electron shell rotates, there is a circular flow of electric current, which corresponds to the appearance of a magnetic moment. The Larmor theorem also states that the moment is so aligned that the field is weakened. The process is equivalent to the well-known case of electromagnetic induction. If a closed conducting loop is brought into a magnetic field, or equivalently if a magnetic field is switched on in the presence of a conducting loop, then a current is induced, which is arranged according to Lenz's rule such that the generating field is weakened.

The magnetic moment of an electron shell which is rotating at the Larmor frequency can be calculated. Its value amounts to

$$m = I \cdot A = -Ze\frac{eB}{2m_{\mathrm{e}}2\pi} \cdot \pi \left\langle r_{\perp}^2 \right\rangle \quad . \tag{2.140}$$

This is the general equation for the magnetic moment of an electric current which encompasses an area A and is of strength I. In determining the current strength, it is to be remembered that there are Z electrons in the shell. In order to evaluate A, it is necessary to consider the distance of the electrons from the rotation axis, $r_{\perp}$, and take the average squared value. For an isotropic electron density distribution, which is always found for closed shells, the following is true when the rotation axis is chosen to be in z direction

$$\left\langle r_{\perp}^2 \right\rangle = \left\langle x^2 \right\rangle + \left\langle y^2 \right\rangle \quad . \tag{2.141}$$

Since

$$\left\langle r^2 \right\rangle = \left\langle x^2 \right\rangle + \left\langle y^2 \right\rangle + \left\langle z^2 \right\rangle \tag{2.142}$$

and

$$\left\langle x^2 \right\rangle = \left\langle y^2 \right\rangle = \left\langle z^2 \right\rangle \tag{2.143}$$

it follows that

$$\langle r_\perp^2 \rangle = \frac{2}{3} \langle r^2 \rangle \quad . \tag{2.144}$$

The magnetic moment is therefore

$$m = -\frac{Ze^2}{6m_e} \langle r^2 \rangle B \quad . \tag{2.145}$$

The minus sign in Eqs. (2.140) and (2.145) expresses the weakening character of diamagnetism. The following is then obtained for the diamagnetic susceptibility:

$$\kappa_{\mathrm{dia}} \approx \mu_0 \frac{M}{B} = -\mu_0 \frac{\rho Z e^2}{6m_e} \langle r^2 \rangle \quad . \tag{2.146}$$

The equation shows that κ_{dia} depends on the number of electrons in the atom and the size of the electron orbitals, as well as, of course, the number of atoms per unit volume ρ.

In this form, the equation is only valid for atoms with isotropic electron shells, i.e., for the contribution of closed shells in the atom. Aligned bonding electrons, which are found in molecules, give rise to a different contribution. For molecules, it has been found to be empirically valid to calculate the total diamagnetic susceptibility as an additive sum of the different parts, namely those due to, first, the electrons directly at the atoms ($\kappa_{\mathrm{a}j}$) and, second, the electrons in valence bonds ($\kappa_{\mathrm{b}j}$):

$$\kappa_{\mathrm{dia}} = \sum_j \kappa_{\mathrm{a}j} + \sum_j \kappa_{\mathrm{b}j} \quad . \tag{2.147}$$

Paramagnetism. Paramagnetism is always found for materials made up of atoms which have a non-vanishing total angular momentum J. Such atoms necessarily have a permanent magnetic dipole moment m. The magnetic dipole moment is proportional to the total angular momentum with the constant of proportionality being termed the gyromagnetic ratio γ:

$$m = \gamma J \tag{2.148}$$

where

$$\gamma = -\frac{g\mu_B}{\hbar} \quad . \tag{2.149}$$

μ_B denotes the Bohr magneton

$$\mu_B = \frac{e\hbar}{2m_e} \quad . \tag{2.150}$$

As a rule, atoms or ions in a bonded state, i.e., the state found in molecules or crystals, have closed electron shells, and, therefore, have a zero total angular momentum and, thus, no permanent magnetic moment. Paramagnetism is found in exceptional cases, for example always in the case of free radicals, where their chemical reactivity is a consequence of the presence of a single valence electron. The value of the g-factor, which determines the value of the gyromagnetic ratio, depends on a number of special factors such as the type of coupling between the spin and orbital angular momenta, and, in a crystal, the effective crystal field at the location of the atom. Crystal field effects can cause the orbital angular moment to be fixed such that only the spin can readjust in an external field. If no such influences are acting upon an atom, then the spin and orbital angular momenta are usually linked by the **LS-coupling**. Then, g can be calculated using the Landé equation

$$g = 1 + \frac{J(J+1) + S(S+1) - L(L+1)}{2J(J+1)} \quad . \tag{2.151}$$

L, S and J are the quantum numbers for the total orbital angular momentum of all electrons, their total spin, and the total angular momentum, respectively. For a pure orbital ($\boldsymbol{J} = \boldsymbol{L}$) or a pure spin ($\boldsymbol{J} = \boldsymbol{S}$) angular momentum, g takes the value 1 or 2, respectively. The latter value expresses the anomalous magneto-mechanical behaviour of electron spins. Equation (2.151) is particularly applicable to the case of rare earth ions, which are paramagnetic since they possess an unpaired electron. This is found in an inner shell and, thus, cannot be influenced by the crystal field.

Atomic magnetic moments are only observable if an external magnetic field is applied. The component of the total angular momentum in the selected direction, J_z, is then quantised:

$$J_z = m_J \hbar \quad , \tag{2.152}$$

with

$$m_J : J, J-1, \ldots, -J \quad . \tag{2.153}$$

The interaction with the magnetic field leads to a splitting of the initially $(2J+1)$-fold degenerate levels. The interaction energy depends on the quantum number m_J according to

$$\epsilon = -\boldsymbol{m} \cdot \boldsymbol{B} = g m_J \mu_B B \quad . \tag{2.154}$$

If the presence of a field only affects the spin we have $g = 2$, and the interaction energy is given by

$$\epsilon = 2 m_S \mu_B B \quad , \tag{2.155}$$

where

$$m_S : S, S-1, \ldots, -S \quad . \tag{2.156}$$

The magnetisation which arises upon applying a magnetic field to a paramagnetic material can be calculated. Consider atoms with an angular momentum quantum number J: By applying Boltzmann statistics, the following expression is obtained for the average value $\langle m_z \rangle$ of the magnetic moment in the field direction

$$\langle m_z \rangle = \left(\sum_{m_J = -J}^{J} \exp - \frac{g m_J \mu_B B}{k_B T} \right)^{-1} \sum_{m_J = -J}^{J} -g m_J \mu_B \exp - \frac{g m_J \mu_B B}{k_B T} \; .$$

$$(2.157)$$

The sums can be directly evaluated to give

$$\langle m_z \rangle = g J \mu_B \Phi_J \left(x = \frac{g J \mu_B B}{k_B T} \right) \; , \qquad (2.158)$$

where the 'Brillouin functions' Φ_J depend on both x and J:

$$\Phi_J(x) = \frac{2J+1}{2J} \coth \left(\frac{(2J+1)x}{2J} \right) - \frac{1}{2J} \coth \frac{x}{2J} \; . \qquad (2.159)$$

Normally it is the case that

$$x \ll 1$$

and the $\Phi_J(x)$ can be expressed in the linear approximation

$$\Phi_J(x) \approx \frac{(J+1)x}{3J} \; . \qquad (2.160)$$

This leads to

$$\langle m_z \rangle = g^2 J(J+1) \mu_B^2 \frac{B}{3 k_B T} = \frac{n_B^2 \mu_B^2 B}{3 k_B T} \; . \qquad (2.161)$$

The variable n_B is defined according to

$$n_B^2 = g^2 J(J+1) \qquad (2.162)$$

and corresponds to an **effective magneton number**. The following expression for the magnetic susceptibility of paramagnetic substances is thus obtained:

$$\kappa_{par} \approx \mu_0 \frac{\rho \langle m_z \rangle}{B} = \mu_0 \rho \frac{n_B^2 \mu_B^2}{3 k_B T} \; , \qquad (2.163)$$

where ρ again denotes the number of paramagnetic atoms per unit volume.

In the final part of this section, we mention an interesting application of paramagnetism. The presence of paramagnetic atoms in a sample can be used to achieve further cooling at very low temperatures. The procedure to be adopted is shown in Fig. 2.21. The angular momentum in a paramagnetic sample makes a contribution to the entropy of the system. The left part of the figure shows the change in this contribution with increasing temperature for two different cases. In the presence of an external field, here 0.05 Tesla, the angular momentum is completely aligned up to a temperature of about 15 mK. Only at this point does thermal motion begin to occur; this introduces disorder and the entropy rises correspondingly. If no external field is present, it is naturally found that the disorder is bigger for all temperatures. Only at very low temperatures, below 5 mK, is the dipole–dipole interaction then sufficient to finally achieve a complete ordering. The arrow shows how a lowering

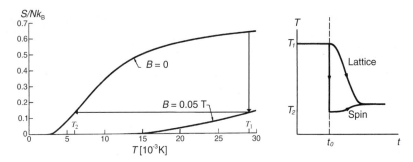

Fig. 2.21. Cooling by means of adiabatic demagnetisation. The influence of a magnetic field on the temperature dependence of the entropy S of a system of N spins (*left*). Temperature exchange between the spin system and the lattice leads to a cooling of the lattice.

of temperature can be achieved in such a system by means of an **adiabatic demagnetisation**. The process begins with the switching on of a magnetic field at an initial temperature of T_1. Switching on the field increases the order in the spin system and correspondingly lowers the entropy. Now the sample is thermally isolated and the field is switched off. At this point, there is a situation in which an external field is no longer acting but the order of the spin system is still unchanged, i.e., it possesses an ordered state characteristic of a much lower temperature T_2. The diagram on the right-hand side shows this and illustrates that a temperature difference has arisen between the spin system of the paramagnetic ions and the other degrees of freedom of the crystal, i.e., all the vibrational degrees of freedom. Over time, this balances itself out and the final temperature for both the lattice and the spin system together lies below the initial value of T_1. An energy exchange between the lattice and the spin system has occurred via a heat flow from the lattice to the spin system such that the crystal has been cooled down.

2.3.3 Magnetic Resonance

By means of magnetic resonance experiments, the paramagnetic properties of liquid or solid materials can be used to gain insight into microscopic structures and dynamic processes. The basic principle of these experiments is that the magnetic moments present in a sample are aligned using a stationary magnetic field and then transitions are excited by irradiating with electromagnetic waves.

In **electron spin resonance** (ESR), the unpaired electrons present in some materials are re-oriented. Equation (2.155) gives the expression for the energy ϵ of a spin, which is carried in the general case by several electrons and which interacts with an external field of strength B_0:

$$\epsilon = 2m_S \mu_B B_0 \ . \tag{2.164}$$

Transitions can only take place between neighbouring states and are, therefore, associated with a change in energy

$$\Delta\epsilon = 2\mu_{\mathrm{B}}B_0 \ . \tag{2.165}$$

To excite this transition, a photon of energy

$$\hbar\omega = \Delta\epsilon \tag{2.166}$$

is required, i.e., its frequency is given by

$$\omega = \frac{e}{m_{\mathrm{e}}}B_0 \ . \tag{2.167}$$

Magnets with field strengths of a few Tesla are usually employed, corresponding to frequencies in the microwave region. A comparison with Eq. (2.139) shows that ω corresponds to the Larmor frequency; the absence of the factor of 2 in the denominator is a consequence of the magneto-mechanical anomaly of the electron spin.

Up until now, no consideration has been taken of the fact that in addition to the electron shell the atomic nucleus can also possess paramagnetic properties. In general, nuclei have a non-vanishing total angular momentum and they, thus, possess a magnetic moment. Bringing a material into an external magnetic field therefore also leads to an interaction of the nuclei with the field. Making the analogy to the electron spin system and taking into account the sign change associated with the positive nuclear charge leads to the following expression for the interaction energy:

$$\epsilon = -g_{\mathrm{n}}m_I\mu_{\mathrm{n}}B_0 \ . \tag{2.168}$$

m_I is the quantum number of the nuclear angular momentum in the field direction, the constant μ_{n} is given by

$$\mu_{\mathrm{n}} = \frac{e\hbar}{2m_{\mathrm{p}}} \ , \tag{2.169}$$

and is referred to as the **nuclear magneton**, and g_{n} is the g-factor of the nucleus. The expression for μ_{n} contains the mass m_{p} of the proton; μ_{n} is, thus, three orders of magnitude smaller than μ_{B}. Transitions in the nuclear spin system require energies given by

$$\Delta\epsilon = g_{\mathrm{n}}\mu_{\mathrm{n}}B_0 \ . \tag{2.170}$$

In **nuclear magnetic resonance (NMR)** experiments, this is achieved using photons of frequency

$$\omega = \frac{g_{\mathrm{n}}e}{2m_{\mathrm{p}}}B_0 \ . \tag{2.171}$$

NMR frequencies lie in the radiowave range.

The magnetisation generated for a system of electron spins in an external field B_0 is given, according to Eqs. (2.162) and (2.163), by

$$M_z = \mu_0 \rho \frac{4S(S+1)\mu_B^2}{3k_B T} \frac{B_0}{\mu_0} = \kappa \frac{B_0}{\mu_0} \quad . \tag{2.172}$$

For a system of nuclear spins with a total angular momentum quantum number I, the corresponding expression is

$$M_z = \mu_0 \rho \frac{g_n^2 I(I+1)\mu_n^2}{3k_B T} \frac{B_0}{\mu_0} = \kappa_n \frac{B_0}{\mu_0} \quad . \tag{2.173}$$

The susceptibility κ_n is six orders of magnitude smaller than the susceptibility of a paramagnetic substance, and can thus be neglected for static experiments. A different situation is found, however, in frequency-dependent experiments. In this case, the excitation of electron spins and atomic nuclei can be easily separated from each other, since the associated frequency ranges are completely different. The separability goes even one step further. The magnetic moments of different nuclei such as hydrogen H, the carbon isotope ^{13}C, and the nitrogen isotope ^{15}N are sufficiently well separated from each other such that it is possible to excite them with different frequencies and hence distinguish between them. Magnetic resonance experiments on these and other nuclei have become today an important and irreplaceable analytical tool with a wide range of applications. NMR spectroscopy uses these nuclei as probes to scan their environment in a well-defined way, and so provide a detailed insight into structural and dynamic properties on a microscopic scale.

Magnetic resonance experiments follow how the spin system, electrons or nuclei, react to irradiation by electromagnetic waves. To describe this, macroscopic equations of motion can be formulated for the time dependence of the magnetisation. In the first instance, let us simply write

$$\frac{1}{\gamma} \frac{d\boldsymbol{M}}{dt} = \boldsymbol{M} \times \boldsymbol{B} \quad . \tag{2.174}$$

The left-hand side is the change with respect to time in the total angular momentum associated with the magnetisation, while the right-hand side describes the total torque exerted by the field $\boldsymbol{B}$. The gyromagnetic ratio for an ESR experiment is given by Eq. (2.149), while that for a NMR experiment is

$$\gamma = \frac{g_n \mu_n}{\hbar} \quad . \tag{2.175}$$

The equation of motion in this initial form is incomplete. For a magnetic resonance experiment, $\boldsymbol{B}$ is a sum of the strong static field and the additional much weaker oscillating field. If the oscillating field is switched off after a certain time, the magnetisation returns to its static equilibrium value after a certain delay. If the static field is applied in the z direction, at equilibrium,

$$M_x = M_y = 0 \tag{2.176}$$

while in the longitudinal direction

$$M_0 = \kappa \frac{B_0}{\mu_0} \tag{2.177}$$

for the electron spin system and

$$M_0 = \kappa_n \frac{B_0}{\mu_0} \tag{2.178}$$

for the nuclear spins. The simplest way of taking account of the requirement that the system returns to equilibrium is to extend Eq. (2.174) in the following way:

$$\frac{dM_x}{dt} = \gamma(\boldsymbol{M} \times \boldsymbol{B})_x - \frac{M_x}{T_2} \tag{2.179}$$

$$\frac{dM_y}{dt} = \gamma(\boldsymbol{M} \times \boldsymbol{B})_y - \frac{M_y}{T_2} \tag{2.180}$$

$$\frac{dM_z}{dt} = \gamma(\boldsymbol{M} \times \boldsymbol{B})_z + \frac{M_0 - M_z}{T_1} \quad . \tag{2.181}$$

These equations of motion for the magnetisation are referred to as the **Bloch equations**. They contain two additional parameters, T_1 and T_2, which are referred to as the **longitudinal** and **transverse relaxation times**, respectively.

If the angular momentum is to be rotated, the correct set up must be used. To achieve this, the electromagnetic wave, in a magnetic resonance experiment, is chosen to propagate perpendicular to the static field, with a polarisation direction which is likewise perpendicular to the static field. From the sum of a static field in the z direction with a strength B_0 and a oscillating field of strength B_1 which rotates in the xy plane there results a total field $\boldsymbol{B}$ with components

$$B_x = B_1 \exp(-\mathrm{i}\omega t) \tag{2.182}$$
$$B_y = -B_1 \mathrm{i} \exp(-\mathrm{i}\omega t)$$
$$B_z = B_0 \quad .$$

In magnetic resonance experiments, it is always true for the amplitudes of both fields that

$$B_1 \ll B_0 \quad .$$

The irradiated linear-polarised electromagnetic field can be broken into two fields with a frequency ω which rotate in opposite directions in the xy plane.

Equation (2.182) contains only that component which rotates in the same sense as the spins. Inserting the field from Eq. (2.182) into the Bloch equations gives

$$\frac{dM_x}{dt} = \gamma M_y B_0 + \gamma M_z B_1 i \exp(-i\omega t) - \frac{M_x}{T_2} \tag{2.183}$$

$$\frac{dM_y}{dt} = -\gamma M_x B_0 + \gamma M_z B_1 \exp(-i\omega t) - \frac{M_y}{T_2} \tag{2.184}$$

$$\frac{dM_z}{dt} = -\gamma M_x i B_1 \exp(-i\omega t) - \gamma M_y B_1 \exp(-i\omega t)$$
$$+ \frac{M_0 - M_z}{T_1} \quad . \tag{2.185}$$

We are looking for a stationary solution and try here

$$M_x = M_\perp \exp(-i\omega t) \tag{2.186}$$

$$M_y = -iM_\perp \exp(-i\omega t) \tag{2.187}$$

$$M_z = M_0 \quad . \tag{2.188}$$

This satisfies Eq. (2.185), and leads to Eqs. (2.183) and (2.184) becoming

$$-i\omega M_\perp = -\gamma B_0 i M_\perp + i\gamma M_0 B_1 - \frac{M_\perp}{T_2} \tag{2.189}$$

$$i\omega i M_\perp = -\gamma B_0 M_\perp + \gamma M_0 B_1 + \frac{iM_\perp}{T_2} \quad . \tag{2.190}$$

The two equations are equivalent, which proves that the trial solution is correct. The following expression relating the amplitude $M_\perp$ of the transverse magnetisation to the amplitude B_1 of the excitation field is then obtained:

$$M_\perp = \frac{\gamma M_0 B_1}{\omega_0 - \omega - i/T_2} = \frac{\gamma M_0 T_2}{T_2(\omega_0 - \omega) - i} B_1 = \frac{\kappa_\perp(\omega)}{\mu_0} B_1 \tag{2.191}$$

with

$$\omega_0 = \gamma B_0 \quad . \tag{2.192}$$

ω_0 is again the Larmor frequency. The result describes the frequency dependence of a susceptibility, namely the susceptibility $\kappa_\perp$ which is measured in the transverse direction under the conditions of a magnetic resonance experiment. $\kappa_\perp$ is complex and can be broken up into its real and imaginary parts:

$$\kappa_\perp(\omega) = \kappa'_\perp + i\kappa''_\perp$$
$$= \frac{T_2^2(\omega_0 - \omega)\gamma M_0}{1 + T_2^2(\omega_0 - \omega)^2} + i\frac{\gamma M_0 T_2}{1 + T_2^2(\omega_0 - \omega)^2} \quad . \tag{2.193}$$

The frequency dependence of these two variables is shown in Fig. 2.22. As always with susceptibilities, the imaginary part describes to what degree the

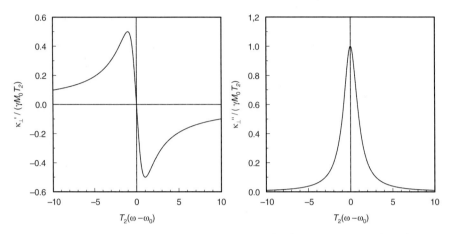

Fig. 2.22. The frequency dependence of the real (*left*) and imaginary (*right*) parts of the susceptibility $\kappa_\perp$ of a system of (electron- or nuclear-) spins upon excitation by electromagnetic waves, which are irradiated perpendicular to the static magnetic field.

energy of the irradiated electromagnetic field is dissipated by the interaction with the system. This absorption signal is a Lorentzian curve with its maximum at the position of the Larmor frequency ω_0 and its width $\Delta\omega$ being approximately the inverse of the transverse relaxation time:

$$\Delta\omega \simeq \frac{1}{T_2} \quad . \tag{2.194}$$

In an ESR experiment $\kappa_\perp''$ is determined. Rather than performing a frequency-dependent experiment, for technical reasons, the Larmor frequency is varied by changing the external magnetic field B_0. This is usually achieved by superimposing a modulation

$$B_0(t) = \overline{B}_0(t) + \Delta B \cos \Omega t \tag{2.195}$$

with a fixed amplitude ΔB onto a continuous rise in $\overline{B}_0$. The signal measured in this way corresponds to

$$\frac{\mathrm{d}}{\mathrm{d}B_0}\kappa_\perp'' \propto \frac{\mathrm{d}}{\mathrm{d}\omega_0}\kappa_\perp'' \quad .$$

Figure 2.23 shows a typical example, namely an ESR spectrum of Mn^{2+} ions, which have a non-vanishing spin and which have been incorporated at low concentration into a crystal of $KMgF_3$. It is immediately apparent that the form of the signal does not correspond to the derivative of the absorption curve in Fig. 2.22: A superimposed fine structure is evident. The experiment is obviously providing more information than that which is to be expected

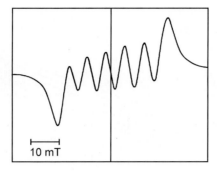

Fig. 2.23. An ESR spectrum, exhibiting fine structure, of Mn^{2+} ions which were incorporated at a concentration of 1% into $KMgF_3$ (from Dormann et al. [16]).

on the basis of the Bloch equations which only contain the parameter T_2. The observed structure is referred to as the **hyperfine structure** of the ESR signal and results from an interaction between the free electron spin and the spin of the ionic nuclei. A quantum-mechanical analysis for the energy levels of an electron spin, which is both exposed to the external field and in interaction with the nuclear spin, yields the following expression:

$$\epsilon(m_S, m_I) = 2m_S\mu_B B_0 + \beta_{\mathrm{hf}} m_S m_I \ . \tag{2.196}$$

The hyperfine structure arises from the second term. The strength of the electron spin-nuclear spin interaction is determined by the coefficient β_{hf}, whose magnitude is given by, for the case of a single free electron $(S = 1/2)$,

$$\beta_{\mathrm{hf}} \propto \mu_n \mu_B |\psi(\boldsymbol{r} = 0)|^2 \ , \tag{2.197}$$

where $\psi(\boldsymbol{r})$ is the electron wavefunction. The equation expresses the fact that the interaction between the electron and nuclear spins via the **Fermi contact energy** is very localised and proportional to the probability of finding an electron at the position of the nucleus.

Figure 2.24 illustrates how the energy levels of the Mn^{2+} ion $(S = 5/2)$ in an external magnetic field B_0 experience an additional splitting due to the interaction with the nucleus $(I = 5/2)$. The transition selection rules are

$$\Delta m_S = 1 \quad , \quad \Delta m_I = 0 \ .$$

These allowed transitions are represented by arrows in the figure.

In principle, NMR experiments could be carried out in the same way. Actually, this is not the case and a different way is chosen. As will be described at the end of this section, it is possible using short strong pulses of electromagnetic radiation to rotate the magnetisation which is initially oriented with the field direction by 90° such that it is in the xy plane. The time-dependent change in the transverse magnetisation can then be observed. The detection

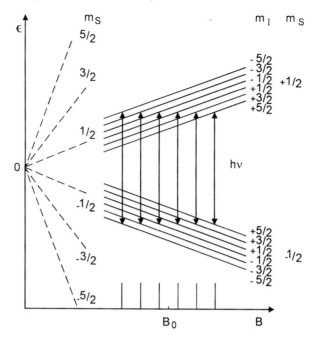

Fig. 2.24. The explanation of the spectrum in Fig. 2.23. The splitting of the ESR resonance due to the interaction between the electron and nuclear spins.

is achieved using an induction coil, and since the signal decays to zero, one talks about a **free induction decay**. This can be described using the Bloch equations. For the initial state, which is described by

$$M_x(t = 0) = M_\perp(0) \ , \tag{2.198}$$
$$M_z(t = 0) = 0 \ , \tag{2.199}$$

the following solution is obtained:

$$M_x = M_\perp(0) \exp\left(-\mathrm{i}\omega_0 t - \frac{t}{T_2}\right) \ , \tag{2.200}$$

$$M_z = M_0\left(1 - \exp-\frac{t}{T_1}\right) \ . \tag{2.201}$$

These equations describe a dying away of the transverse magnetisation within a time T_2 and a recovery of the equilibrium state in the longitudinal direction within a time T_1; this is completely what is to be expected from the meanings of the two relaxation times.

Figure 2.25 shows the result of an experimental measurement of the time dependence of the transverse magnetisation for an organic liquid. As was the case for the ESR experiment, it is evident that the Bloch equations provide

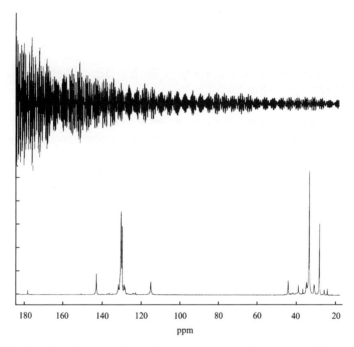

Fig. 2.25. The free decay of the transverse magnetisation measured for an organic liquid after a 90° pulse (*top*). The NMR spectrum obtained from a Fourier transformation of the signal (*bottom*) (from McBriety and Packer [17]).

only a rough description of the time dependence. The decay is only approximately described by an exponential function and there is again clearly a fine structure. The physical basis for this becomes clear when a Fourier analysis is carried out on the free induction decay signal. The result is shown in the lower part of the figure. It is evident that the magnetic moments in the sample do not all precess with the same frequency in the external field. If the nuclei are in different chemical environments, they experience different local magnetic fields. Although these changes in the field are small – they lie in the ppm (*parts per million*) range – they are of great importance.

The value of NMR spectroscopy as an analytical method in chemistry lies in such a spectrum. Concerning the physical origin for the shift of the precession frequency in liquids, there are two mechanisms. First, the external field B_0 is modified by the diamagnetic effect of the electron shell in accordance with the construction of the electronic environment. In addition, there is a second effect referred to as the **spin–spin interaction**. It can be demonstrated that the nuclear spins of atoms which are chemically bonded to each other interact reciprocally by means of the bonding electrons. The corresponding energy –

although it is weak, it is detectable in the ppm range – can be described as follows:

$$\Delta\epsilon_{ss} = \beta_{ss}\boldsymbol{I}_j \cdot \boldsymbol{I}_k \quad . \tag{2.202}$$

The special feature is that the energy splitting $\Delta\epsilon_{ss}$ is independent of the orientation of the bonding vector between the nuclei j and k with respect to the external field.

In this respect, the spin–spin interaction is completely different to the fundamental coupling of the nuclear spins via the magnetic dipole field, which is also present. The dipole–dipole interaction is considerably stronger than the spin–spin interaction, but it, and in particular the sign of the prefactor, depends on the orientation of the internuclear vector with respect to the external field. The average value of the dipole–dipole interaction over all possible orientations is zero. In liquids, the molecules can rearrange themselves very quickly and experience all possible directions within a short time τ_{or}, such that the dipole–dipole interaction is not observed in the NMR experiment. The condition for this is that

$$\tau_{or} \ll \frac{1}{\Delta\omega_{dd}} \quad . \tag{2.203}$$

The variable $\Delta\omega_{dd}$ denotes the change in the precession frequency due to the dipole–dipole interaction, which would be obtained for a rigid isotropic distribution of molecules.

In solids, there is no fast reorientational motion and the dipole–dipole interaction between the nuclei is completely active. Since its magnitude exceeds considerably the chemical shift and the spin–spin interaction, it is not possible, in the first instance, to record a signal like that in Fig. 2.25 for a solid. The dipole–dipole interaction can, however, be removed in this case by means of an experimental method: A sufficiently fast rotation of the sample about an axis inclined at an angle of 54.7° to the direction of the external field causes the averaging out of the dipole–dipole interaction. This **magic angle** is the solution of the equation:

$$3\cos^2\vartheta - 1 = 0 \quad . \tag{2.204}$$

It can be shown that this is the expression for the orientational dependence of the dipole–dipole interaction.

Finally, it will be briefly discussed how a short pulse can cause the magnetisation which is initially aligned with the static magnetic field to be rotated by 90°. Figure 2.26 describes the situation and what happens. The drawings on the left and in the middle show the equilibrium state before the application of the pulse. The nuclear spins precess at the Larmor frequency about the direction of the field, with there being a uniform distribution over all azimuthal angles. As shown in the middle, there is only the longitudinal magnetisation M_0. Radiowaves are now irradiated along the x direction with a polarisation in the y direction. There is a component of the linearly polarised wave which precesses with the same direction of rotation as that of the spins in the

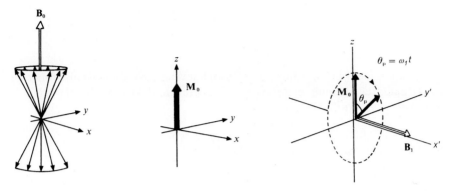

Fig. 2.26. System of spins in a static magnetic field $\mathbf{B}_0$. Precession with the Larmor frequency (*left*) and resulting longitudinal magnetisation (*centre*). Rotation of the spin system by a transverse magnetic field $\mathbf{B}_1$ oscillating with the Larmor frequency. Precession in the rotating coordinate system x', y' (*right*).

xy plane. If the frequency of the irradiated field is chosen to agree with the precession frequency of the spins, the irradiated field is able to cause the spins to rotate. This can be understood by considering a coordinate system which rotates at the Larmor frequency – this is denoted by x' and y' in the drawing on the right-hand side. In this rotating frame, the field with an amplitude B_1 is again responsible for a precessional motion, this time about the direction of the varying field. The change in the angle due to this precessional motion in the rotating coordinate frame is denoted by θ_P and grows proportionally with time. The pulse duration can be chosen to ensure that θ_P equals exactly 90° such that a purely transverse initial state is generated, whose free induction decay can subsequently be detected.

2.4 General Properties of Susceptibilities

The subject of this chapter has been the effects which external fields can cause in matter. It has been seen that linear dependences are observed over a wide range of conditions: Deformations are proportional to the applied tensile or shear stress, polarisations are proportional to the electric field which is acting, while magnetisations are proportional to the magnetic field which exists. We are always dealing with linear responses which can be quantitatively described by means of response functions such as the compliance and susceptibility. For these response functions, there exists a series of generally applicable laws, some of which have already been introduced at the end of Sect. 2.1.3. Equations were derived which link different response functions together. The relationship between the creep compliance $\alpha_c(t)$ and the time-dependent modulus $a(t)$ is given by the integro-differential equation

$$1 = \int_0^t \frac{d\alpha_c}{dt}(t - t')a(t')dt'$$

(Eqs. (2.66) and (2.68)), while the relation between the pulse response function $\alpha_p(t)$ and the dynamic compliance or **mechanical susceptibility** $\alpha(\omega)$ is expressed by (Eq. (2.73)):

$$\alpha(\omega) = \alpha'(\omega) - i\alpha''(\omega) = \int_0^\infty \alpha_p(t'') \exp(i\omega t'')dt'' \quad .$$

An important result was that the real and imaginary parts of $\alpha(\omega)$ have a well-defined physical meaning. The real part expresses for a periodic load the part of the external power which is reversibly introduced and removed again, while the imaginary part describes the loss, i.e., the part of the power which is dissipated and lost as heat. The proof of this was given for the dynamic-mechanical experiment; the statement, though, is, as already mentioned, valid for all discussed susceptibilities. In the mechanical case, the work per unit volume is given by

$$dw = \sigma_{zz}de_{zz} \quad . \tag{2.205}$$

In the dielectric case, the following is valid

$$dw = EdP \tag{2.206}$$

while we have the following for the application of a magnetic field:

$$dw = \mu_0 HdM \quad . \tag{2.207}$$

For variables which are related to each other by means of a susceptibility, one is dealing with a pair of energy-conjugated variables (see Eq. (A.4) in Appendix A). Therefore, it is always true that the imaginary part describes the energy dissipation, i.e., the transfer of work or field energy into disordered thermal motion.

A further one will now be added to these previously stated general laws. It may sound amazing at first but the real and imaginary parts of a susceptibility are actually not independent, but rather they are coupled with each other. This may be surprising since both, as described above, have completely different meanings. The relationship is expressed by means of the **Kramers–Kronig relations**.

The Kramers–Kronig Relations. In order to describe the relations, fundamental laws from the mathematical theory of complex functions, in particular the theorem of residues, are used. Consider as a starting point Eq. (2.73): It gives the frequency dependence of the susceptibility on the basis of pulse response functions. If the expression is considered purely mathematically, it can be generalised and extended to be applicable to complex frequencies

$$z = \omega' + i\omega'' \quad . \tag{2.208}$$

Inserting z in the place of the previous real variable ω leads to

$$\alpha(z) = \int\limits_0^\infty \alpha_p(t'') \exp(i\omega't'') \exp(-\omega''t'')dt'' \quad . \tag{2.209}$$

The result of the generalisation is a complex-valued function over the plane of the complex frequency z. Which properties does this function possess and in particular, where is it analytic? One can be immediately recognised: $\alpha(z)$ is analytic for the whole upper half of the complex plane, i.e., for $\omega'' \geq 0$. Here, only multiplication by an exponentially decaying function is added to the original integral. To be analytic in the upper half of the complex plane means that $\alpha(z)$ has no singularities. Singularities definitely exist, but they are all in the lower half. It is instructive here to consider two examples. The susceptibility associated with a simple relaxation process was given in Eq. (2.80). Extending the applicability to complex frequencies leads to

$$\alpha(z) = \frac{\alpha(0)}{1 - iz\tau} \quad . \tag{2.210}$$

This function has a singularity at

$$z = -\frac{i}{\tau} \tag{2.211}$$

which lies in the lower half of the complex plane. The same is found for the generalized susceptibility of a damped harmonic oscillator. Starting from Eq. (2.101), it is given by

$$\alpha(z) = \frac{\alpha(0)}{1 - z^2\tau'^2 - iz\tau} \quad . \tag{2.212}$$

$\alpha(z)$ now has two singularities at

$$z_{1/2} = \pm \left(\frac{1}{\tau'^2} - \frac{\tau^2}{4\tau'^4} \right)^{1/2} - i\frac{\tau}{2\tau'^2} \quad , \tag{2.213}$$

which also lie in the lower half of the z-plane.

In the first instance, the extension of the range of definition of the susceptibility has been treated as a purely mathematical procedure. There is, in fact, an important link to physics, which becomes clear upon considering the positions of the singularities in the lower half of the z-plane. It is only necessary to examine which time dependencies result from inserting the specified z value in place of a real frequency. The singularity for a simple relaxation process gives

$$X(t) \propto \exp(-izt) = \exp -\frac{t}{\tau} \ , \tag{2.214}$$

while the following results for a damped harmonic oscillator

$$X(t) \propto \exp \left[\pm i \left(\frac{1}{\tau'^2} - \frac{\tau^2}{4\tau'^4} \right)^{1/2} t - \frac{\tau}{2\tau'^2} t \right] \ . \tag{2.215}$$

These are, however, exactly the time dependences of the eigenmodes of the systems which become visible during the subsequent decay process after switching off an external force. The singularities in $\alpha(z)$ occur at the positions of the system eigenmodes. It is also understandable, for it is exactly the eigenmodes which can exist without excitation by an external force. An infinitely large susceptibility expresses this. Why can no singularity appear in the lower half of the complex plane? The answer is that the associated modes would be described by processes with an exponentially increasing amplitude, and these cannot exist for stable linear systems.

In order to derive the Kramers–Kronig relation, consider now not $\alpha(z)$ but rather the function

$$\frac{\alpha(z)}{z - \omega_0} \ . \tag{2.216}$$

It is also analytic in the whole upper half of the plane including the ω' axis with one exception: There is a singularity at the location $z = \omega_0$ on the ω' axis. Consider the calculation of a special integral over a closed path, namely that shown in Fig. 2.27. According to a basic theorem valid for complex functions the integral must vanish since no singularity is included. The integral can be expressed as a sum of four pieces, namely two straight lines and two semicircles:

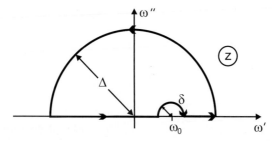

Fig. 2.27. The integration route in the plane of the complex frequency $z = \omega' + i\omega''$, as selected in the derivation of the Kramers–Kronig relations.

$$0 = \int_{-\infty}^{\omega_0 - \delta} \frac{\alpha(z)}{z - \omega_0} dz + \int_{\pi}^{2\pi} \alpha(\omega_0 + \delta \exp i\vartheta) i d\vartheta \qquad (2.217)$$

$$+ \int_{\omega_0 + \delta}^{+\infty} \frac{\alpha(z)}{z - \omega_0} dz + \lim_{\Delta \to \infty} \int_{0}^{\pi} \alpha(\omega_0 + \Delta \exp i\vartheta) i d\vartheta \quad . \qquad (2.218)$$

The large semicircle with radius Δ should run into infinity. The first term runs from infinity to a distance δ from the singularity at ω_0, the second term corresponds to going round the singularity via the small semi-circle with radius δ, the third term goes from there to infinity, while the fourth term gives the contribution of the large semi-circle with the radius $\Delta \to \infty$. The following was used for the small circle:

$$z = \omega_0 + \delta \exp i\vartheta \quad , \qquad (2.219)$$

where ϑ is a variable angle. It, thus, follows that

$$dz = i\delta \exp i\vartheta d\vartheta \qquad (2.220)$$

and therefore

$$\frac{dz}{z - \omega_0} = i d\delta \quad . \qquad (2.221)$$

This expression is equally valid for the large semi-circle and was used in the second and fourth terms. The theorem of residues can now be applied to determine the value of the integral which goes round the singularity ω_0:

$$\lim_{\delta \to 0} \int_{\pi}^{2\pi} \alpha(\omega_0 + \delta \exp i\vartheta) i d\vartheta = -i\pi\alpha(\omega_0) \quad . \qquad (2.222)$$

The fourth term makes no contribution since the susceptibility disappears at infinity in the lower half of the space. The following is thus obtained

$$0 = -i\pi\alpha(\omega_0) + \lim_{\delta \to 0} \left[\int_{-\infty}^{\omega_0 - \delta} \frac{\alpha(z) dz}{z - \omega_0} + \int_{\omega_0 + \delta}^{\infty} \frac{\alpha(z) dz}{z - \omega_0} \right] \quad . \qquad (2.223)$$

The expression in square brackets, which formulates a coupled simultaneous approach from both sides to the singularity, is referred to as the 'principal value'; it can be written in the following short-hand form:

$$P \int_{-\infty}^{\infty} \frac{\alpha(\omega)}{\omega - \omega_0} d\omega = \lim_{\delta \to 0} \left[\int_{-\infty}^{\omega_0 - \delta} \frac{\alpha(\omega)}{\omega - \omega_0} d\omega + \int_{\omega_0 + \delta}^{\infty} \frac{\alpha(\omega)}{\omega - \omega_0} d\omega \right] \quad . \qquad (2.224)$$

In this way, the result has already been obtained. The breaking up of the susceptibility into real and imaginary parts and the comparison to the real and imaginary parts in Eq. (2.223) leads to the two equations

$$\alpha'(\omega_0) = \frac{1}{\pi} P \int_{-\infty}^{\infty} \frac{\alpha''(\omega)}{\omega - \omega_0} d\omega \quad, \tag{2.225}$$

$$\alpha''(\omega_0) = -\frac{1}{\pi} P \int_{-\infty}^{\infty} \frac{\alpha'(\omega)}{\omega - \omega_0} d\omega \quad. \tag{2.226}$$

These equations link the real and imaginary parts of the susceptibility in a one-to-one way and are referred to as the Kramers–Kronig relations.

A consideration of the susceptibilities which are measured in dynamic-mechanical experiments on viscoelastic materials such as polymers reveals that the Kramers–Kronig relations in the above form require the addition of a small supplementary term. The contribution of the plastic component, which is always found for polymers, hasn't yet been included. The eigenmode of a plastic body is at $z = 0$. This can be recognised from the fact that the above-stated singularity associated with a simple relaxation process approaches $z = 0$ for $\tau \to \infty$, i.e., upon the transition into a plastic body. Alternately, it can be seen by solving the equation of motion for a plastic body

$$\frac{dX}{dt} \propto \xi \quad. \tag{2.227}$$

Inserting

$$\xi(t) = \xi_0 \exp(-i\omega t) \tag{2.228}$$

and

$$X(t) = X_0 \exp(-i\omega t) \tag{2.229}$$

leads to the following expression for the susceptibility: $\alpha(\omega)$

$$\frac{X_0}{\xi_0} = \alpha(\omega) \propto \frac{i}{\omega} \quad, \tag{2.230}$$

and the corresponding expression for $\alpha(z)$

$$\alpha(z) \propto \frac{i}{z} \quad. \tag{2.231}$$

Such a singularity at the origin was not taken into account in the above calculation of the integral. A later consideration of such processes is possible. As is implied in Eq. (2.230), plastic bodies only make a contribution to the imaginary part of the susceptibility,

$$\alpha''(\omega) = \frac{A_v}{\omega} \quad, \tag{2.232}$$

since they cannot store energy. The coefficient A_v characterises the viscous forces included in the plastic component. This term must be subtracted before Eqs. (2.225) and (2.226) are written, i.e., the imaginary part $\alpha''(\omega)$ in the equations has to be replaced by

$$\alpha''(\omega) - \frac{A_v}{\omega} \quad . \tag{2.233}$$

Then, instead of Eq. (2.226), the following is obtained

$$\alpha''(\omega_0) = -\frac{1}{\pi} P \int_{-\infty}^{\infty} \frac{\alpha'(\omega)}{\omega - \omega_0} d\omega + \frac{A_v}{\omega} \quad . \tag{2.234}$$

What is actually the origin of the Kramers–Kronig relations? The answer is surprisingly simple: They are simply based on the **causality principle**. This is expressed here in terms of a simple mathematical property: The pulse response function $\alpha_p(t)$ only exists for positive times. It seems evident that a pulse can only have an effect after an event, but it is, indeed, an expression of the causality principle. If the pulse response function did not have a vanishing value for negative times, then the analytical behaviour in the upper half of the complex plane would be lost for the function defined in Eq. (2.209), $\alpha(z)$, and with it the validity of the Kramers–Kronig relations.

2.5 Exercises

1. Viscoelastic behaviour can be modelled by combinations of purely elastic, spring-like elements (e.g., described by the shear modulus $G = \sigma_{zx}/\tan\gamma$) and purely viscous damped elements (described by the viscosity $\eta = \sigma_{zx}/(d\tan\gamma/dt)$). The connection in series of an elastic and a viscous element – this is how shear deformations add together – is termed a Maxwell element. A Kelvin element refers to the connection of an elastic and a viscous element in parallel – this is how stresses, by contrast, add up together. Describe the time dependencies of the following reactions:

    (a) a Maxwell and a Kelvin element in a creep experiment (the shear stress changes from 0 at time $t = 0$ to a constant value σ_0),
    (b) a Maxwell element in a stress relaxation experiment (the shear deformation changes from 0 at time $t = 0$ to a constant value γ_0),
    (c) a Maxwell and a Kelvin element in a dynamic experiment with an oscillating stress $\sigma_{zx}(t) = \sigma_0 \exp(-i\omega t)$,

    for small perturbations $\tan\gamma \approx \gamma$.

2. A cone-plate rheometer consists of a cone, whose top point touches a plate. The purely viscous liquid with viscosity η, which is to be investigated, is found in the space between the cone and the plate. The cone can rotate relative to the plate about the perpendicular axis (see Fig. 2.28).

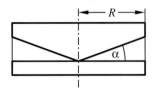

Fig. 2.28. The construction of a cone-plate rheometer.

(a) How large must the externally applied torque be such that the cone rotates at a constant angular velocity ω?

(b) The cone rotates in an oscillatory fashion $\varphi(t) = \varphi_0 \exp(-i\omega t)$. Calculate the required time-dependent torque under neglect of the rotational inertia of the cone.

3. The dielectric susceptibility in the case of a Debye relaxation is given by

$$\chi(\omega) = \frac{\chi_0}{1 - i\omega\tau} \quad .$$

Show that a plot of the imaginary part of $\chi(\omega)$ against the real part gives a semi-circle.

4. Calculate the molar diamagnetic susceptibility of atomic hydrogen. The probability of finding an electron in the ground state is given by $|\psi(r)|^2$ as calculated for the wave function

$$\psi = \left(\pi a_0^3\right)^{-1/2} \exp -\frac{r}{a_0} \quad \text{with} \quad a_0 = \hbar^2/m_e^2 = 0.529 \cdot 10^{-10}\,\text{m} \quad .$$

Show that $\langle r^2 \rangle = 3a_0^2$ is true and use it to calculate

$$\kappa_{\text{dia}} = -\mu_0 N_A \frac{Ze^2}{6m_e} \langle r^2 \rangle \quad (N_A \text{ is the Avogadro number}) \quad .$$

5. A paramagnetic salt contains 10^{22} ions per cm^3 with a magnetic moment of one Bohr magneton.

a) A magnetic field of strength $B = 1$ Tesla is applied at a temperature of 300 K. Calculate the excess part of the moment which is parallel to the field.

b) How large is the magnetisation which results?

6. In solids, the proton resonance is split, because of the dipole–dipole interaction, into a large number of close together neighbouring components. Let us examine as the simplest example the resonance behaviour of a proton pair.

a) For what angle between the inter-proton vector and the magnetic field is the maximum splitting observed?

b) How large then is the splitting $\Delta\nu$ of the resonance frequency for a proton distance of 0.2 nm?

c) At what angle does the splitting disappear?

3

Molecular Fields and Critical Phase Transitions

It was shown in the previous chapter that fields can be amplified because of the reaction which is caused in matter. The polarisation caused by an electric field provides a contribution itself to the field, and the same happens for paramagnetic materials with the magnetic field via the magnetisation. It is usually found that this amplification is weak, i.e., the additional contribution is small in comparison to the external field which causes the reaction. This, however, is not always the case. For certain materials, conditions can exist for which the additional contribution to the field dominates and that, upon a further increase, a situation is reached where the polarisation or the magnetisation is so large that a self-stabilisation is achieved. Exactly this occurs for ferroelectrics and ferromagnets. In such systems, a permanent polarisation or magnetisation is found for temperatures below a critical point, the **Curie temperature**. In most cases – always for ferromagnets and also very frequently for ferroelectrics – the transition into the self-stabilised state occurs in a continuous manner in the form of a **second-order phase transition**; when crossing the Curie temperature on cooling, the magnetisation or polarisation increases from zero. This contrasts with the normally encountered first-order phase transitions, where there are jump-like changes in the structure. Although second-order phase-transitions are less common than first-order ones, they are very important as far as the fundamental understanding of the properties of condensed matter is concerned because of their universal characteristics. The basic situation is always the same: It is always found that there is a certain degree of freedom in the structure generating a **molecular field** which can exert in return a stabilising effect on itself. In this way, order can arise spontaneously. An **order parameter** describes how large the order is. The reorientation of electric and magnetic dipoles represent such degrees of freedom and P and M are the related order parameters.

Two further examples will be considered in this chapter. The structural properties of nematic liquid crystals were discussed in the first chapter. They possess a structural anisotropy in the liquid state. Indeed, the existence of the nematic phase is also based on a self-stabilisation process. The orientation-dependent interaction forces favour a parallel arrangement of the rod-like molecules in a nematogen material and, thus, create a molecular field which can maintain itself. The spontaneous self-organisation begins at the transition from the isotropic liquid into the nematic phase. The order parameter here is the variable S_2 which characterises the degree of orientation.

The fourth example is of a completely different nature and deals with the phase behaviour of binary fluid polymer mixtures. If the composition of the mixture is suitably chosen, a transition from the mixed homogeneous state into a two-phase structure occurs in a continuous fashion. The difference in the composition of the two phases which form on crossing a critical temperature is initially infinitely small and then gradually grows. It is this difference which assumes, in this case, the role of the order parameter. Order is generated through the separation of the homogenous mixture into two different phases.

By means of these examples, we will recognise the common features in second-order phase transitions, or as is also said **critical phase transitions**; the changes in properties near the transition temperature occur in a characteristic manner. Second-order phase transitions are theoretically very well understood. A large proportion of observations can be described using a simple thermodynamic theory which was developed by Landau. It will be described and correspondingly applied.

3.1 The Ferroelectric State

Equation (2.119), which was obtained in the derivation of the Clausius–Mosotti equation, shows immediately that a polarisation of matter can stabilise itself and gives the criterion for this. The dielectric susceptibility diverges for

$$1 = \frac{\rho\beta}{3\varepsilon_0} \ . \tag{3.1}$$

A divergence means that the external field which must be applied in order to generate a definite polarisation becomes ever smaller and finally disappears. The basis of this phenomenon is the Lorentz field which is in the same direction as the external field and always amplifies it. For the condition described by Eq. (3.1), it is exactly strong enough for a self-stabilisation. This follows immediately from the equation

$$P = \rho\beta E_{2z} = \rho\beta\frac{P}{3\varepsilon_0} \ . \tag{3.2}$$

The condition for the transition into the ferroelectric state is thus solely a sufficiently high polarisability or atom density in the solid.

3.1.1 Transition Scenarios

Figure 3.1 shows as an example the structural changes which arise for $BaTiO_3$ – a crystal with the perovskite structure – upon the transition into

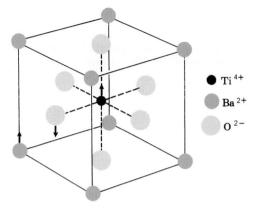

Fig. 3.1. BaTiO$_3$: The atomic displacements associated with the transition into the ferroelectric phase.

the ferroelectric phase. The permanent polarisation P_s arises here because of a relative displacement of the positive and negative parts of the lattice which produces in each individual cell a dipole moment. The transition occurs, upon cooling, at a temperature $T_c = 127\,°C$, as is apparent from Fig. 3.2. It is further apparent that there are two other ferroelectric phases of BaTiO$_3$ which appear at lower temperatures.

A diverging susceptibility means that the internal forces, which arise upon a relative displacement of the two parts of the lattice and keep the electric field forces in equilibrium, become ever weaker upon approaching the Curie temperature. This decrease in the restoring forces can be directly observed.

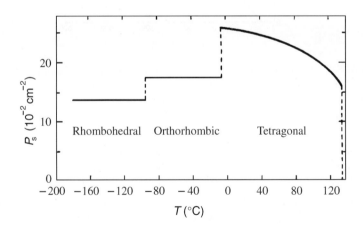

Fig. 3.2. BaTiO$_3$ below the Curie temperature: The spontaneous polarisation P_s in the different ferroelectric phases (from Burns [18]).

A displacement of the two parts of the lattice from each other occurs, even without an external field, solely because of the thermal energy. This excites a vibration whose frequency ν is in the infra-red region, and which can be measured. ν is as always determined by a force constant and a mass:

$$(2\pi\nu)^2 = \frac{a}{m_\mathrm{r}} \ . \tag{3.3}$$

m_r is the reduced mass of the two components of the lattice, calculated per unit cell, and a is the effective force constant, which is likewise given with respect to one unit cell. Figure 3.3 shows, using the example of $SrTiO_3$ – another crystal with the perovskite structure – the temperature dependence of the frequency of this 'transverse-optical' mode (see Sect. 5.1.1). The frequency becomes zero at the Curie temperature and, thus, reflects the disappearance of the restoring force.

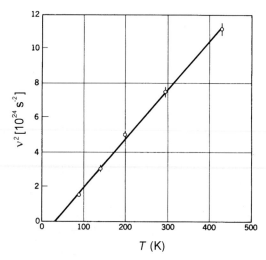

Fig. 3.3. $SrTiO_3$ upon approaching the Curie temperature: Disappearance of the restoring force of the transverse-optical mode (the vibration frequency, ν, was determined by Cowley [19] using inelastic neutron scattering).

It is apparent from the figure that the restoring force changes linearly with the difference in temperature from T_c. Figure 3.4 shows for the example of a third Perovskite-crystal, $LiTaO_3$, that the inverse of the dielectric susceptibility behaves in the same way, i.e., it increases linearly as a function of the difference in temperature from T_c. It is not difficult to describe theoretically this dependence. It is simply necessary to choose the following form for the temperature dependence of the polarisability in the region of the Curie temperature:

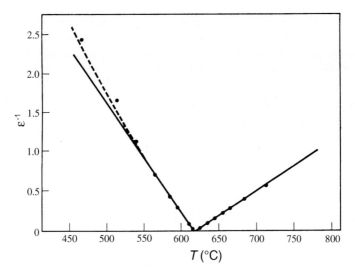

Fig. 3.4. LiTaO$_3$ in the region around the Curie temperature: Temperature dependence of the (inverse) dielectric constant (from Lines [20]).

$$\frac{1}{3\varepsilon_0} \sum_j \rho_j \beta_j = 1 - c(T - T_c) \ . \tag{3.4}$$

This starting equation suggests itself. Above T_c, the expression on the left-hand side becomes less than the critical value of one and this is expressed here in the simplest possible way, by using a linear term. Using now the Clausius–Mosotti equation, we obtain

$$\frac{\varepsilon - 1}{\varepsilon + 2} = \frac{1}{3\varepsilon_0} \sum_j \rho_j \beta_j = 1 - c(T - T_c) \ . \tag{3.5}$$

It, thus, follows that

$$-\frac{3}{\varepsilon + 2} = -c(T - T_c) \tag{3.6}$$

and, therefore,

$$\varepsilon + 2 \propto \frac{1}{T - T_c} \ . \tag{3.7}$$

For ε, $\chi \gg 1$ this becomes

$$\varepsilon \approx \chi \propto \frac{1}{T - T_c} \ . \tag{3.8}$$

This is exactly the behaviour shown in Fig. 3.4 by the dielectric constant for $T > T_c$.

For the first example of BaTiO$_3$, a finite spontaneous polarisation immediately establishes itself at the Curie temperature, which must of course be associated with a jump to some distinct displacement of the two parts of the lattice. The other two examples show that this is not always the case. For SrTiO$_3$ and LiTaO$_3$, the polarisation sets in continuously at the Curie temperature, beginning with $P_s = 0$.

3.1.2 The Landau Theory of Critical and Nearly-Critical Phase Transitions

Figure 3.4 shows that a divergence of the dielectric susceptibility upon approaching the Curie temperature is also observed for the low temperature side, i.e., in the ferroelectric phase. Landau presented a way by which this behaviour and also other properties of ferroelectrics can be understood using a simple thermodynamic theory. The polarisation is the variable which controls the phase transition; P disappears above T_c, while below T_c, P begins to increase. The value of P represents, for every temperature, an equilibrium value in the sense of thermodynamics and the equilibrium condition can be directly formulated. The requirement is a knowledge of the Helmholtz free energy as a function of the polarisation and the temperature, i.e., of the expression

$$f(P, T) \ .$$

The equilibrium value of the polarisation at a particular temperature then follows from the minimum condition

$$\left. \frac{\partial f}{\partial P} \right|_{eq} = 0 \ . \tag{3.9}$$

Landau suggested that the Helmholtz free energy density in the region of the Curie temperature be expressed as a series expansion:

$$f = f_0 + \sum_j c_j(T) P^j \ . \tag{3.10}$$

Since the following must of course be valid

$$f(P) = f(-P) \ , \tag{3.11}$$

only even-order terms appear in the expansion:

$$f = f_0 + c_2 P^2 + c_4 P^4 + c_6 P^6 + \dots \ . \tag{3.12}$$

In the discussion of the transverse-optical mode, the force constant a was introduced to describe the restoring forces, which become active upon a relative shift of the two parts of the lattice. This force constant determines the value of the second-order term in the series expansion for the Helmholtz free energy density:

$$\frac{\partial^2 f}{\partial P^2}(P = 0) = 2c_2 \propto a \ . \tag{3.13}$$

Since a becomes zero at the Curie temperature and then increases linearly above it the following can be written:

$$c_2 = b(T - T_c) \quad , \quad \text{with} \quad b > 0 \ . \tag{3.14}$$

Limiting the expansion to the next term then leads to the following expression for the Helmholtz free energy density

$$f = f_0 + b(T - T_c)P^2 + c_4 P^4 \ . \tag{3.15}$$

Figure 3.5 shows this dependence for a fixed, positive coefficient c_4 for three different temperatures, namely one above T_c, one at the Curie temperature and one below T_c. The consequences are immediately recognisable: Above the Curie temperature, the equilibrium is, as observed, at $P = 0$, while below T_c, a non-vanishing value for P establishes itself – this is likewise in agreement with what is observed. The latter corresponds to the permanent polarisation P_s.

Evaluating the expression yields the temperature dependence of the polarisation below T_c. The minimum condition for the Helmholtz free energy leads to:

$$\frac{\partial f}{\partial P} = 0 = 2b(T - T_c)P + 4c_4 P^3 \ . \tag{3.16}$$

It follows that the equilibrium value of the permanent polarisation is given by

$$P_{\text{eq}}^2 = P_s^2 = \frac{b(T_c - T)}{2c_4} \quad , \tag{3.17}$$

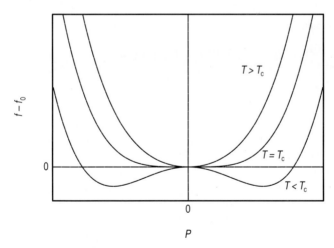

Fig. 3.5. The Landau expansion for the Helmholtz free energy density at the transition into the ferroelectric phase ($T = T_c$) as well as for temperatures above and below T_c. The case of a second-order transition.

i.e.,

$$|P_s| \propto (T_c - T)^{1/2} \quad . \tag{3.18}$$

Hence, the polarisation which arises obeys a square root law.

For the calculation of the dielectric susceptibility, equilibrium states are sought which arise in the presence of a given electric field. These are also determined by a minimum condition, but this time with respect to the Gibbs free energy. We choose the reduced form $\hat{g}$, which is given by Eq. (A.11) in Appendix A:

$$\hat{g} = f - EP = f_0 + b(T - T_c)P^2 + c_4 P^4 - EP \quad . \tag{3.19}$$

The minimum condition

$$\frac{\partial \hat{g}}{\partial P} = 0 = -E + 2b(T - T_c)P + 4c_4 P^3 \tag{3.20}$$

leads to

$$\frac{1}{\varepsilon_0 \chi} = \frac{dE}{dP}(E = 0) = 2b(T - T_c) + 12c_4 P_{eq}^2 \quad , \tag{3.21}$$

where P_{eq} denotes the equilibrium polarisation in the absence of an electric field. Two cases can now be distinguished. For $T > T_c$, there is $P_{eq} = 0$, which leads to

$$\varepsilon_0 \chi = \frac{1}{2b(T - T_c)} \quad . \tag{3.22}$$

This result was already obtained, via a different route, in Eq. (3.8). The Landau expansion yields now, though, an additional expression for the dielectric susceptibility in the ferroelectric phase, i.e., for $T < T_c$. Using the value for P_{eq} given by Eq. (3.17) gives

$$\frac{1}{\varepsilon_0 \chi} = 2b(T - T_c) + 12\frac{b(T_c - T)}{2} \tag{3.23}$$

and thus

$$\varepsilon_0 \chi = \frac{1}{4b(T_c - T)} \quad . \tag{3.24}$$

It can be seen that the expressions for the dielectric susceptibilities above and below the Curie temperature differ by a factor of 2. This agrees with the experiment as is apparent from a look at Fig. 3.4.

The square root law for the emergence of a polarisation upon going below the Curie temperature, as predicted by Landau's treatment, describes second-order phase transitions. The law is obviously not applicable to the transition of $BaTiO_3$ into the ferroelectric phase, where a finite polarisation is immediately established. However, a change of the parameters in the power series expansion for the Helmholtz free energy also allows this transition to be described in a qualitatively correct fashion. This is achieved when the following form is chosen:

$$f - f_0 = b(T - T^*)P^2 + c_4 P^4 + c_6 P^6 \quad . \tag{3.25}$$

T^* is a temperature which lies somewhat below the temperature T_c of the phase transition. If, in addition,

$$c_4 < 0 \qquad \text{and} \qquad c_6 > 0$$

are selected, plots of the Helmholtz free energy as a function of P have the forms shown in Fig. 3.6. The consequences here are also immediately recognisable. For $T > T_c$, the equilibrium value of the polarisation is zero, while for $T < T_c$, the polarisation has a finite value, which is determined by the minimum. The phase transition occurs at the Curie temperature T_c. As is evident, a non-vanishing polarisation arises immediately at the phase transition. The non-polar and ferroelectric phase co-exist at this point; both have the same Helmholtz free energy.

It can be easily realised that the susceptibility is generally linked to the curvature of $f(P)$ at the minimum, i.e., that

$$\varepsilon_0 \chi \propto \left(\frac{\partial^2 f}{\partial P^2}(P_{\text{eq}}) \right)^{-1} \tag{3.26}$$

is valid. A consideration of the form of the curves in Fig. 3.6 reveals that there is now no longer a divergence. For $BaTiO_3$, the transition into the ferroelectric phase is associated with a jump-like change and thus corresponds to a first-order phase transition. The course of the transition deviates, however, from the normal case. The distinctive feature is that the phase transition is

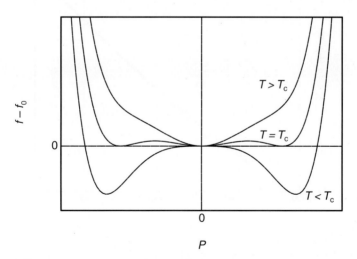

Fig. 3.6. The Landau expansion of the Helmholtz free energy density at the transition into the ferroelectric phase (T_c) as well as for temperatures above and below T_c. The case of a 'weakly first order' transition.

already signalled previously by means of a susceptibility which, although not diverging as would be the case for a second-order phase transition, shows an anomalously sharp rise. The transition, thus, also possesses properties which are characteristic of second-order phase transitions. In order to express this, the BaTiO$_3$ phase transition and other similar types are referred to as **nearly critical** or **weakly first-order transitions**.

3.2 The Ferromagnetic State

As was the case with ferroelectrics, it is also found for ferromagnetics that the transition into the ordered state is already clearly visible previously. As is shown in Fig. 3.7 for the example of nickel, the paramagnetic susceptibility diverges when the temperature of the phase transition – again referred to as the Curie temperature T_c – is approached from above. After passing through T_c, a permanent magnetisation establishes itself. Figure 3.8 shows that this happens for nickel in a continuous fashion, beginning with a vanishing magnetisation, without there being a jump-like change. A similar continuous build-up is also found for all other ferromagnetic materials. The transition from the paramagnetic into a ferromagnetic phase is, thus, in the strict sense, a second-order phase transition.

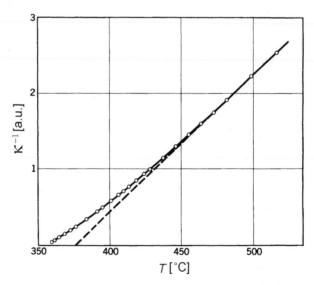

Fig. 3.7. The temperature dependence of the inverse of the paramagnetic susceptibility of nickel (from Kouvel and Fischer [21]).

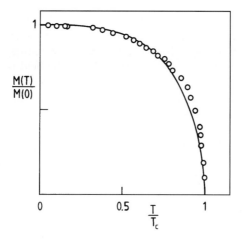

Fig. 3.8. The temperature dependence of the magnetisation of nickel in the ferromagnetic state. Comparison with the prediction of the Landau theory $M \propto (T_c - T)^{1/2}$ (from Weiss [22]).

3.2.1 Weiss Domains

Actually, the emergence of a permanent magnetisation is not directly observed when a sample is cooled below the Curie temperature. A structure made up of many individual domains with different magnetisation orientations develops in the sample. Figure 3.9 shows, again for the example of nickel, this internal structure made up of **Weiss domains**. The ferromagnetic state only becomes

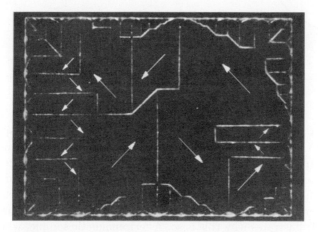

Fig. 3.9. Weiss domains in nickel, observed on the surface of a single crystal. The magnetisation directions are indicated (from de Blois in [23]).

visible externally when a magnetic field is applied. For a changing external
field H_z, the total magnetisation in the sample then passes through the hysteresis curve shown in Fig. 3.10. After the initial switching on and build up of
the field, a permanent value for M_{tot} remains if the field is then switched off.
This magnetisation can only be made to vanish again if a field with a characteristic **coercive field strength** H_k is applied in the opposite direction. The
right-hand side of the figure shows the changes in the internal structure of the
sample which occur upon magnetising the sample for the first time. The size
of the individual Weiss domains is not fixed, but rather they can change by
shifting the boundaries. In this way, the domains which are magnetised in the
direction of the external field can grow at the expense of those magnetised
in the opposite direction. These boundary shifts do not require large forces.
A second mechanism comes into play at higher field strengths. Initially, the
magnetisation directions of all the Weiss domains are aligned with the lattice;
the internal crystal forces fix a set of preferred directions (two in Fig. 3.9). At
higher field strengths, the magnetisation of the individual Weiss domains can
be rotated away from these preferred directions to the field direction, which
leads to a further continuous increase in the total magnetisation of the sample.

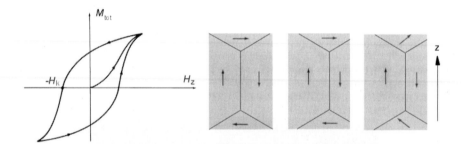

Fig. 3.10. The change in the domain structure of a ferromagnet upon applying an
external magnetic field for the first time (*right*) and the total hysteresis curve (*left*).

It could at first sight be thought that the domain structure simply forms
because of the fact that there is no favoured direction for a sample to which
no field is applied and, with the phase transition setting in at different points
at the same time, one of the directions is chosen by chance in each case.
Although this is true, it doesn't completely describe the effect. Actually, there
are also energetic reasons which support the formation of a domain structure.
Figure 3.11 illustrates why this is so. A ferromagnetic sample, which consists
only as a single domain, creates an extended magnetic stray field around it
which carries magnetic field energy. The field lines run from the magnetic
north pole (N) to the magnetic south pole (S) via the external region. If the

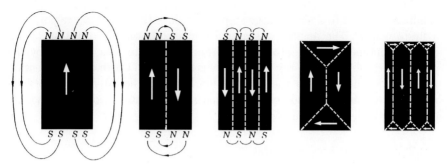

Fig. 3.11. The decrease in the stray field outside a ferromagnetic sample via the formation of Weiss domains.

sample is divided into two domains, there is a shorter route by which the field lines can return into the material. This shortening increases ever more as the domain structure becomes more sophisticated and optimised, as is shown in the two examples on the right-hand side. In the end, there is no longer any stray field; this, of course, implies, to a first consideration, a lowering of the total energy of the system.

Such a consideration could allow one to assume that the finest possible domain structure would be advantageous. This is, however, not the case since the formation of boundaries between two domains is not achieved without a cost in energy. Figure 3.12 shows the structure of the boundary between two domains with opposite magnetisation directions – such a boundary is referred to as a **Bloch wall**. The figure indicates that the change in the orientation of the magnetic dipole does not occur abruptly between two neighbouring atoms, but rather takes place in a continuous fashion stretched out over a transition region. This shows, in the first instance, that it is energetically more favourable for the acting interaction mechanisms to achieve the reorientation in many

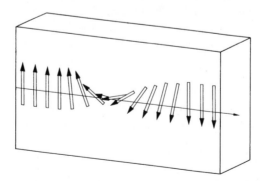

Fig. 3.12. The structure of a Bloch wall. The dipole direction changes in a continuous fashion.

small as opposed to a single large step. Why does the transition region have a finite width? The answer was actually given above: The rotation of the magnetic moments of the atoms away from the orientations favoured by the crystal lattice requires energy; thus, there is an optimal width of the transition region which depends on the competition between the interaction between magnetic dipoles on neighbouring atoms and the interaction of individual dipoles with the crystal field.

3.2.2 Exchange Force Fields

By analogy to the case of ferroelectrics, it could be assumed that the paramagnetic susceptibility in ferromagnetics is carried by atomic magnetic dipoles which adjust themselves in the magnetic field according to Boltzmann statistics, and that κ attains a critical value at the phase transition which is so large that a self-stabilisation is achieved. However, a short consideration shows that this cannot be the case. If the paramagnetic susceptibility is given by Eq. (2.163), the magnetisation is:

$$M = \rho \frac{n_B^2 \mu_B^2}{3 k_B T_c} B \quad . \tag{3.27}$$

Self-stabilisation would require that

$$B = \mu_0 M \quad , \tag{3.28}$$

thus, leading to the condition

$$\mu_0 \rho \frac{n_B^2 \mu_B^2}{3 k_B T_c} = 1 \quad . \tag{3.29}$$

An estimation shows that this condition cannot be achieved. If typical values for the atomic density ρ are used, then the left-hand side of the equation has a value on the order of 10^{-2}. Therefore, ferromagnetism obviously cannot be due to a classical interaction between the magnetic dipoles. Heisenberg recognised the real cause. The basis of ferromagnetism is an interaction mechanism of a different kind which is based solely on quantum-mechanical effects. **Exchange forces** between the spins of neighbouring atoms become active and these are so strong that they lead to a cooperative stabilisation of the spin alignment. The details of how these forces arise cannot be described here, and only a general hint is given. A consequence of the fermion character of the electron is that it is not possible to change the relative orientation of a pair of spins without influencing the wavefunctions. A change in the relative orientation always leads, therefore, to a change in the overlap of the wavefunctions of the two involved electrons, and this has energetic consequences.

In ferromagnetics, the resulting energy changes are large and so arranged that there is a preference for a parallel orientation of the spins. This can be formally expressed as follows:

$$u_{jk} = -\beta_{ex} \boldsymbol{S}_j \cdot \boldsymbol{S}_k \quad . \tag{3.30}$$

The interaction energy u_{jk} between two neighbouring spins $\boldsymbol{S}_j$ and $\boldsymbol{S}_k$ is written here using a scalar product, with the strength being defined by the coefficient β_{ex}. β_{ex} increases with decreasing temperature because of the temperature dependence of the lattice constants. At the Curie temperature, the limiting value is reached at which a self-stabilisation sets in. Below T_c, the spins are no longer uniformly distributed over all the crystal field orientations, but rather there is a preferred orientation in each domain.

In order to describe how this preferred orientation and with it the build-up of a permanent magnetisation sets in upon going through the Curie temperature, Landau theory can again be used. As far as the Landau formalism is concerned, the question as to how the interaction comes into existence, whether it be via classical interactions between magnetic dipoles or quantum-mechanical exchange forces, is irrelevant. The order parameter which controls the transition into the ferromagnetic phase is always the magnetisation. It always determines the molecular field which acts, in always the same fashion, upon the spins, independent of whether a field of exchange forces or of dipole forces is being considered. We now need to know the value of the magnetisation which leads to a minimum in the Gibbs or Helmholtz free energy. With the more general case of the presence of an external magnetic field in mind, we choose the (reduced) density of the Gibbs free energy $\hat{g}$ (Eq. (A.11) in Appendix A) as the thermodynamic potential and specify it here as

$$\hat{g} = g_0 + b(T - T_c)M^2 + c_4 M^4 - \mu_0 H M \quad , \tag{3.31}$$

in complete analogy to Eq. (3.19) for ferroelectrics. Plots of this function, in the absence of a field, for temperatures above and below the transition point are given in Fig. 3.5, and the same conclusion can be made now referring to ferromagnetics: The equilibrium value of the magnetisation disappears for temperatures above T_c,

$$M_{eq}(T > T_c) = 0 \quad , \tag{3.32}$$

while upon cooling below T_c, the build-up of magnetisation follows the square root law

$$M = \left(\frac{b}{2c_4}\right)^{1/2} (T_c - T)^{1/2} \quad . \tag{3.33}$$

Expressions for the magnetic susceptibility above and below T_c can also be obtained from the corresponding result for ferroelectrics. Transferring Eqs. (3.22) and (3.24) to the case of ferromagnetics gives for temperatures above T_c

$$\frac{\kappa}{\mu_0} \approx \frac{dM}{dB} = \frac{1}{2b(T - T_c)} \quad , \tag{3.34}$$

while in the ferromagnetic phase one finds

$$\frac{\kappa}{\mu_0} \approx \frac{1}{4b(T_c - T)} \quad . \tag{3.35}$$

Critical Fluctuations. A further consideration of Figs. 3.7 and 3.8 shows that while the prediction of Landau's theory reproduces the curves very well overall, upon a closer examination, systematic differences are observed. Differences of this type are always found for second-order phase transitions. Sometimes, they exist, as in the example, over a wider temperature range, while, in other cases, they are focused on a very small region near T_c. Landau's **molecular field theory** actually loses its validity here. The origin for this is, in principle, easy to understand, and follows as a consequence of a fundamental property of critical transitions. The trigger for the phase transition is the fact that the restoring forces in the material, which make the polarisation or magnetisation generated by an external field disappear again after switching off the external field, become ever smaller upon approaching the Curie temperature. In this temperature range, a magnetisation or polarisation can temporarily emerge without an external field, simply via thermal fluctuations. Local regions form, in which there is a preferred orientation of the electric dipoles or the spins. They are not stable, but vanish again. Their lifetime depends on the weak but still present restoring forces; the weaker they are, the larger and longer-lasting the temporarily ordered regions become. In Landau's theory, the stabilising action is attributed solely to the average value of the molecular field, the latter being expressed via the polarisation P or the magnetisation M. If the fluctuations are very strong, the average field is no longer the decisive parameter, and Landau's theory which is based on this loses its validity.

Actually, it is also possible to describe theoretically the region very close to the Curie temperature, which is marked by strong fluctuations. The basis for this is a symmetry property which **critical fluctuations** possess near to T_c, namely their self-similarity. In this range, magnetised or polarised regions of all sizes appear, without there being a definite characteristic length. As a consequence images of the varying magnetic or polarization structure are invariant upon changing the spatial resolution, i.e., they always have a similar appearance. It is exactly this property which is used in a **renormalisation group theory** due to Wilson in order to make precise statements about the changes in the order parameter and susceptibility in the vicinity of T_c. It is found that power laws are always obtained for the magnetisation,

$$M \propto (T_c - T)^\nu \ , \tag{3.36}$$

and also for the susceptibilities below the Curie temperature,

$$\kappa \propto (T_c - T)^{-\gamma} \tag{3.37}$$

and above T_c,

$$\kappa \propto (T - T_c)^{-\gamma'} \ . \tag{3.38}$$

Renormalisation group theory provides algorithms by which the **critical coefficients** ν, γ and γ' can be calculated. The results of these calculations are in general in very good agreement with the observations.

In spite of this, Landau's theory maintains its validity within the proper limits: Its applicability begins when the fluctuation effects no longer play a large role and the molecular field takes over control in terms of the average field.

3.3 The Nematic Liquid-Crystalline State

In the first discussion of the properties of nematic liquid-crystalline phases in Sect. 1.3, it was already indicated that the existence of these special ordered states is to be understood as the expression of a self-stabilisation process. If the temperature dependence of the nematic order parameter in Fig. 1.20 is considered again, the similarity with Fig. 3.2, which shows the temperature dependence of the permanent polarisation for $BaTiO_3$, is obvious. Both cases involve phase transitions of the same type. The only difference is that the determining parameter is, in one case, the polarisation, and, in the other, the nematic order parameter S_2. Considering the temperature dependence of P and S_2, an accelerating decrease is observed for the ordered phase upon approaching the transition temperature, while there is a disappearance of order in the high-temperature phase; these are typical indications of a nearly-critical or weakly first-order phase transition.

Typical characteristics of this type of transition are also already found in the isotropic phase of nematogen substances. As an example, Fig. 3.13 shows the result of light-scattering experiments on heptyl-phenyl-cyclohexane (PCH7). 'HV-measurements' were performed, in which light scattering is initiated by a horizontally polarised laser beam, with the scattered radiation being measured by a vertically positioned analyser. In such an experiment, only scattering processes which change the polarisation direction are visible, and this is only possible if the sample contains optically anisotropic elements. The appearance of scattered light where the polarisation direction has been altered is, in the first instance, not especially surprising for the rod-like PCH7-molecule, which has an anisotropic polarisability. The special feature of the experimental result is the sharp increase in the intensity upon approaching the transition temperature T_{ni}. This observation suggests that the molecules have a tendency, already in the isotropic phase, to form temporary aggregates, whose number and size rapidly increase upon approaching T_{ni} in the fashion of a critical fluctuation. Therefore, in this way, the nematic phase announces itself already in the isotropic liquid state. The intensity increase can be described by a simple power law,

$$I \propto (T - T^*)^{-1} \ , \tag{3.39}$$

as is shown in Fig. 3.13. The limiting temperature T^* lies a few degrees below the transition point $T_{ni} = 58\,°C$, as can also be seen from the figure.

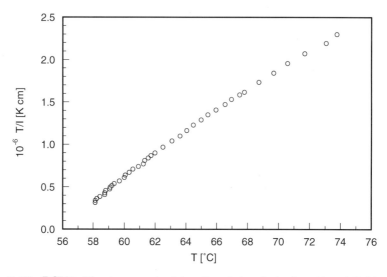

Fig. 3.13. PCH7: The increase in intensity of depolarised scattered light upon approaching the temperature T_{ni} of the transition into the nematic phase.

3.3.1 The Landau–de Gennes Expansion

What would be the form of a Landau expansion which describes the isotropic-nematic phase transition in a qualitatively correct fashion? The answer to this question was given by de Gennes. In order to describe the Helmholtz free energy density as a function of the nematic order parameter and the temperature

$$f(S_2, T)$$

the following power series expansion is to be used

$$f - f_0 = b(T - T^*)S_2^2 - c_3 S_2^3 + c_4 S_2^4 \quad . \tag{3.40}$$

A difference as compared to the power series expansions for ferroelectrics and ferromagnetics is the inclusion of a third-order term, which appears because the symmetry has changed. While previously a sign reversal of the order parameter left the Helmholtz free energy unchanged, there is a change for a nematic liquid crystal. This becomes immediately clear upon considering an example. $S_2 = -1/2$ corresponds to an orientation distribution, where all molecules have their long axes perpendicular to the director, with their orientations being uniformly distributed in this plane. By comparison, the structure is completely different for an order parameter $S_2 = +1/2$: Here, there is a wide distribution with a maximum in the director direction. Under these conditions, it is clear that a third-order term must be included in the Landau expansion for the Helmholtz free energy.

The consequences are shown in Fig. 3.14, which shows the dependence of the Helmholtz free energy on S_2 for three temperatures. The middle curve corresponds to the transition temperature, where the isotropic liquid and a nematic phase with a well-defined order parameter co-exist. At higher and lower temperatures, there is of course only the isotropic state and only the nematic state, respectively.

The Landau expansion in Eq. (3.40) can be evaluated to determine the position of the transition point. The following applies for the equilibrium value of S_2:

$$\frac{\mathrm{d}f}{\mathrm{d}S_2} = 0 = 2b(T - T^*)S_2 - 3c_3 S_2^2 + 4c_4 S_2^3 \ . \tag{3.41}$$

The co-existence condition is given by

$$f(S_2 = 0) - f_0 = f(S_2 = S_2(T_{\mathrm{ni}})) - f_0 = 0 \ , \tag{3.42}$$

which means that

$$0 = b(T - T^*)S_2(T_{\mathrm{ni}})^2 - c_3 S_2(T_{\mathrm{ni}})^3 + c_4 S_2(T_{\mathrm{ni}})^4 \ . \tag{3.43}$$

Combining Eqs. (3.43) and (3.41) leads immediately to an expression for the order parameter at the phase transition:

$$S_2(T_{\mathrm{ni}}) = \frac{c_3}{2c_4} \ . \tag{3.44}$$

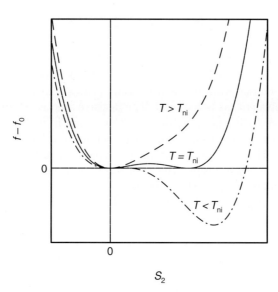

Fig. 3.14. The Landau–de Gennes expansion for the Helmholtz free energy density of a nematic liquid crystal at the clearing point T_{ni} as well as for temperatures where only the isotropic ($T > T_{\mathrm{ni}}$) or only the nematic phase ($T < T_{\mathrm{ni}}$) exists.

The situation at the temperature T^* is given by

$$T_{\mathrm{ni}} - T^* = \frac{c_3^2}{4bc_4} \quad . \tag{3.45}$$

Finally, it follows that the temperature dependence of the order parameter in the nematic phase is described by

$$S_2 = \frac{3c_3}{8c_4} + \left(\left(\frac{3c_3}{4c_4} \right)^2 - \frac{2b}{4c_4}(T - T^*) \right)^{1/2} \quad . \tag{3.46}$$

As a thermodynamic theory, the Landau approach works with phenomenologi-cal coefficients. If experimental data exist, these can be determined by a fitting procedure. Further experimental results, for example, the dependence of S_2 on T, can then be predicted.

3.3.2 The Maier–Saupe Theory

A specific theory allows the order parameter at the transition temperature to be calculated. It was developed by Maier and Saupe in 1958 and is briefly described here. The theory deals directly with the self-stabilisation in the nematic phase. The starting point is the following expression for the molecular field which acts to orientate the rod-like molecules:

$$u(\vartheta) = -u_0 S_2 \frac{3\cos^2 \vartheta - 1}{2} \quad . \tag{3.47}$$

$u(\vartheta)$ is the 'nematic potential' appearing in this phase, which each molecule experiences upon a rotation. The angle $\vartheta = 0$ corresponds to the director direction. The chosen form of the potential ensures that the positions ϑ and $180°\text{-}\vartheta$ are equivalent, as is required. The decisive step is the inclusion of the order parameter S_2 in the description of the strength of the molecular field. The chosen form means that it becomes ever more difficult for a molecule to deviate from the director direction, the higher the order parameter in the nematic phase is. In addition, the expression contains a variable which is material dependent, namely the coefficient u_0.

The orientation distribution function $w(\vartheta, \varphi)$ of the molecules can be cal-culated, for a given potential $u(\vartheta)$, using Boltzmann statistics. It is given as

$$w(\vartheta, \varphi) = \frac{1}{Z} \exp \left(\frac{u_0 S_2}{k_{\mathrm{B}} T} \cdot \frac{3\cos^2 \vartheta - 1}{2} \right) \tag{3.48}$$

or

$$w(\vartheta, \varphi) = \frac{1}{Z'} \exp \left(\frac{3 u_0 S_2}{2 k_{\mathrm{B}} T} \cos^2 \vartheta \right) \quad . \tag{3.49}$$

We introduce a new variable,

$$x = \frac{3u_0 S_2}{2k_B T} \quad , \tag{3.50}$$

and determine the partition function $\mathcal{Z}'$ using

$$\frac{1}{\mathcal{Z}'} \int\limits_{\vartheta=0}^{\pi} \int\limits_{\varphi=0}^{2\pi} \exp(x\cos^2\vartheta) \cdot \sin\vartheta d\vartheta d\varphi = 1 \quad . \tag{3.51}$$

$\mathcal{Z}'(x)$ can be expressed as a function of x:

$$\mathcal{Z}'(x) = 2\pi \int\limits_{\vartheta=0}^{\pi} \exp(x\cos^2\vartheta) d(-\cos\vartheta) = 4\pi \int\limits_{0}^{1} \exp(xy^2) dy \quad . \tag{3.52}$$

If the orientation distribution function is known, the order parameter follows generally as:

$$S_2 = \int\limits_{\vartheta=0}^{\pi} \int\limits_{\varphi=0}^{2\pi} w(\vartheta,\varphi) \frac{3\cos^2\vartheta - 1}{2} \sin\vartheta d\vartheta d\varphi \quad . \tag{3.53}$$

The equation expresses here a self-consistency condition: S_2 is not only the result on the left-hand side, but it is also contained in the distribution function w on the right. The equation can be solved – i.e., self-consistency is achieved – and gives the value of the order parameter. Equation (3.53) is re-expressed as

$$S_2 = -\frac{1}{2} \int\limits_{\vartheta=0}^{\pi} \int\limits_{\varphi=0}^{2\pi} w(\vartheta,\varphi) d\varphi \sin\vartheta d\vartheta \tag{3.54}$$

$$+\frac{3}{2} \cdot \frac{1}{\mathcal{Z}'} \int\limits_{\vartheta=0}^{\pi} \int\limits_{\varphi=0}^{2\pi} \exp(x\cos^2\vartheta) \cos^2\vartheta \sin\vartheta d\vartheta d\varphi$$

$$= -\frac{1}{2} + \frac{3}{2} \frac{4\pi}{\mathcal{Z}'} \int\limits_{0}^{1} \exp(xy^2) y^2 dy \quad . \tag{3.55}$$

This leads to

$$S_2 = -\frac{1}{2} + \frac{3}{2} \frac{1}{\mathcal{Z}'} \frac{d\mathcal{Z}'}{dx} \quad . \tag{3.56}$$

Denoting the right-hand side of Eq. (3.56) as $\Phi(x)$ and expressing S_2 as a function of x yields

$$\frac{2k_B T}{3u_0} x = \Phi(x) \quad . \tag{3.57}$$

We find here functions of the variable x on both sides of the equation. Figure 3.15 shows the dependence of both functions and points to the solution: It is necessary to make the left-hand side of the equation intersect with $\Phi(x)$. For high temperatures, i.e., the gradient of the line is large, the only intersection point is at the origin of the co-ordinate system. This changes at the temperature T_{ni}, where there is a second solution: It is associated with a definite value of the order parameter, namely $S_2 = 0.44$. This actually agrees well with experimental observations. The Maier–Saupe theory also gives the temperature dependence of the order parameter in the nematic phase. The figure contains a straight line which corresponds to this case – it can be shown that the highest intersection point corresponds to the equilibrium value.

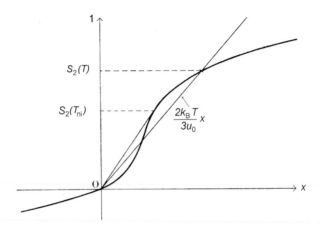

Fig. 3.15. The graphical solution of the self-consistent Eq. (3.57) for S_2.

3.4 Phase Separation in Binary Polymer Melts

The discussion of the properties of polymer materials has up to now only dealt with single component systems. In fact, a large part of, in particular, application-oriented research is devoted to mixtures or blends of different polymers. The basis for this is the observation that mixtures often have better mechanical properties than the pure substances. It is possible in this way to reduce the brittleness of materials and hence increase their fracture resistance. In order to optimise the properties of polymer blends, a good basic understanding of the mixing behaviour is important. It would be desirable to know

- whether two particular polymers are miscible or not, and whether this can be predicted,

- what is the composition of the two separate phases in the demixed state,
- how the transition of a homogeneous melt into a two-phase structure occurs and which morphology then emerges.

At first, it may seem surprising that these questions should be discussed in a chapter which is dealing with critical transitions and molecular fields. The following observations can explain this.

3.4.1 Binodals and Critical Concentrations

Figure 3.16 shows the phase diagrams of different mixtures of polystyrene and polybutadiene, which differ in the molecular weights of the two partners. The variables in the diagram are the composition of the blend, which is described by means of the volume fraction ϕ of polystyrene, and the temperature; the pressure is constant at normal conditions. Both partners are liquid in their pure states for the shown temperature range. A consideration of the phase diagram reveals, for a given ratio of polystyrene and polybutadiene at a particular temperature, whether there is a homogeneous melt or a phase-separated two-component liquid. All three mixtures are found to be homogeneous at high temperature, while upon cooling there is a transition into a two-phase structure. The line at which the transition occurs and which separates the two regions in the phase diagram is referred to as a **binodal**.

Figure 3.17 shows a schematic representation of such a binodal which sheds light on the structural changes which occur upon the transition from the homogeneous into the two-phase region. The binodal generally reveals the

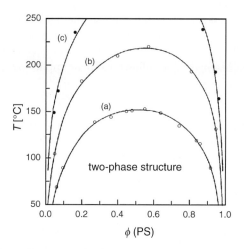

Fig. 3.16. The phase diagrams of polystyrene-polybutadiene blends with different molecular weights M. (a) $M(\mathrm{PS}) = 2250$, $M(\mathrm{PB}) = 2350$; (b) $M(\mathrm{PS}) = 3500$, $M(\mathrm{PB}) = 2350$; (c) $M(\mathrm{PS}) = 5200$, $M(\mathrm{PB}) = 2350$ (from Roe and Zin [24]).

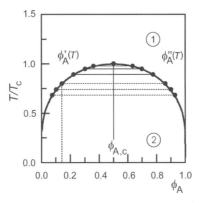

Fig. 3.17. A schematic representation of a binodal with a single- and a two-phase region at high and low temperature, respectively. The continuously changing phase separation at the critical point (*solid lines*) and the discontinuous transition into the two-phase state in the general case (*dotted lines*).

composition of the two phases which exist when the system is at equilibrium at a particular temperature. Both are mixed phases with different proportions, ϕ'_A and ϕ''_A, of the component A. Consider that there is initially a homogeneous melt with the volume fraction ϕ_A – this corresponds to the vertical dotted line. Upon cooling, a second phase is formed upon crossing the binodal, with its composition being given by the horizontal line at the transition point as ϕ''_A. If the system is further cooled such that the distance from the binodal is increased, the composition of the two phases changes, for example, as shown by the two further horizontal dotted lines in the figure. The volume fractions ϕ_1 and $\phi_2 = 1 - \phi_1$ of the two co-existing mixed phases are determined at each temperature by the requirement that the total mass of the two individual components is maintained. The following, thus, applies for polymer A:

$$\phi_A = \phi_1 \cdot \phi'_A + (1 - \phi_1)\phi''_A \ . \tag{3.58}$$

The volume fraction of the A-rich phase, ϕ_1, is then given by

$$\phi_1(T) = \frac{\phi''_A(T) - \phi_A}{\phi''_A(T) - \phi'_A(T)} \ , \tag{3.59}$$

while that of the B-rich phase is

$$\phi_2(T) = 1 - \phi_1(T) = \frac{\phi_A - \phi'_A(T)}{\phi''_A(T) - \phi'_A(T)} \ . \tag{3.60}$$

In the described case, the phase separation sets in in a continuous fashion in so far as the volume fraction ϕ_2 of the B-rich phase begins at zero upon crossing through the binodal; the composition of the newly formed phase is, however, from the outset different to the initial state. There is a specific composition of the melt where this is not the case and for which the phase separation proceeds in a completely continuous fashion. This happens when the **critical composition** $\phi_{A,c}$ is chosen for component A – this is where there is a horizontal tangent in the binodal, i.e., in the case shown in Fig. 3.17 at $\phi_A = 0.5$. The phase separation then occurs at the corresponding critical temperature T_c and with

$$\phi_A' = \phi_A'' = \phi_{A,c} \ .$$

The composition of the two phases only becomes different upon further cooling, and this also happens in a continuous fashion. Exactly these, though, are again the characteristics of a critical transition, which exists here.

As with all critical transitions, it is necessary to identify the associated order parameter. The choice is easy to make here: The difference $\phi_A'' - \phi_A'$ is a suitable candidate for this role. It shows the typical temperature behaviour with a continuous setting in beginning at zero. In addition, it describes the increase in order since it takes the value one when there is a complete separation into the two individual components.

For the polymer blend in Fig. 3.16, the phase separation occurs upon cooling. The opposite case of a phase separation upon heating is also found. Figure 3.18 shows an example of this, namely a mixture of polystyrene and poly(vinylmethylether). The two polymers are completely miscible at a temperature under 100 °C. A separation into two phases occurs upon heating, as is shown by the binodal in the figure. While the critical volume fraction of polystyrene is 0.5 for the mixture in Fig. 3.16, it is now, in Fig. 3.18, at 0.3, i.e., it is shifted away from the centre.

It is a typical feature of critical transitions that the phase transition in the unstructured phase is announced by means of increasing critical fluctuations. This effect can be particularly clearly observed with polymer blends:

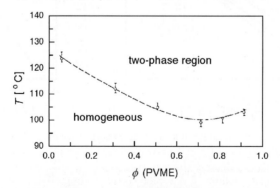

Fig. 3.18. The phase diagram of a polystyrene-poly(vinylmethylether) mixture ($M(\text{PS}) = 2 \times 10^5$, $M(\text{PVME}) = 4.7 \times 10^4$) (from Hashimoto et al [25]).

If the critical composition is chosen then a sharp increase in intensity is observed in scattering experiments upon approaching T_c because of increasing concentration fluctuations. Figure 3.19 shows the results of neutron scattering experiments which were carried out on a polystyrene-poly(vinylmethylether) blend at the critical composition. The inverse of the scattered intensity as a function of the scattering vector squared is plotted on the left-hand side. It is seen that the scattered intensity increases with increasing temperature. The plot on the right-hand side shows the temperature dependence of the inverse of the scattered intensity in the forward direction $S(0)^{-1}$. A divergence in the intensity is observed, and the associated temperature corresponds to T_c. It can be shown that the scattered intensity as a function of q and T can be described by

$$S_c \propto \frac{1}{|T - T_c|} \cdot \frac{1}{1 + \xi_\phi^2 q^2} \tag{3.61}$$

with

$$\xi_\phi \propto \frac{1}{|T - T_c|^{1/2}} \ . \tag{3.62}$$

The variable ξ_ϕ, which diverges like the scattered intensity $S(0)$ upon approaching T_c, denotes the distance over which the concentration fluctuations are correlated with each other, i.e., it represents the diameter of the regions in which a brief transient demixing of the melt occurs. It is easy to appreciate why ξ_ϕ has this meaning. As was discussed in Sect. 1.5.2, there is a general Fourier relationship between the scattering function and the pair distribution

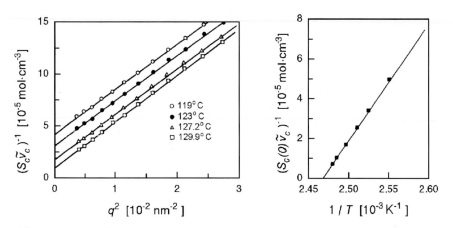

Fig. 3.19. The results of neutron scattering experiments on a blend of (deuterated) polystyrene ($M = 3.8 \times 10^5$) and poly(vinylmethylether) ($M = 6.4 \times 10^4$) at the critical composition. S_c denotes the scattering function with respect to the structural unit with the molar volume $\tilde{v}_c$. A plot corresponding to Eq. (3.61). The linear extrapolation of $S(q \to 0)^{-1}$ to the value zero gives the critical temperature (*right*) (from Schwahn [26]).

function. In the case of a polymer blend, the pair distribution function denotes the probability of finding, within a distance r of a monomer unit, a unit of the same type. The pair distribution function provides, because of its spreading out in space, information about the size of the demixing regions. Let the general Fourier relationship

$$g_2(\boldsymbol{r}) \propto \int \exp(\mathrm{i}\boldsymbol{q}\boldsymbol{r})S_{\mathrm{c}}(\boldsymbol{q})\mathrm{d}^3\boldsymbol{q} \tag{3.63}$$

be applied to the scattering function of Eq. (3.61). A direct calculation then shows that

$$g_2(\boldsymbol{r}) \propto \frac{1}{r}\exp-\frac{r}{\xi_\phi} \quad . \tag{3.64}$$

It is seen that it is the parameter ξ_ϕ in the expression for the pair distribution function which determines the spreading out.

It is also possible to distinguish whether the phase separation in a binary polymer melt occurs as a critical transition or, for compositions further away from the critical value, as a first-order phase transition, just by considering the adopted structure. Figure 3.20 shows two characteristic images obtained for blends of polystyrene and partially brominated polystyrene with a polarisation microscope using a phase contrast method. The image on the left-hand side with spherical precipitates is typical for a composition far away from the critical value, while that on the right-hand side, which shows two inter-penetrating continuous phases, is found for critical or almost critical transitions. The different formation kinetics will be discussed later. Here, it will only be noted that spherical precipitates arise via a process of nucleation and subsequent growth, while the structure on the right-hand side is typical

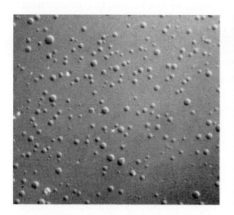 

Fig. 3.20. Structure formation during phase separation for two blends of polystyrene and partially brominated polystyrene. *left*: nucleation with subsequent growth of spherical precipitates ($\phi(\mathrm{PS}) = 0.8$); *right*: spinodal decomposition ($\phi(\mathrm{PS}) = 0.5$) [27].

for a **spinodal decomposition**. Nucleation requires an activation step, while spinodal decomposition can occur spontaneously without such a step.

3.4.2 Flory–Huggins Theory

Flory and Huggins developed a thermodynamic theory which describes the mixing properties of a pair of polymers in the liquid phase. The theory provides a fundamental understanding as to why different types of phase diagrams appear and, in particular, deals with the effect of the molecular weight of the partners in the blend.

When discussing the miscibility of two components, it is necessary, for polymers as also in the case of mixtures of low-molecular weight substances, to investigate the change in the Gibbs free energy upon mixing. A schematic representation of the situation is given in Fig. 3.21, which also introduces the relevant thermodynamic variables. In the initial state, there are $\tilde{n}_A$ moles of polymer A in a volume $\mathcal{V}_A$ and $\tilde{n}_B$ moles of polymer B in a volume $\mathcal{V}_B$. The mixing of the two components is made possible by removing the dividing wall between the two compartments such that both polymers can expand to fill the total volume $\mathcal{V} = \mathcal{V}_A + \mathcal{V}_B$. In order to determine whether such a mixing takes place, it is necessary to consider the change in the Gibbs free energy. This change, which is referred to as the **Gibbs free energy of mixing** and is denoted by $\Delta\mathcal{G}_{\mathrm{mix}}$, is given by

$$\Delta\mathcal{G}_{\mathrm{mix}} = \mathcal{G}_{\mathrm{AB}} - (\mathcal{G}_{\mathrm{A}} + \mathcal{G}_{\mathrm{B}}) \ . \tag{3.65}$$

$\mathcal{G}_{\mathrm{A}}, \mathcal{G}_{\mathrm{B}}$ and $\mathcal{G}_{\mathrm{AB}}$ refer to the Gibbs free energy of the pure components A and B in the initial state and the Gibbs free energy of the homogeneous mixture.

The Flory–Huggins theory describes $\Delta\mathcal{G}_{\mathrm{mix}}$ as the sum of two components:

$$\Delta\mathcal{G}_{\mathrm{mix}} = -T\Delta\mathcal{S}_{\mathrm{t}} + \Delta\mathcal{G}_{\mathrm{loc}} \ . \tag{3.66}$$

They represent the two aspects of the mixing process. Firstly, a mixing always leads to an increase in the entropy associated with the motion of the centres of gravity of the molecules. Secondly, the change in the local environment, generally, leads to changes in the intermolecular interactions between

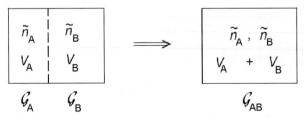

Fig. 3.21. The variables used to describe the mixture of two polymers A and B.

the monomers. $\Delta\mathcal{G}_{\text{loc}}$ represents the second part, while $\Delta\mathcal{S}_t$ is the increase in the translational entropy. $\Delta\mathcal{S}_t$ and the resulting decrease $-T\Delta\mathcal{S}_t$ in the Gibbs free energy always promote mixing. On the other hand, $\Delta\mathcal{G}_{\text{loc}}$ can also work against a mixing especially if the average interaction energy between the monomers is increased upon mixing. In fact, it can be shown for the van der Waals interactions, which are dominant in polymer systems, that the attractive forces between like monomers are always stronger than those between unlike monomers. Since it represents a Gibbs free energy, $\Delta\mathcal{G}_{\text{loc}}$ can also include positive or negative changes in the local entropy due to changes in the chain packing, such that mixing may be further promoted or hindered, respectively.

The separation of $\Delta\mathcal{G}_{\text{mix}}$ into two components alone has not achieved much. In order to be able to calculate phase diagrams, explicit expressions for $\Delta\mathcal{S}_t$ and $\Delta\mathcal{G}_{\text{loc}}$ are required, so that the sum of the two contributions can be evaluated. The Flory–Huggins theory provides such expressions in the following forms:

1. The increase in the translational entropy is described by

$$\frac{\Delta\mathcal{S}_t}{\tilde{R}} = \tilde{n}_A \ln \frac{\mathcal{V}}{\mathcal{V}_A} + \tilde{n}_B \ln \frac{\mathcal{V}}{\mathcal{V}_B} \quad . \tag{3.67}$$

Let the volume fractions ϕ_A and ϕ_B be introduced according to

$$\phi_A = \frac{\mathcal{V}_A}{\mathcal{V}} \quad \text{and} \quad \phi_B = \frac{\mathcal{V}_B}{\mathcal{V}} \quad . \tag{3.68}$$

Equation (3.67) then becomes

$$\frac{\Delta\mathcal{S}_t}{\tilde{R}} = -\tilde{n}_A \ln \phi_A - \tilde{n}_B \ln \phi_B \quad . \tag{3.69}$$

2. The following expression serves to describe the change in the local interactions:

$$\Delta\mathcal{G}_{\text{loc}} = \tilde{R}T \frac{\mathcal{V}}{\tilde{v}_c} \chi_F \phi_A \phi_B \quad . \tag{3.70}$$

It contains two parameters. $\tilde{v}_c$ denotes the volume of a reference unit, which is the same for both the polymers; normally, the volume of the monomer unit of one of the partners is chosen. The decisive parameter is the **Flory–Huggins parameter** χ_F. It is dimensionless and expresses – in units of $\tilde{R}T$ – the change in the local Gibbs free energy of a reference unit that would occur upon a transition into the mixed state.

The physical basis of Eq. (3.70) is easy to appreciate. The interaction energy can be determined as a statistical average, since each individual polymer chain is in contact with a very large number of others. There is a probability ϕ_B that a monomer unit of polymer A makes contact with units of the B chains, and there is vice versa a probability ϕ_A that B units make contact with A units.

The total number of contacts between unlike monomers is, thus, proportional to the product $\phi_A\phi_B$. Each contact of this type leads to an energy increase $\chi_F \check{R} T$ per mole of reference units. Multiplication by the number of moles of reference units in the sample, $\mathcal{V}/\tilde{v}_c$, gives $\Delta\mathcal{G}_{loc}$.

The expression which describes the increase in the translational entropy upon mixing corresponds to standard expressions, which are also used for the mixing of ideal gases. In both cases, the gain in entropy is based on the fact that the volume, in which the centres of gravity of the molecules are contained, is expanded by the factor $\mathcal{V}/\mathcal{V}_A$, or $\mathcal{V}/\mathcal{V}_B$.

Using Eqs. (3.69) and (3.70) leads to the following expression for the Gibbs free energy of mixing

$$\Delta\mathcal{G}_{mix} = \tilde{R}T\mathcal{V}\left(\frac{\phi_A}{\tilde{v}_A}\ln\phi_A + \frac{\phi_B}{\tilde{v}_B}\ln\phi_B + \frac{\chi_F}{\tilde{v}_c}\phi_A\phi_B\right) \tag{3.71}$$

$$= \tilde{R}T\tilde{n}_c\left(\frac{\phi_A}{N_A}\ln\phi_A + \frac{\phi_B}{N_B}\ln\phi_B + \chi_F\phi_A\phi_B\right) . \tag{3.72}$$

In the first equation, the molar volumes of the two polymers $\tilde{v}_A$ and $\tilde{v}_B$ were introduced, based on:

$$\tilde{n}_A = \mathcal{V}\frac{\phi_A}{\tilde{v}_A} \quad\text{and}\quad \tilde{n}_B = \mathcal{V}\frac{\phi_B}{\tilde{v}_B} , \tag{3.73}$$

while, in the second equation, the number of moles of reference units as given by

$$\tilde{n}_c = \frac{\mathcal{V}}{\tilde{v}_c} \tag{3.74}$$

was used. The numbers N_A and N_B denote the molecular weights of the two partners, expressed in the common reference unit of volume $\tilde{v}_c$:

$$N_A = \frac{\tilde{v}_A}{\tilde{v}_c} \quad\text{and}\quad N_B = \frac{\tilde{v}_B}{\tilde{v}_c} . \tag{3.75}$$

The Flory–Huggins equation provides the basis for discussing in a transparent way the question of the miscibility of a pair of polymers. For reasons of simplicity, a 'symmetrical' polymer blend, in which both components have the same degree of polymerisation, is considered here:

$$N_A = N_B = N . \tag{3.76}$$

It then follows that

$$\frac{\tilde{n}_c}{N} = \tilde{n}_A + \tilde{n}_B , \tag{3.77}$$

and the Flory–Huggins equation thus becomes

$$\Delta\mathcal{G}_{mix} = \tilde{R}T(\tilde{n}_A + \tilde{n}_B)(\phi_A\ln\phi_A + \phi_B\ln\phi_B + N\chi_F\phi_A\phi_B) . \tag{3.78}$$

An inspection of Eq. (3.78) reveals that it only contains one relevant parameter, namely the product $N\chi_{\mathrm{F}}$. Figure 3.22 shows the dependence of $\Delta\mathcal{G}_{\mathrm{mix}}$ on ϕ_{A} for different values of $N\chi_{\mathrm{F}}$. For $N\chi_{\mathrm{F}} = 1.6$ and likewise for smaller and negative values of this parameter, it is found that $\Delta\mathcal{G}_{\mathrm{mix}}$ is negative for all compositions ϕ_{A}, with the minimum being at $\phi_{\mathrm{A}} = 0.5$. This means that mixing is always observed here. As $N\chi_{\mathrm{F}}$ increases to larger values, changes are observed. The curve changes its shape, and for values of the parameter $N\chi_{\mathrm{F}}$ above a critical value,

$$(N\chi_{\mathrm{F}}) > (N\chi_{\mathrm{F}})_{\mathrm{c}} \ ,$$

a maximum, rather than a minimum, is now observed at $\phi_{\mathrm{A}} = 0.5$. This change leads to a qualitatively new situation. Even if $\Delta\mathcal{G}_{\mathrm{mix}}$ is negative, this does not mean that a homogeneous mixture always forms. Consider as an example the curve at $N\chi_{\mathrm{F}} = 2.4$ for a mixture with $\phi_{\mathrm{A}} = 0.45$. Two arrows are drawn here in the figure. The first arrow shows that the formation of a homogeneous mixture of A and B from an initially separated state would result in a decrease in the Gibbs free energy. The second arrow shows, however, that there is the possibility of a further decrease upon the formation of a two-phase structure

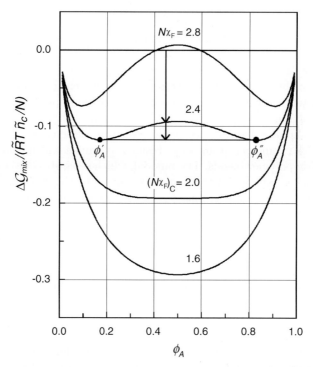

Fig. 3.22. The Gibbs free energy of mixing of a binary polymer melt ($N_{\mathrm{A}} = N_{\mathrm{B}} = N$), calculated using the Flory–Huggins theory.

made up of two mixed phases with compositions ϕ'_A and ϕ''_A. The special property of the curve which leads to this result is the presence of two minima at exactly both these compositions.

The critical value $(N\chi_F)_c$, which separates the region of complete miscibility from that in which two-phase structures can form, can be calculated: It is at this value that the curvature is zero for $\phi_A = 0.5$:

$$\frac{\partial^2 \Delta \mathcal{G}_{mix}(\phi_A = 0.5)}{\partial \phi_A^2} = 0 \quad . \tag{3.79}$$

Since the first and second derivatives of $\Delta \mathcal{G}_{mix}$ are given by

$$\frac{1}{(\tilde{n}_A + \tilde{n}_B)\tilde{R}T} \frac{\partial \Delta \mathcal{G}_{mix}}{\partial \phi_A} = \ln \phi_A + 1 - \ln(1 - \phi_A) - 1 + N\chi_F(1 - 2\phi_A) \quad , \tag{3.80}$$

$$\frac{1}{(\tilde{n}_A + \tilde{n}_B)\tilde{R}T} \frac{\partial^2 \Delta \mathcal{G}_{mix}}{\partial \phi_A^2} = \frac{1}{\phi_A} + \frac{1}{1 - \phi_A} - 2N\chi_F \quad , \tag{3.81}$$

it follows that the critical value is given as

$$(N\chi_F)_c = 2 \quad . \tag{3.82}$$

A comparison of the curves in Fig. 3.22 with those in Fig. 3.5, which explain thermodynamically the critical transition in ferroelectrics, immediately reveals the analogous behaviour. In both cases, there are two minima in the plots of the Gibbs or Helmholtz free energy for the structured phase. Their position determines in both cases the order parameter, which, beginning at zero at the critical temperature, rises in a continuous fashion upon entering into the structured phase.

A general phase diagram can be derived from these curves. By using the parameters $N\chi_F$ and ϕ_A, a universal result is obtained which is applicable to all symmetric polymer blends. The binodal, which separates the homogeneous state from the two-phase structures, is determined by the dependence on $N\chi_F$ of the values of the compositions ϕ'_A and ϕ''_A in the two equilibrated mixed phases. ϕ'_A and ϕ''_A follow from the minimum condition

$$\frac{\partial \Delta \mathcal{G}_{mix}}{\partial \phi_A} = 0 \quad . \tag{3.83}$$

Using Eq. (3.80), the following analytical expression for the binodal is obtained:

$$N\chi_F = \frac{1}{1 - 2\phi_A} \cdot \ln \frac{1 - \phi_A}{\phi_A} \quad . \tag{3.84}$$

This function is plotted in Fig. 3.23; it is seen again that the critical point is at $\phi_A = 0.5$ and $N\chi_F = 2$.

Phase diagrams for particular polymer blends, as for example in Figs. 3.16 and 3.18, always have T and ϕ_A as the variables. Such phase diagrams can be

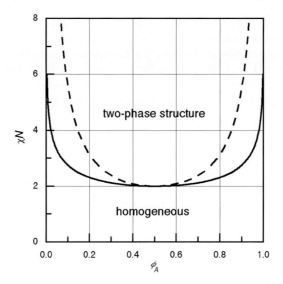

Fig. 3.23. $\chi_F \phi_A$ phase diagram of a symmetric polymer blend ($N_A = N_B = N$). The spinodal (*dashed line*) is shown in addition to the binodal.

derived from the universal ($N\chi_F, \phi_A$) phase diagram if the temperature dependence of χ_F is known and then used. The dependence $\chi_F(T)$, thus, determines the appearance of the phase diagram; different temperature dependencies lead to different types of (ϕ_A, T) phase diagrams. It can be assumed, for many polymer blends, that the contact energy between unlike monomers, given by $\tilde{R} T \chi_F$, is positive and remains largely unchanged over the temperature range of interest. In such systems, χ_F changes with respect to temperature according to

$$\chi_F \propto \frac{1}{T} \; .$$

It is immediately evident which type of mixing behaviour results. A homogeneous melt is found at high temperatures if the molecular weight of both components is not too high. The increase in χ_F upon decreasing temperature then leads to a phase separation. For a symmetric blend of two polymers with the same molecular weight at the critical composition 1:1, it occurs when the critical value $\chi_F = 2/N$ is reached. Since this happens also at the critical temperature T_c, the following can then be written

$$\chi_F = \frac{2}{N} \cdot \frac{T_c}{T} \; . \tag{3.85}$$

The resulting phase diagram is that shown in Fig. 3.17. The combination of Eqs. (3.85) and (3.84) gives the form of the binodal as

$$\frac{T}{T_c} = \frac{2(1 - 2\phi_A)}{\ln\left((1 - \phi_A)/\phi_A\right)} \; . \tag{3.86}$$

The system in Fig. 3.16 corresponds approximately to this case. The different type of phase diagram in Fig. 3.18 shows that a different temperature dependence is observed for the PS-PVME system. It is evident here, for low temperatures, that the Flory–Huggins parameter initially takes a negative value. Upon heating, it increases through the zero point and into the positive region. Phase separation again also occurs here at the critical point, i.e., when

$$N\chi_{\mathrm{F}}(T_{\mathrm{c}}) = 2 \ .$$

This particular mixing process behaviour occurs because of the effect of two competing factors. It was mentioned above that both the contact interaction between neighbouring monomers as well as local entropy effects contribute to the Flory–Huggins parameter. The latter can lead to a de-mixing if mixing results in a volume decrease and, thus, to a restriction in the mobility of the monomers. For the polystyrene-poly(vinylmethylether) system, an energy gain, which is found in this special case for unlike monomers, has, in the first instance at lower temperatures, the upper hand. Upon increasing temperature, entropic effects assume an ever greater importance and lead, in the end, to de-mixing.

While the Flory–Huggins theory was developed for polymer blends, it is interesting to ask whether it can be applied to the case of liquid mixtures of substances with normal low molecular weights by simply setting $N = 1$. In fact, it is usually found that no useful results are obtained by doing this; the main reason for this lies in the use of Eq. (3.70) for $\Delta\mathcal{G}_{\mathrm{loc}}$, which is specific for the theory. Since a polymer makes a very large number of contacts with other chains, it is true that the number of AB contacts corresponds to the statistical average, and it is exactly this upon which Eq. (3.70) is based. This is no longer true for mixtures of small molecules, where there is always a preference for choosing a like or an unlike molecule as a contact partner such that Eq. (3.70) loses its validity. In contrast to the situation with small molecules, each polymer chain, when considering the sum over all contacts, experiences the same local field – this is exactly the same as how in the molecular field description each spin in a ferromagnet, each dipole in a ferroelectric or each rod-like molecule in a nematic phase experiences the same field, as given by M, P, or S_2. The Flory–Huggins theory uses the fact that such a molecular field exists and thus belongs, like the Landau- and the Maier–Saupe theory, to the class of molecular field theories.

3.4.3 Spinodal Decomposition

Phase separation sets in if, starting from a homogeneous phase, the binodal is crossed upon decreasing or increasing the temperature. Structure formation can be easily followed for polymers, since, first, it occurs slowly, because of the high viscosity, and, second, it can always be stopped by a rapid freezing to below the glass temperature. The images in Fig. 3.20 were obtained in this way. Both samples consist of polystyrene and partially-brominated polystyrene, both with the same degree of polymerisation; they differ only in their composition. The melt was, in both cases, homogeneous for temperatures above 220 °C. Phase separation was, then, caused, in both cases, by a temperature jump from 230 °C to 200 °C. The sample, which gave the structure on the right-hand side, has with $\phi(\text{PS}) = 0.5$, the critical composition, while the other sample with $\phi(\text{PS}) = 0.8$ is far away from this condition. The different appearance of the two structures indicates that the phase separation follows a different mechanism in the two cases. Why this is so, indeed why this must be the case becomes clear upon taking another look at the form of the curves $\Delta\mathcal{G}_{\text{mix}}(\phi_A)$ in Fig. 3.22 and upon asking the question as to how a homogeneous mixture reacts to thermally excited local concentration fluctuations after a temperature jump into the two-phase region. Figure 3.24 shows a schematic representation of such a local fluctuation, which starts from a value ϕ_0 and involves an increase $\delta\phi_A$ in the concentration of the A chains in one half of a volume element d^3r and a corresponding decrease in the other volume half. There is a change in the Gibbs free energy associated with the fluctuation, which can be written as

$$\delta\mathcal{G} = \frac{1}{2}(g(\phi_0 + \delta\phi_A) + g(\phi_0 - \delta\phi_A))\mathrm{d}^3r - g(\phi_0)\mathrm{d}^3r \quad . \tag{3.87}$$

$g(\phi_A)$ denotes the composition-dependent Gibbs free energy per unit volume of the homogeneous polymer blend. Expanding $g(\phi_A)$ as a power series up to second order yields the following expression for $\delta\mathcal{G}$:

$$\delta\mathcal{G} = \frac{1}{2}\frac{\partial^2 g}{\partial\phi_A^2}(\phi_0)\delta\phi_A^2\mathrm{d}^3r \quad . \tag{3.88}$$

We use the Flory–Huggins equation to calculate $\partial^2 g/\partial\phi_A^2$ and write

$$\frac{\partial^2 g}{\partial\phi_A^2} = \frac{1}{\mathcal{V}}\frac{\partial^2 \Delta\mathcal{G}_{\text{mix}}}{\partial\phi_A^2} \quad , \tag{3.89}$$

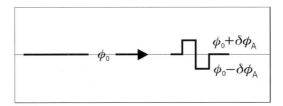

Fig. 3.24. A local concentration fluctuation.

where $\Delta\mathcal{G}_{mix}$ is given by Eq. (3.71). The change in the Gibbs free energy associated with a local concentration fluctuation is thus given as

$$\delta\mathcal{G} = \frac{1}{2}\frac{1}{V}\frac{\partial^2\Delta\mathcal{G}_{mix}}{\partial\phi_A^2}(\phi_0)\delta\phi_A^2 d^3\boldsymbol{r} \ . \tag{3.90}$$

This is a very interesting result since it tells us that it is only the sign of the curvature $\partial^2\Delta\mathcal{G}_{mix}/\partial\phi_A^2$ which determines whether a concentration fluctuation leads to an increase or a decrease in the Gibbs free energy. For a state to be stable, it is necessary that $\delta\mathcal{G}$ is positive, since only then is it sure that a spontaneous local concentration increase of species A is not long lived, but decays again. This can still be the case even after the temperature jump, when the homogeneous phase is fundamentally unstable. The homogeneous melt then remains largely stable with respect to small concentration fluctuations. The curves in Fig. 3.22 always possess, near to the minima, regions with a positive curvature; but, near to the critical concentration there are also regions where the curvature is negative. This has drastic consequences. A negative curvature means that each still small concentration fluctuation leads to a decrease in the Gibbs free energy, and there are no restoring forces. Indeed, there now exists a tendency for a further increase in the concentration fluctuation. It is clear that phase separation, under these conditions, occurs very easily and in a different manner. Such a behaviour is referred to as **spinodal decomposition** and exactly this leads to the structure shown on the right-hand side of Fig. 3.20 which consists of two interpenetrating continuous phases. The lower part of Fig. 3.25 shows schematically how such a spinodal decomposition occurs, namely, starting from very small values, the amplitude of a concentration fluctuation increases continuously, such that two equilib-

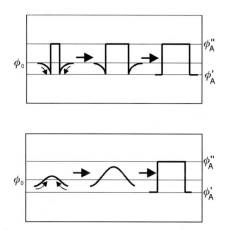

Fig. 3.25. Phase separation mechanisms: nucleation and subsequent growth (*above*) and spinodal decomposition (*below*). The *curved arrows* indicate the direction of the diffusive motion of polymer A.

rium phases with the compositions ϕ'_A and ϕ''_A are finally formed. The two small upwards-pointing arrows indicate that, in this case, the A chains, whose concentration is shown, flow, in contrast to the usual behaviour, in the direction of increasing concentration.

The upper part of the figure shows what happens when the system always removes small concentration fluctuations. Phase separation can then only occur if the thermally-driven fluctuation exceeds a certain size and, thus, a seed with the concentration ϕ''_A of the new equilibrium phase forms. Once this has been created, it can grow further. The growth then occurs in the normal way by means of the diffusion of chains to the growing surface; this is now, as shown in the figure, in the direction of decreasing concentration. Nucleation takes place statistically at a few positions in the sample; this can still be recognised later in the individual precipitations. By contrast, spinodal decomposition involves the whole sample at once, since overall it is unstable to the same degree with respect to concentration fluctuations. The structure is continuous and shows no individual precipitations – this is still clear when the equilibrium phases have formed.

A curve can be drawn in a phase diagram to indicate whether a binary polymer melt, when brought into the two-phase region by a temperature change, is stable or unstable with respect to local fluctuations, i.e., whether phase separation occurs via nucleation and growth or via a spinodal decomposition. As we have seen, a positive curvature in the $\Delta \mathcal{G}_{mix}(\phi_A)$ curve changes into a negative curvature at the boundary. This transition takes place when

$$\frac{\partial^2 \Delta \mathcal{G}_{mix}}{\partial \phi_A^2} = 0 \ . \tag{3.91}$$

The so-defined curve is referred to as the **spinodal**. For the discussed symmetric mixing, it can be directly given. Equation (3.81) is used again, and the following expression is obtained for the dependence of the spinodal:

$$N\chi_F = \frac{1}{2\phi_A(1 - \phi_A)} \ . \tag{3.92}$$

This is the dashed line in Fig. 3.23.

3.5 Exercises

1. A dipole generates an electric field according to

$$\boldsymbol{E}(\boldsymbol{r}) = \frac{3\,(\boldsymbol{pr})\,\boldsymbol{r} - r^2\boldsymbol{p}}{4\pi\varepsilon_0 r^5} \ .$$

   (a) Consider a system consisting of two atoms with a separation a; let the polarisability of the atoms be β. What condition for β and a must be satisfied for the system to be ferroelectric? How are the dipoles arranged with respect to the interatomic vector?

(b) Can the system also adopt an anti-ferroelectric state?

(c) Generalise the case in (a) to that of a linear chain. Show that a sponta-neous polarisation can arise in this case when $\beta \geq \pi \varepsilon_0 a^3 / \sum n^{-3}$. The sum here runs over all natural numbers and equates to $\sum n^{-3} = 1.202$.

2. Consider two electric and magnetic dipoles, p_0 and m_0, positioned on an axis. Calculate the energy difference between a parallel and an anti-parallel arrangement. The distance between the dipoles is 5 Å. The following expressions apply

$$u = -\boldsymbol{p}\boldsymbol{E}_{\text{dipole}} \quad \text{resp.} \quad u = -\boldsymbol{m}\boldsymbol{B}_{\text{dipole}}$$

$$E_{\text{dipole}} = \frac{2p_0}{4\pi\varepsilon_0 r^3} \quad ; \quad B_{\text{dipole}} = \frac{2\mu_0 m_0}{4\pi r^3} \quad .$$

The electric dipole moment is $1\text{D} = 3.34 \times 10^{-30}$ Cm, while the magnetic dipole moment is $\mu_B = 9.27 \times 10^{-24}$ J T^{-1}.

3. Consider the transition from paramagnetic to ferromagnetic behaviour in Landau's molecular field approximation. The Gibbs free energy density is given by

$$g = g_0 + b(T - T_c)M^2 + c_4 M^4 \quad .$$

Determine M_{eq} by applying the equilibrium condition $\partial g/\partial M|_{\text{eq}} = 0$. Calculate the equilibrium value of g, the entropy density $s = -\partial g/\partial T$ and the heat capacity $c_v = T\partial s/\partial T$ as functions of the temperature near the critical temperature T_c.

4

Charges and Currents

Chapter 2 described how condensed matter reacts to the application of an external field, namely via a deformation in a mechanical field, a polarisation in an electric field, or a magnetisation upon applying a magnetic field. Generally, it is structural changes which are involved here, in which case it was established that there is, under normal conditions, a linear dependence with respect to the particular field strength. In addition to these structural consequences, there is a second type of reaction, namely fields can cause currents to flow. A first example of this has already been discussed: If a mechanical shear field is applied to a liquid, it begins to flow, with the flow rate depending on the viscosity. We come now to the electric currents which are made to flow in many materials by electric fields. The condition for this is the existence of free charges, i.e., changes which are mobile over large distances. These can be electrons or ions, as is the case in, e.g., metals or electrolytes, as well as quasi-particles, such as the positively charged 'holes' in semiconductors. For the large majority of encountered conditions, there is also a linear relationship here between the magnitude of the current and the field strength, with the electrical conductivity being the constant of proportionality.

While it is the case that the electrical current is caused to flow and carried by an electric field, it is also true that it can be influenced by magnetic fields via the Lorentz force. These can change the course of charges and also lead to separation of charges. For metals, the presence of a magnetic field can induce loops of current, which make a contribution to the magnetic susceptibility.

In Chap. 3, it was explained that a magnetisation or polarisation can exist via a self-stabilisation process even without the presence of an external field, i.e., it can become permanent. This is, in fact, also true for electric currents: Permanent, constant macroscopic currents can establish themselves without the action of an electric field – this is what happens in superconductors. It was seen that the transition from a para- to a ferromagnetic state occurs in the form of a critical phase transition. Similarly, it is found that the change from a normal conducting to a superconducting state with a permanent current occurs by means of a second-order phase transition – this can be observed for certain crystals at low temperatures. The conditions for the appearance of such a phase transition are of a special nature. Currents in the superconducting state are carried by a special type of quasi-particle, namely Cooper pairs, which are two coupled electrons with opposite spins. The binding together completely changes the character of the charge carrier from that of a fermion to a particle which resembles a spinless boson, and this enables a stationary electric current to flow without any external driving force.

All this will be discussed in this chapter under the heading of 'charges and currents'. The discussion will focus first of all on crystalline solids, not only because the physical understanding is most developed here, but also because the vast majority of technical applications are associated with this class of materials. In addition, the flow of current in liquid electrolytes will be discussed in a shorter section at the end of the chapter.

4.1 Metallic Electrons

It is well known that metals possess characteristic properties, by which they are clearly distinguished from other materials. They have the highest electrical conductivities and at the same time very high heat conductivities. Metals are impenetrable to electromagnetic radiation over a wide range of frequencies up to the 'plasma frequency', which lies in the ultra-violet region. These characteristic specific properties are a consequence of the existence of free electrons. Metallic crystals exist as a lattice of positively charged ion cores in which the electrons can to a large extent move freely like in a gas. Over a century ago and indeed before the introduction of quantum mechanics, Drude created a classical theory based on this picture which can satisfactorily describe some main characteristics of the electrical and heat conductivity of metals.

4.1.1 The Classical Drude Model

Electrical Conductivity. The Drude model considers the electrons to be like an ideal gas which is held in the crystal because of the attractive forces of the ions which, considered altogether, create an unstructured potential well. The average kinetic energy per electron is then given by

$$\frac{3}{2}k_{\mathrm{B}}T = \frac{m_{\mathrm{e}}}{2}\langle v^2\rangle \ . \tag{4.1}$$

The electrons undergo, like the particles in an ideal gas, a free translational motion, which is interrupted by collisions. It is only by means of these collisions that the Maxwell velocity distribution, which is found in the equilibrium state of an ideal gas, can be established. The distance λ_{f} – the **mean free path** – and the average time of free flights, τ_{f}, specify the distance travelled and the time between collisions, respectively. The relation between λ_f and τ_f can be described by

$$\tau_f = \frac{\lambda_f}{\sqrt{\langle v^2 \rangle}} \; . \tag{4.2}$$

If an electric field is applied to the metal, all electrons, regardless of their direction of flight, experience, in the time τ_f between two collisions, an acceleration in the field direction. For a field in the x direction, there is an average increase in the corresponding velocity component of

$$\langle \Delta v_x \rangle = \frac{-eE_x}{m_e} \frac{\tau_f}{2} \; . \tag{4.3}$$

The effect of the field is, thus, to superimpose upon the initially isotropic motion a uniform motion in the x direction with the average velocity $\langle \Delta v_x \rangle$. This means that there is a flow of charge in the x direction with a current density, j_η, equal to

$$j_\eta = -e\rho_e \frac{-eE_x\tau_f}{2m_e} = \frac{\rho_e e^2 \tau_f}{2m_e} E_x \tag{4.4}$$

(ρ_e is the number of electrons per unit volume). The current has a constant value because the additional kinetic energy gained by the acceleration in the field is completely lost again in the collisions. In this way, the electron gas and, thus, the sample heats up and an Ohmian resistance is generated in the sample. **Ohm's law** expresses the proportionality between the current density and the electric field as

$$j_\eta = \sigma_{el} E_x \; , \tag{4.5}$$

where σ_{el} is the **electrical conductivity**. The **Drude theory** therefore yields for σ_{el} the equation

$$\sigma_{el} = \frac{\rho_e e^2 \tau_f}{2m_e} \; , \tag{4.6}$$

which expresses the fact that, in addition to the electron density, there is only one more parameter which determines the size of the electrical conductivity, and that is the time τ_f between two collisions.

Heat Transport and the Wiedemann–Franz Law. The second characteristic property of metals, namely their high heat conductivity, can be treated by adopting the same approach as used for gases. This is illustrated in Fig. 4.1 which also introduces the variables which are used in the description. There is a linear fall in temperature in the x direction, and we then ask: What is the resulting flow of heat through a surface perpendicular to x? This is given by the difference in particle currents, which flow from the left- and right-hand sides through the plane at $x = 0$ and have different average kinetic energies according to the temperature difference. A qualitative approximation can be

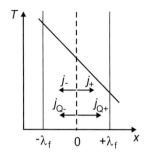

Fig. 4.1. Heat currents in an (electron) gas upon applying a temperature gradient.

made based simply upon average values. As the first variable, the average velocity in the x direction is introduced as

$$v_x = \sqrt{\langle v_x^2 \rangle} \; . \tag{4.7}$$

Since the average value of the total velocity is correspondingly given by

$$v = \sqrt{\langle v^2 \rangle} \; , \tag{4.8}$$

the relationship between the two variables is

$$v_x = \frac{v}{\sqrt{3}} \; . \tag{4.9}$$

The following expression can be used to describe the density of the particle currents which run through the plane of the sample from the left- and right-hand sides:

$$j_+ = -j_- = \beta \rho_e v_x \; , \tag{4.10}$$

where β denotes a dimensionless statistical coefficient on the order of one. It is thus taken into account that both particle currents, firstly, must be equal because the mass distribution in the gas doesn't change, and, secondly, are proportional to the particle density ρ_e and the average velocity in the x direction. In contrast to the particle currents, the heat currents j_{Q+} and j_{Q-}, which run through the reference plane in both directions, have different values, since they have different associated kinetic energies. The average kinetic energy is fixed at the position of the last collision before passing through the sample plane where, through the collision, the electrons took up the local thermal energy. Since the last collisions took place at a distance of about λ_f, the following can be written for the two heat currents

$$j_{Q+} = \beta \rho_e v_x \langle u_{\text{kin}} \rangle (x = -\lambda_f) \tag{4.11}$$

$$j_{Q-} = -\beta \rho_e v_x \langle u_{\text{kin}} \rangle (x = \lambda_f) \; . \tag{4.12}$$

The total heat current flowing through the plane of the sample is given by the sum of the two individual currents and, thus, has a density

$$j_Q = j_{Q+} + j_{Q-} = \beta \rho_e v_x [\langle u_{\text{kin}} \rangle (x = -\lambda_f) - \langle u_{\text{kin}} \rangle (x = \lambda_f)]$$
$$= -2 \beta \rho_e v_x \frac{\mathrm{d} \langle u_{\text{kin}} \rangle}{\mathrm{d} T} \frac{\mathrm{d} T}{\mathrm{d} x} \lambda_f \quad . \tag{4.13}$$

It can be seen that the heat current is proportional to the temperature gradient. Let the **mean free path** be replaced by the collision time τ_f according to

$$\lambda_f = v \tau_f \tag{4.14}$$

and let the **heat conductivity** λ_Q be introduced by writing

$$j_Q = -\lambda_Q \frac{\mathrm{d} T}{\mathrm{d} x} \quad . \tag{4.15}$$

If all numerical coefficients for λ_Q are neglected, the following expression is obtained

$$\lambda_Q \simeq \rho_e \tau_f \frac{\langle u_{\text{kin}} \rangle}{m_e} \frac{\mathrm{d}}{\mathrm{d} T} \langle u_{\text{kin}} \rangle \quad . \tag{4.16}$$

Using

$$\langle u_{\text{kin}} \rangle = \frac{3}{2} k_B T \tag{4.17}$$

leads to the end result

$$\lambda_Q \simeq \frac{\rho_e \tau_f k_B^2 T}{m_e} \quad . \tag{4.18}$$

It was established above in Eq. (4.6) that the electrical conductivity depends only, with the exception of the density ρ_e, upon the average time between collisions, τ_f. The same behaviour is also found here for the heat conductivity. This leads to an interesting conclusion: Since σ_{el} and λ_Q depend in the same way on the variables ρ_e and τ_f, it can be expected that the following constant condition applies for all metals:

$$\frac{\lambda_Q}{\sigma_{\text{el}} T} \simeq \frac{k_B^2}{e^2} = 0.745 \times 10^{-8} \frac{\mathrm{J}\Omega}{\mathrm{s K}^2} \quad . \tag{4.19}$$

The **Wiedemann–Franz law** expresses this expectation which is found to be experimentally valid. In fact, the experimental value is slightly different to that in Eq. (4.19), but this is not surprising given the qualitative nature of the derivation. For most metals, values in a narrow range are found:

$$\frac{\lambda_Q}{\sigma_{\text{el}} T} = (2.4 \pm 0.2) \times 10^{-8} \frac{\mathrm{J}\Omega}{\mathrm{s K}^2} \quad . \tag{4.20}$$

The explanation of the Wiedemann–Franz law in this way at the end of the 19th century was seen as a confirmation of the picture provided by the Drude model. Today, it is known that electrons in metals do not behave at all like particles in a classical ideal gas. While this was not apparent from the treatment of conductivity, it becomes so when considering the heat capacity of metals. There are two contributions to the heat capacity per unit volume:

$$c_v = c_{v,\text{vib}} + c_{v,\text{e}} \ , \tag{4.21}$$

where the first term, $c_{v,\text{vib}}$, and the second term, $c_{v,\text{e}}$, are due to the vibration of the ionic lattice and the electrons, respectively. If the electrons constitute an ideal gas, the following should apply:

$$c_{v,\text{e}} = 3\rho_\text{e}\frac{k_\text{B}}{2} \ . \tag{4.22}$$

As will be explained later in Sect. 5.1.1, the take-up of heat by a lattice is due to the excitation of lattice vibrations. Since, first, the number of independent lattice vibrations corresponds to the number of degrees of freedom of the atoms in the crystal, and, second, each lattice vibration is equivalent to a harmonic oscillator, there is the following contribution to the heat capacity

$$c_{v,\text{vib}} = 3\rho_\text{c}k_\text{B} \tag{4.23}$$

(Eq. (5.38); ρ_c denotes the atomic density in the crystal). The Drude model would thus predict that the specific heat of a metal is

$$c_v = 3\rho_\text{c}k_\text{B} + 3\rho_\text{e}\frac{k_\text{B}}{2} \ . \tag{4.24}$$

Experimentally, it is rather found that, and this is expressed by the 'Dulong–Petit law',

$$c_v \approx 3\rho_\text{c}k_\text{B} \ , \tag{4.25}$$

i.e., a result which is practically solely explained by the lattice vibrations, with there being no contribution which can be compared in the least with that in Eq. (4.22) due to the electrons. The explanation is given by quantum mechanics, and here, in the simplest treatment, by the Fermi Gas Model.

4.1.2 The Fermi Gas Model

In the Fermi gas model, it is assumed, as in the classical Drude model, that the ionic cores of the metal create a homogeneous attractive potential, which prevents the freely mobile electrons from leaving the sample. For the quantum-mechanical treatment, which the Fermi gas model uses, it is necessary to know the eigenstates of the electrons in such a potential well. This requires the solution of the time-independent Schrödinger equation for a free electron

$$-\frac{\hbar^2}{2m_\text{e}}\Delta\psi = \epsilon\psi \tag{4.26}$$

and the determination of the eigenvalues of the energy ϵ. The potential energy at the bottom of the potential well has been set to zero. An appropriate 'ansatz' for the solution is as always for free particles the wavefunction

$$\psi \propto \exp(\mathrm{i}\boldsymbol{k}\boldsymbol{r}) \tag{4.27}$$

with $\boldsymbol{k}$ denoting the wavevector. Inserting ψ into the Schrödinger equation leads to

$$-\frac{\hbar^2}{2m_{\mathrm{e}}}(-k^2)\psi = \epsilon\psi \tag{4.28}$$

and thus the following equation linking the modulus of the wavevector, k, and the energy of the state:

$$\epsilon = \frac{\hbar^2}{2m_{\mathrm{e}}}k^2 \quad . \tag{4.29}$$

For free electrons, $\boldsymbol{k}$ plays the role of a quantum number, i.e., the wavevector defines in a clear way a definite eigenstate and its eigenenergy. The physical meaning of $\boldsymbol{k}$ lies in the fact that the momentum of the eigenstate is defined as $\hbar\boldsymbol{k}$; the chosen eigenfunctions have a momentum with a sharp value.

The discussion has so far not considered the walls of the potential well, i.e., the volume of the metallic body in which the electrons are contained. Upon a first consideration, it would seem that the consideration of the existence of walls means that the choice of a wave-like function can no longer be maintained. The electron waves are reflected at the walls, and standing waves would result as the stationary solutions. The situation can be managed by a superposition of the travelling waves with wavevectors in the opposite direction, however, this leads to a problem: The flow of currents, a central property of metals, cannot be described by standing waves. For a one-dimensional system, i.e., electrons in a very narrow wire, there is a way out. Let the wire be closed into a loop such that a system with a finite length but without surfaces exists. If the wire has a length $\mathcal{N}_{\mathrm{c}}a$, which corresponds to $\mathcal{N}_{\mathrm{c}}$ cells with a lattice constant a, the wavefunction must simply satisfy the condition

$$\psi(x) = \psi(x + \mathcal{N}_{\mathrm{c}}a) \quad . \tag{4.30}$$

Born and von Karman suggested that this **cyclic boundary condition** could be generalised in a formal mathematical way, such that it is applied in the form

$$\psi(\boldsymbol{r}) = \psi(\boldsymbol{r} + \mathcal{N}_1\boldsymbol{a}_1) = \psi(\boldsymbol{r} + \mathcal{N}_2\boldsymbol{a}_2) = \psi(\boldsymbol{r} + \mathcal{N}_3\boldsymbol{a}_3) \tag{4.31}$$

to a three-dimensional crystal. Hereby, a crystal is considered which has lattice constants $\boldsymbol{a}_1, \boldsymbol{a}_2, \boldsymbol{a}_3$, with there being $\mathcal{N}_1, \mathcal{N}_2, \mathcal{N}_3$ atoms along these three directions, respectively. In contrast to the one-dimensional case, there is of course no realisation of this Born–von Karman boundary condition in three dimensions. On the other hand, they do allow, and this is the matter of importance, the selection of discrete eigenfunctions which have the same density of levels in energy space as would result from, for example, standing waves.

The great advantage of the Born–von Karman boundary conditions is that the stationary solutions, Eq. (4.27), associated with travelling waves

$$\psi(\mathbf{r}, t) \propto \exp\left(-\mathrm{i}\frac{\epsilon}{\hbar}t + \mathrm{i}\mathbf{k}\mathbf{r}\right) \tag{4.32}$$

are maintained as solutions; as mentioned above, this is essential for the description of transport phenomena. The three-dimensional cyclic boundary conditions take into account the finite nature of the sample body without introducing surface effects. As long as it is bulk properties of the crystal which are being considered, such surface effects are of no importance, and their complete suppression, as is achieved by the boundary conditions, is actually desired.

The requirement to be satisfied in Eq. (4.31) is equivalent to

$$\exp(\mathrm{i}\mathbf{k}\mathcal{N}_1\mathbf{a}_1) = \exp(\mathrm{i}\mathbf{k}\mathcal{N}_2\mathbf{a}_2) = \exp(\mathrm{i}\mathbf{k}\mathcal{N}_3\mathbf{a}_3) = 1 \ . \tag{4.33}$$

In this case, a solution can be directly stated using the reciprocal lattice: The conditions are met for all wavevectors $\mathbf{k}$ of the series

$$\mathbf{k}_{i_1 i_2 i_3} = i_1\frac{\widehat{\mathbf{a}}_1}{\mathcal{N}_1} + i_2\frac{\widehat{\mathbf{a}}_2}{\mathcal{N}_2} + i_3\frac{\widehat{\mathbf{a}}_3}{\mathcal{N}_3} \quad \text{where} \quad i_1, i_2, i_3 \quad \text{are integers} \ . \tag{4.34}$$

The validity of this statement is immediately apparent upon inserting the above solutions into Eq. (4.33), remembering the definition in Eq. (1.129) of the unit vectors of the reciprocal lattice $\widehat{\mathbf{a}}_1, \widehat{\mathbf{a}}_2$ and $\widehat{\mathbf{a}}_3$.

Although discrete wavevectors have been chosen, they lie very close to each other such that only their density in $\mathbf{k}$ space needs to be considered. This can be directly stated. Since there are altogether $\mathcal{N}_c = \mathcal{N}_1\mathcal{N}_2\mathcal{N}_3$ wavevectors $\mathbf{k}$ which satisfy the cyclic boundary conditions in a unit cell of the reciprocal lattice with volume $\widehat{V}_c$, the following is obtained, using Eq. (1.137), for the density:

$$\frac{\mathcal{N}_c}{\widehat{V}_c} = \frac{\mathcal{N}_c V_c}{(2\pi)^3} = \frac{V}{(2\pi)^3} \ . \tag{4.35}$$

An important parameter in the calculation of derived variables is the **density of levels** $\mathcal{D}$:

$$\mathcal{D}(\epsilon)\mathrm{d}\epsilon \tag{4.36}$$

denotes how many eigenstates are found in the energy interval $\mathrm{d}\epsilon$ at the energy ϵ. The density of levels can be derived from the density of levels in $\mathbf{k}$ space using the equations

$$\mathcal{D}(\epsilon)\mathrm{d}\epsilon = 2\frac{V}{(2\pi)^3}4\pi k^2\frac{\mathrm{d}k}{\mathrm{d}\epsilon}\mathrm{d}\epsilon = 2\frac{V}{(2\pi)^3}4\pi k^2\frac{1}{\hbar^2 k/m_e}\mathrm{d}\epsilon \ , \tag{4.37}$$

which leads to

$$\mathcal{D}(\epsilon) = \frac{V(2m_e)^{3/2}}{2\pi^2\hbar^3}\sqrt{\epsilon} \ . \tag{4.38}$$

For the electron spin, there are two possible orientation directions, and, correspondingly, for the electron, there are two independent eigenstates for each $\boldsymbol{k}$. This is taken into account by the factor two on the right-hand side of Eq. (4.37).

Next the electronic ground state must be considered. Electrons are Fermi particles, and thus each eigenstate can only be occupied by one electron. In the same way that the ground state of an atom involves the successive occupation of the orbitals starting at the lowest energy level, it is found that the electronic ground state of a Fermi gas in the potential well of a metal is likewise given by the occupation of all eigenstates starting from the state with zero momentum, $\boldsymbol{k} = 0$. The highest energy which is reached upon filling all eigenstates with the total number, $\mathcal{N}_e$, of electrons is referred to as the **Fermi energy**, which is denoted here as ϵ_F. Using the expression for the density of levels we obtain

$$\mathcal{N}_e = \int_0^{\epsilon_F} \mathcal{D}(\epsilon)d\epsilon = \frac{\mathcal{V}(2m_e)^{3/2}}{2\pi^2\hbar^3}\frac{2}{3}\epsilon_F^{3/2} \quad , \tag{4.39}$$

and it follows that

$$\epsilon_F = \frac{\hbar^2}{2m_e}(3\pi^2)^{2/3}\rho_e^{2/3} \quad . \tag{4.40}$$

The result shows that the Fermi energy which determines the electronic ground state is solely dependent on the density of electrons in the metal

$$\rho_e = \frac{\mathcal{N}_e}{\mathcal{V}} \quad . \tag{4.41}$$

Using Eqs. (4.38) and (4.39), the following expression for the density of levels at the Fermi level is obtained:

$$\frac{\mathcal{D}(\epsilon_F)}{\mathcal{N}_e} = \frac{3}{2\epsilon_F} \quad . \tag{4.42}$$

The total energy U_{e0} of the electronic ground state can also be calculated:

$$U_{e0} = \int_0^{\epsilon_F} \mathcal{D}(\epsilon)\epsilon d\epsilon = \frac{\mathcal{V}(2m_e)^{3/2}}{2\pi^2\hbar^3}\frac{2}{5}\epsilon_F^{5/2} \quad , \tag{4.43}$$

or, using Eq. (4.39),

$$\frac{U_{e0}}{\mathcal{N}_e} = \frac{3}{5}\epsilon_F \quad . \tag{4.44}$$

The electronic ground state of a Fermi gas is only adopted at absolute zero, because of the fact that the thermal energy at finite temperatures leads to the excitation and occupation of states above the Fermi energy. The occupation of the energy levels at non-zero temperatures is given in the following way by **Fermi–Dirac statistics**:

$$w(\epsilon_i) = \frac{1}{\exp\frac{\epsilon_i - \mu_e}{k_B T} + 1} \tag{4.45}$$

describes the probability that an eigenstate with energy ϵ_i is occupied by a Fermi particle. The expression implies that

$$0 \leq w \leq 1 \ .$$

The variable μ_e is referred to as the **chemical potential** of the electrons and determines, together with the temperature, the distribution. μ_e depends on the number of electrons in the metal and follows from the requirement

$$\sum_i w(\epsilon_i) = \mathcal{N}_e \ , \tag{4.46}$$

or

$$\int_0^\infty \mathcal{D}w\mathrm{d}\epsilon = \mathcal{N}_e \ . \tag{4.47}$$

It is immediately apparent that at $T = 0\,\mathrm{K}$ the chemical potential agrees with the Fermi energy

$$\mu_e = \epsilon_F \tag{4.48}$$

because Eq. (4.45) gives here the result

$$w(\epsilon_i < \epsilon_F) = 1$$

$$w(\epsilon_i > \epsilon_F) = 0 \ ,$$

which exactly reproduces the electronic ground state adopted at absolute zero.

Electronic Heat Capacity. The question as to why the electrons make a much smaller contribution to the heat capacity as compared to that which was initially expected from the classical treatment can now be addressed. Figure 4.2 shows the energy distribution in a Fermi gas. Specifically, the number of electrons per energy interval is shown for the case of absolute zero and a non-zero temperature. It is apparent that only electrons with energies near to the **Fermi surface** at ϵ_F can be excited. Electrons in the region

$$\epsilon_F - \epsilon \simeq k_B T$$

take up thermal energy and are transported into the region

$$\epsilon - \epsilon_F \simeq k_B T \ .$$

All electrons which occupy lower lying energy levels remain completely inactive as if they did not exist at all. The consequence is a dramatic reduction in the electronic contribution to the heat capacity to values, which are very small as compared to the contribution of the lattice vibrations.

The magnitude of the contribution can be calculated. The following expression for the increase in the electronic energy relative to the energy U_{e0} of the electronic ground state can be written

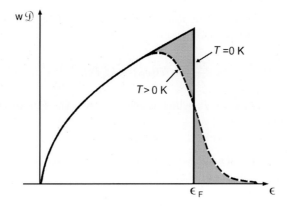

Fig. 4.2. The number of electrons per energy interval for the ground state at $0\,\mathrm{K}$ (all levels are filled up to the Fermi energy ϵ_F) and at a somewhat higher temperature (thermal excitation of the electrons near the Fermi surface according to Fermi–Dirac statistics) (*dashed*).

$$\mathcal{U}_e - U_{e0} = \int\limits_0^{\epsilon_F} (\epsilon_F - \epsilon)[1 - w(\epsilon)]\mathcal{D}(\epsilon)\mathrm{d}\epsilon + \int\limits_{\epsilon_F}^{\infty} (\epsilon - \epsilon_F)w(\epsilon)\mathcal{D}(\epsilon)\mathrm{d}\epsilon \quad . \qquad (4.49)$$

Here, the first term represents the energy which is required to transport all excited electrons, in the first instance, to the Fermi surface, while the second term corresponds to the energy needed to lift, in a second step, the electrons from ϵ_F into the final higher levels. The following then applies for the change with respect to temperature in the electronic energy, that is, for the electronic heat capacity:

$$\frac{\mathrm{d}\mathcal{U}_e}{\mathrm{d}T} = \int\limits_0^{\infty} (\epsilon - \epsilon_F)\mathcal{D}(\epsilon)\frac{\mathrm{d}w}{\mathrm{d}T}\mathrm{d}\epsilon \quad . \qquad (4.50)$$

Since the redistribution occurs in the vicinity of the Fermi surface, this can be approximated by

$$\frac{\mathrm{d}\mathcal{U}_e}{\mathrm{d}T} \approx \mathcal{D}(\epsilon_F)\int\limits_0^{\infty} (\epsilon - \epsilon_F)\frac{\mathrm{d}w}{\mathrm{d}T}\mathrm{d}\epsilon \quad . \qquad (4.51)$$

Making the substitution

$$x = \frac{\epsilon - \epsilon_F}{k_B T} \qquad (4.52)$$

leads to

$$\frac{\mathrm{d}\mathcal{U}_e}{\mathrm{d}T} = \mathcal{D}(\epsilon_F)\frac{(k_B T)^2}{T}\int\limits_{x=-\epsilon_F/(k_B T)}^{\infty} \frac{x^2\exp x}{(\exp x + 1)^2}\mathrm{d}x \quad . \qquad (4.53)$$

At low temperatures the limit for $T \to 0$ can be used for the integral, which is $\pi^2/3$. The result is, thus,

$$\frac{\mathrm{d}\mathcal{U}_e}{\mathrm{d}T} = \mathcal{D}(\epsilon_F)\frac{\pi^2}{3}k_B^2 T \quad . \tag{4.54}$$

Introducing a temperature T_F which is referred to as the **Fermi temperature** by the definition

$$\epsilon_F = k_B T_F \quad , \tag{4.55}$$

and using Eq. (4.42) leads to the final result for the heat capacity of the electrons per unit volume in a metallic sample, $c_{v,e}$:

$$c_{v,e} = \frac{1}{V}\frac{\mathrm{d}\mathcal{U}_e}{\mathrm{d}T} = \frac{\pi^2}{2}\rho_e k_B \frac{T}{T_F} \quad . \tag{4.56}$$

It is now possible to estimate the extent to which $c_{v,e}$ is reduced as compared to that expected from the classical Drude model

$$c_v = \frac{3}{2}\rho_e k_B \quad . \tag{4.57}$$

Typical Fermi energies are on the order of a few eV. This corresponds to Fermi temperatures of $T_F \simeq 10^4 - 10^5\,\mathrm{K}$ and leads to a reduction of more than two orders of magnitude.

Even though the contribution is very small, it is nevertheless possible to determine it using precise measurements. As will be discussed in Sect. 5.1.1, the Debye T^3-law, Eq. (5.63), applies for the contribution of the lattice vibrations to the heat capacity. The heat capacity of a metal is thus overall described by

$$c_v = \frac{36\pi^4}{15}\rho_c k_B \left(\frac{T}{T_D}\right)^3 + \frac{\pi^2}{2}\rho_e k_B \frac{T}{T_F} \quad . \tag{4.58}$$

Figure 4.3 shows a plot of c_a/T against T^2, where c_a is the heat capacity per mole of potassium. The y-intercept in the plot corresponds to the electronic part, which is thus clearly recognisable and determinable.

The evaluation of the results yields an additional insight, namely, the y-intercept is larger than is to be expected on the basis of Eqs. (4.58), (4.55) and (4.40) and the known density ρ_e. The origin of this effect will be discussed later, namely a periodic fluctuation of the potential, which is neglected in the Fermi gas model, leads to there being an apparent change in the mass of the electron. For the case of potassium, it turns out to be enlarged by 25%

$$\frac{m_{\exp}}{m_e} = 1.25 \quad . \tag{4.59}$$

As a consequence, there is a reduction in the Fermi energy and correspondingly also in the Fermi temperature.

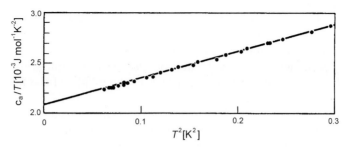

Fig. 4.3. The molar heat capacity of potassium below 1 K. The temperature dependence provides evidence for an electronic contribution in addition to phonon generation (from Lien and Phillips [28]).

Paramagnetism of the Electrons. A large difference as compared to the classical expectation is also found for the magnetic properties of metals. Electrons have together with their spin a magnetic moment. Paramagnetic properties would therefore be expected for metals, whereby the relationship between the strength B of the magnetic field and the induced magnetisation M_z would be given classically by Eq. (2.172)

$$M_z = \mu_0 \rho_e \frac{\mu_B^2}{k_B T} \frac{B}{\mu_0} \quad . \tag{4.60}$$

Actually the fermion character of the electron causes the magnetic susceptibility to be much smaller. The situation for the electronic ground state is shown in Fig. 4.4. The density of levels is plotted as in Fig. 4.2, except that it is rotated by 90° and here shown together with its mirror image, with the right and left sides corresponding to states with spins parallel and anti-parallel to

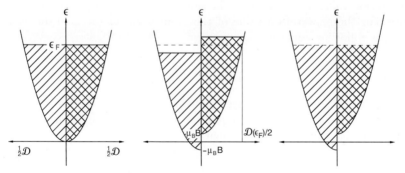

Fig. 4.4. Paramagnetism of metallic electrons at low temperature. *Left*: The starting point in the absence of a field. Occupied energy levels for the two spin orientations are represented separately. *Middle*: The change in the energies upon switching on a magnetic field. *Right*: The lowering of the energy through a redistribution and re-establishment of a uniform Fermi surface.

the field direction, respectively. As long as no magnetic field is present, it is found that both parts are filled up to the Fermi surface at ϵ_F. If a magnetic field of strength B is applied, the energy of all parallel oriented spins is increased by $\mu_B B$, while that of the anti-parallel spins is reduced by the same amount. In the first instance, the situation in the middle of the figure is obtained. This does not, however, correspond to the new equilibrium state since the spin system can gain energy by a redistribution into the situation shown on the right-hand side of the figure. The redistribution occurs at the Fermi surface and thus at a density of levels per unit volume of $3\rho_e/(4\epsilon_F)$ (from Eq. (4.42)). The number of electrons per unit volume for which the spin is inverted equals $\mu_B B 3\rho_e/(4\epsilon_F)$. Each spin inversion is accompanied by a change in the magnetic dipole moment of $2\mu_B$. The reorientation of the spins in the vicinity of the Fermi surface caused by the application of a magnetic field thus leads to a magnetisation

$$M_z = 2\mu_B \frac{3\rho_e}{4\epsilon_F} \mu_B B = \mu_0 \rho_e \frac{3\mu_B^2}{2k_B T_F} \frac{B}{\mu_0} \quad . \tag{4.61}$$

The difference to the classical prediction is immediately apparent: It is – apart from a factor – again the replacement of T by T_F as in the case of the heat capacity.

The Temperature Dependence of the Electrical Conductivity. Are there also important differences in the description of the electrical and heat conductivity resulting between the Drude electron gas and the Fermi gas picture? It will be shown that the changes that exist are rather hidden and not immediately apparent. In order to treat electrical conductivity, it is the equation of motion of the electrons in an applied electric field which is required. Fundamentally, it is true that the equation of motion for the quantum-mechanical expectation value of a variable is the same as for the variable in the classical sense. For the expectation value of the momentum in the field direction x, $\langle p_x \rangle$, the following equation thus applies:

$$\frac{d}{dt}\langle p_x \rangle = -eE_x \quad . \tag{4.62}$$

Since $\langle p_x \rangle = \hbar k_x$ is an exactly determined observable for the wave-like eigenfunctions of the Fermi gas, the equation of motion becomes an equation of motion for the wavevector component k_x:

$$\hbar \frac{d}{dt} k_x = -eE_x \quad . \tag{4.63}$$

An increase in the x component of the wavevector, k_x, which is proportional to time, is thus expected for all electrons, i.e.,

$$\Delta k_x = \frac{-eE_x}{\hbar} t \quad . \tag{4.64}$$

This is illustrated for the $\mathbf{k}$ space in Fig. 4.5. The Fermi surface corresponds in $\mathbf{k}$ space to the surface of a sphere with radius k_F, whose value is given according to Eq. (4.29), as

$$k_F = \frac{(2m_e\epsilon_F)^{1/2}}{\hbar} . \tag{4.65}$$

All states within the **Fermi sphere** would be occupied at absolute zero; for non-zero temperatures there is, as discussed above, simply a redistribution in a narrow region around the Fermi surface. Equation (4.64) seems to imply that the Fermi sphere, as illustrated, would be displaced with a constant velocity along the negative k_x axis. This is, however, not the case. To see this, it is simply necessary to imagine what happens when the electric field is switched off. The Fermi sphere would not remain in the displaced state, but rather it would move back to its original position with its centre at $\mathbf{k} = 0$, since this corresponds to the equilibrium state in the absence of a field. This moving back to the equilibrium state is an irreversible process, and such a process can, for small displacements from equilibrium, be described by a relaxation equation. In this case, it has the form

$$\frac{\mathrm{d}\Delta k_x}{\mathrm{d}t} = -\frac{\Delta k_x}{\tau} , \tag{4.66}$$

where τ denotes the relaxation time. The complete equation of motion for the Fermi sphere is obtained by summing the two effects described by Eqs. (4.64) and (4.66):

$$\frac{\mathrm{d}\Delta k_x}{\mathrm{d}t} = -\frac{eE_x}{\hbar} - \frac{\Delta k_x}{\tau} . \tag{4.67}$$

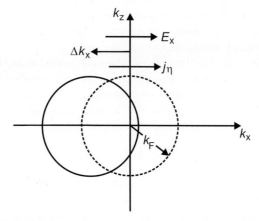

Fig. 4.5. A Fermi sphere in $\mathbf{k}$ space in the absence of a field (*dashed*) together with its displacement under the influence of an electric field. The displaced position is associated with a current density j_η.

It is known that a stationary flow of current results upon applying a field to a metal. Stationarity means that

$$\frac{\mathrm{d}\Delta k_x}{\mathrm{d}t} = 0 \ , \tag{4.68}$$

which leads to the following result for the stationary displacement of a Fermi sphere upon applying an electric field:

$$\Delta k_x = \frac{-eE_x\tau}{\hbar} \ . \tag{4.69}$$

For the electrons, the displacement of the Fermi sphere means that all of them experience the same change in their velocity, which is given by

$$\Delta v_x = \frac{\hbar \Delta k_x}{m_e} = \frac{-eE_x\tau}{m_e} \ . \tag{4.70}$$

This is equivalent to the flow of an electrical current with density

$$j_\eta = -e\rho_e \Delta v_x = \frac{\rho_e e^2 \tau}{m_e} E_x \ . \tag{4.71}$$

It can be seen that this result, which was derived here for a Fermi gas, is the same as that in Eq. (4.4) which was obtained for the Drude model. The following expression is again obtained for the electrical conductivity:

$$\sigma_{\mathrm{el}} = \frac{\rho_e e^2 \tau}{m_e} \ . \tag{4.72}$$

For the Drude model, the time constant, which was written as τ_f, was associated with the time between collisions. Now τ refers to a general relaxation time, but the microscopic basis is the same; a Fermi gas also returns to equilibrium through many collisions.

Quantum effects reveal themselves when the question of the temperature dependence of τ is considered. Figure 4.6 shows, as a representative example, plots of the inverse of σ_{el} against temperature as measured for three different sodium samples. The variable $\sigma_{\mathrm{el}}^{-1}$, which corresponds to the electrical resistance, can be expressed as the sum of two terms

$$\sigma_{\mathrm{el}}^{-1}(T) = \sigma_{\mathrm{el,imp}}^{-1} + \sigma_{\mathrm{el,vib}}^{-1} \tag{4.73}$$

where the first is independent of temperature, while the second increases with increasing temperature. Both terms arise from the interaction of the electrons with the crystal lattice. As will be shown later, a restriction in their free motion always results if the periodic structure of the lattice is disturbed. This can happen in two different ways. Firstly, each crystal has a certain number of lattice defects, which are always present even at low temperature. They provide the first contribution to the resistance. The second contribution is

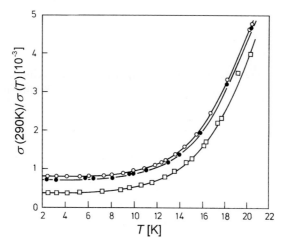

Fig. 4.6. The temperature dependence of the electrical conductivity of three different sodium samples, which differ in the number of perturbing sites (from McDonald and Mendelssohn [29]).

due to lattice vibrations. Their amplitudes increase with temperature, which correspondingly leads to an increase in the deviations from a strict periodicity and, thus, an increase in the perturbation of the free motion of the electrons. The collisions of the electrons with lattice defects and displaced atoms occur with characteristic rates. Typical values are $\tau_{imp}^{-1} \simeq 10^9 \, s^{-1}$ for the collisions with lattice defects and, at room temperature, $\tau_{vib}^{-1} \simeq 10^{14} \, s^{-1}$ for the collisions with vibrating atoms. The basis of Eq. (4.73) is that both rates can be combined in an additive manner

$$\tau^{-1} = \tau_{imp}^{-1} + \tau_{vib}^{-1} \ . \tag{4.74}$$

The part of the resistance due to lattice vibrations increases with temperature, as the amplitude of the vibrations increases in a way which is approximately described by

$$\sigma_{el,vib}^{-1} \propto \tau_{vib}^{-1} \propto T \ . \tag{4.75}$$

Although it is not directly evident from the expression for the conductivity, Eq. (4.72), the physical facts imply that the electrical current in a metal is only carried by electrons in the immediate vicinity of the Fermi surface. In this respect, Fig. 4.5 conveys an unrealistic picture, since in reality the displacement Δk_x is always only tiny in comparison to the radius k_F, and it is only the electrons in the upper, not overlapping part of the states in $\boldsymbol{k}$ space near k_F which carry the current. The acceleration of electrons and changes in momentum by means of collisions only takes place in this narrow region.

The same argument also applies to the conduction of heat. The expression for the heat conductivity λ_Q, which was derived classically in Eq. (4.16), can qualitatively be used for a Fermi gas, if it is considered that the kinetic energy of the active electrons is

$$\langle u_{\mathrm{kin}} \rangle \simeq k_{\mathrm{B}} T_{\mathrm{F}} \quad , \tag{4.76}$$

and that the change with respect to temperature of the kinetic energy, which is only taken up by the electrons in the vicinity of the Fermi surface, is given by

$$\rho_{\mathrm{e}} \frac{\mathrm{d}\langle u_{\mathrm{kin}} \rangle}{\mathrm{d}T} = c_{v,\mathrm{e}} \simeq \rho_{\mathrm{e}} k_{\mathrm{B}} \frac{T}{T_{\mathrm{F}}} \tag{4.77}$$

according to Eq. (4.56). Interestingly, the product of these two expressions, which determines λ_{Q}, no longer depends on the Fermi temperature, and so the same result as in the classical derivation is obtained:

$$\lambda_{\mathrm{Q}} \simeq \frac{\rho_{\mathrm{e}} \tau k_{\mathrm{B}}^2 T}{m_{\mathrm{e}}} \quad . \tag{4.78}$$

Since both the expression for σ_{el} and that for λ_{Q} are identical with the classical result, the Wiedemann–Franz law also remains valid.

The Frequency Dependence of the Conductivity and the Dielectric Function. The above discussion has considered the current which flows in a metal upon applying a constant electric field. What changes are to be expected if the field oscillates with time, with periods of oscillation on the order of standard AC power supplies or with much higher frequencies, corresponding to wavelengths in the IR or UV region? To consider this question, the equation of motion, Eq. (4.67), can again be used as a starting point, but a field which varies periodically with frequency ω is inserted on the right-hand side. The change with respect to time in the wavenumber is equivalent to the change in the velocity of the electrons, such that the following can be written

$$m_{\mathrm{e}} \left(\frac{\mathrm{d}\Delta v_x}{\mathrm{d}t} + \frac{\Delta v_x}{\tau} \right) = -e E_0 \exp(-\mathrm{i}\omega t) \quad . \tag{4.79}$$

The stationary solution of the equation of motion is found by assuming that

$$\Delta v_x(t) = \Delta v_0 \exp(-\mathrm{i}\omega t) \quad . \tag{4.80}$$

Inserting this into Eq. (4.79) leads to an equation for the amplitude Δv_0 of the variation in velocity:

$$m_{\mathrm{e}} \left(-\mathrm{i}\omega + \frac{1}{\tau} \right) \Delta v_0 = -e E_0 \quad . \tag{4.81}$$

The periodically alternating field causes an alternating current with density

$$j_\eta = j_{\eta 0} \exp(-\mathrm{i}\omega t) = -e \rho_{\mathrm{e}} \Delta v_0 \exp(-\mathrm{i}\omega t) \tag{4.82}$$

to flow, whose amplitude $j_{\eta 0}$ is given by

$$j_{\eta 0} = \frac{\rho_e e^2 \tau}{m_e (1 - i\omega\tau)} E_0 \quad . \tag{4.83}$$

The constant of proportionality is a frequency-dependent conductivity which obeys the law

$$\sigma_{el}(\omega) = \frac{\sigma_{el}(\omega = 0)}{1 - i\omega\tau} \quad . \tag{4.84}$$

The parameter in the denominator, $\sigma_{el}(\omega = 0)$, corresponds to the static conductivity of Eq. (4.72). The **frequency-dependent conductivity** is a complex variable with a real and an imaginary part,

$$\sigma_{el}(\omega) = \frac{\sigma_{el}(0)}{1 + \omega^2\tau^2} + i\frac{\sigma_{el}(0)\omega\tau}{1 + \omega^2\tau^2} \quad . \tag{4.85}$$

The complex character means that, in the general case, a phase difference can arise between the current and the field strength.

Deviations from the static value are only found when the product $\omega\tau$ is no longer negligibly small. It was explained above in the discussion of the static conductivity that the relaxation time τ at room temperature is determined by the time between collisions with the vibrating lattice, which is of the order of 10^{-14} s. This means that the static value is still found for field frequencies which are well below 10^{14} Hz. A frequency dependence is only observed for frequencies comparable to the collision frequencies, in which case a rapid falloff in the conductivity is found.

When the metallic electrons move with an electric field, the propagation of electromagnetic waves is affected. In order to understand this effect, it is necessary to determine the frequency dependence of the refractive index or the dielectric function. It can be easily recognised that $\varepsilon(\omega)$ is directly linked to $\sigma(\omega)$. The alternating field causes a spatial displacement of the electrons and, since this occurs relative to the lattice of fixed positively charged cores, a polarisation arises which is proportional to the displacement Δx:

$$P(t) = -e\rho_e \Delta x(t) \quad . \tag{4.86}$$

The current density is also linked to the displacement, via

$$j_\eta(t) = -e\rho_e \Delta v_x(t) = e\rho_e i\omega \Delta x(t) \quad . \tag{4.87}$$

It follows that the ratio of the polarisation to the current density is given by

$$\frac{P(t)}{j_\eta(t)} = \frac{1}{-i\omega} \tag{4.88}$$

and the amplitude P_0 of the polarisation is therefore

$$P_0 = \frac{j_{\eta 0}}{-i\omega} = \frac{\sigma_{el}(\omega)}{-i\omega} E_0 \quad . \tag{4.89}$$

The following expression is, thus, obtained for the dielectric susceptibility χ, as defined by Eq. (2.91):

$$\varepsilon_0 \chi(\omega) = \frac{\sigma_{\text{el}}(\omega)}{-i\omega} \quad . \tag{4.90}$$

Using Eq. (4.84) leads to

$$\varepsilon_0 \chi(\omega) = \frac{\sigma_{\text{el}}(0)}{-i\omega - \omega^2 \tau} \tag{4.91}$$

and for the dielectric constant $\varepsilon(\omega)$ to

$$\varepsilon(\omega) = 1 + \chi(\omega) = 1 + \frac{\rho_e e^2 \tau}{\varepsilon_0 m_e} \frac{1}{-i\omega - \omega^2 \tau} \quad . \tag{4.92}$$

It is well known that metals are not transparent to visible light. Why this is so can be understood using this equation for $\varepsilon(\omega)$. For visible light, we have

$$\omega \tau \gg 1 \quad ,$$

such that the dielectric function can, to a good approximation, be described by

$$\varepsilon(\omega) \approx 1 - \frac{\rho_e e^2}{\varepsilon_0 m_e \omega^2} \quad . \tag{4.93}$$

It is evident that negative values can be obtained for $\varepsilon(\omega)$. This occurs for

$$\omega^2 < \omega_{\text{pl}}^2 = \frac{\rho_e e^2}{\varepsilon_0 m_e} \quad , \tag{4.94}$$

i.e., for all frequencies which are below a certain limiting frequency ω_{pl}. The fact that the dielectric function is negative has drastic consequences for the propagation of waves. This becomes clear upon solving the equation

$$\mu_0 \frac{\partial^2 \boldsymbol{D}}{\partial t^2} = \nabla^2 \boldsymbol{E} \tag{4.95}$$

which comes from the Maxwell equations and describes the propagation of waves in matter. The displacement vector $\boldsymbol{D}$ is related to the electric field $\boldsymbol{E}$ according to

$$\boldsymbol{D} = \varepsilon_0 \varepsilon(\omega) \boldsymbol{E} \quad . \tag{4.96}$$

For a planar monochromatic wave

$$E_x(y, t) = E_0 \exp i(-\omega t + ky) \tag{4.97}$$

it follows, upon inserting into Eq. (4.95), that

$$\mu_0 \varepsilon_0 \varepsilon(\omega) \omega^2 = k^2 \quad . \tag{4.98}$$

Introducing the velocity of light c_l as defined by

$$c_l^{-2} = \mu_0 \varepsilon_0 \quad , \tag{4.99}$$

leads to, making use of Eqs. (4.93) and (4.94), the result:

$$\left(1 - \frac{\omega_{pl}^2}{\omega^2} \right) \frac{\omega^2}{c_l^2} = k^2 \quad . \tag{4.100}$$

The left-hand side becomes negative when $\omega < \omega_{pl}$, with the result that the wavenumber k becomes imaginary

$$k = ik'' \quad .$$

The wavenumber being imaginary means that

$$E_x(y, t) = E_0 \exp(-i\omega t - k'' y) \quad , \tag{4.101}$$

i.e., the wave can no longer propagate with constant amplitude, but rather it decays exponentially. Since propagation of light in metals is not possible for frequencies below ω_{pl}, an electromagnetic wave with such a frequency which hits a metal surface is reflected. A light field with an exponentially decaying intensity, which is referred to as an **evanescent wave**, only forms near the surface; the depth of penetration is on the order of the wavelength of the incident radiation. For metals, the limiting frequency ω_{pl} is in the UV region; metals are thus impenetrable to visible light.

Plasma Oscillations. The following dispersion relation follows from Eq. (4.100) for frequencies above ω_{pl}:

$$\omega^2 = \omega_{pl}^2 + c_l^2 k^2 \quad . \tag{4.102}$$

This dispersion relation has unusual properties. Firstly, it can be established that the phase velocity ω/k is greater than the velocity of light, whereas the group velocity $d\omega/dk$ is less than the velocity of light. Both statements are a consequence of the fact – in contrast to the situation with electromagnetic waves in a vacuum – that the frequency for $k \to 0$ does not go to zero but assumes a finite value ω_{pl}. Obviously, restoring forces are even acting here in this case of a homogeneous field distribution. Figure 4.7 illustrates how these forces arise. Imagine a disc-like metallic sample and consider that all the electrons are shifted upwards by Δx as shown. In this way, a negative and positive layer of charge forms on the upper and lower surfaces, respectively, and there is, thus, as in a capacitor, a homogeneous electric field E_x, whose strength is given by

$$\varepsilon_0 E_x = e \rho_e \Delta x \quad . \tag{4.103}$$

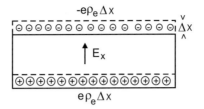

Fig. 4.7. The fundamental mode of the plasma vibrations: The surface charges which form upon a collective displacement of all electrons generate a restoring electric field.

The field acts, in turn, on the electrons and leads to an acceleration which is described by the equation of motion

$$\rho_e m_e \frac{\mathrm{d}^2 \Delta x}{\mathrm{d}t^2} = -e\rho_e E_x = -\frac{1}{\varepsilon_0} e^2 \rho_e^2 \Delta x \quad . \tag{4.104}$$

The solution of this equation is a vibration with eigenfrequency

$$\omega^2 = \frac{\rho_e e^2}{\varepsilon_0 m_e} \quad . \tag{4.105}$$

Exactly the same expression was obtained in Eq. (4.94) for $\omega_{\rm pl}$. It can now be recognised why the known limiting frequency is referred to as the **plasma frequency**: It is the frequency of the eigenvibration of the electron gas, which can undergo this vibration without an external field.

A further theoretical analysis shows that $\omega_{\rm pl}$ is the fundamental mode of a large number of vibrations, which the electronic plasma can undergo in a metallic body. Every local accumulation generates electric fields which cause restoring forces and hence vibrations. Eigenmodes with well-defined frequencies result from wave-like modulations of the electron density. Time- and spatially variable electric fields generate magnetic fields, such that vibrations occur in which charge displacements and electromagnetic fields are coupled. These can only exist above the frequency of the fundamental vibration of the plasma, $\omega_{\rm pl}$; below this limiting value, electromagnetic waves cannot propagate.

4.2 Electrons and Holes in Semiconductors

Metals are clearly different to other materials because of their high electrical conductivity. Other crystals also contain displaceable electrical charges, and the question arises as to why their contribution to current flow is much less. In semiconductors, the currents are much less than in metals, but they are still measurable and can be used. There are, however, also materials which act as electric insulators, i.e., there is indeed no measurable flow of current upon applying an electric field. It is necessary to explain the origin of these enormous differences in electrical conductivity – they amount to over 20 orders of magnitude.

The answer will be given in this section, namely, it is a consequence of the effect on the electron states of variations of the potential which are periodic with the lattice. In the previous discussion of metals, such variations were not considered and a homogeneous potential was assumed. Under these conditions, the electrons behave as free particles and can thus have all energy values. Variations of the potential which are periodic with the lattice change the situation, in that certain regions in the energy are no longer accessible to the electrons. **Energy bands** are formed, i.e., energy intervals, which are occupied by eigenstates, together with energetically inaccessible band gaps in between them. This becomes apparent from a theoretical perturbation treatment of the question as to how the energies of free electrons in a homogeneous potential well change upon adding a weak lattice periodic potential.

4.2.1 Electrons in a Periodic Crystal Field

Figure 4.8 presents in advance the result of the theoretical analysis carried out for a one-dimensional system with period a. The solid and partially dashed line represents the parabolic dependence of the energy on the wavenumber of the free electron as described by Eq. (4.29). Gaps are formed in the accessible energies if a weak potential, which varies with the lattice period a, is switched on. The positions k, at which they appear, are found at all half values of the reciprocal lattice points

$$G_h = h\frac{2\pi}{a} \quad , \quad h \text{ is an integer} \ , \tag{4.106}$$

of the system. The different energy bands occur in k space in different **Brillouin zones**, with the band with the lowest energy being in the first Brillouin zone $|k| < \pi/a$, the next lowest being in the second Brillouin zone $\pi/a < |k| < 2\pi/a$, etc. The band gaps are always at the boundaries of the Brillouin zones.

In order to explain this, it is necessary to solve the Schrödinger equation

$$\left[-\frac{\hbar^2}{2m_e}\frac{\mathrm{d}^2}{\mathrm{d}x^2} + u(x) \right]\psi(x) = \epsilon\psi(x) \ . \tag{4.107}$$

The potential $u(x)$ varies with the period of the lattice:

$$u(x) = u(x + la) \ . \tag{4.108}$$

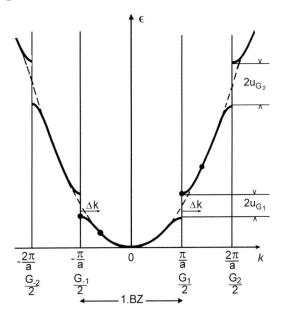

Fig. 4.8. The changes in the eigenvalues of the energy of the electrons upon the transition from a flat potential to one with a structure which is periodic with the lattice. The formation of energy bands and gaps. The main effects are at the boundaries of the Brillouin zones.

$u(x)$ can thus be represented as a Fourier series

$$u(x) = \sum_{G_h} u_{G_h} \exp(iG_h x) \quad . \tag{4.109}$$

The coefficients are assigned, in this way, to the points of the reciprocal lattice, Eq. (4.106). u_0 is chosen to be

$$u_0 = 0 \quad .$$

Since the potential is real, it follows that

$$u_{G_h} = u_{-G_h} \quad .$$

Right from the start, let us specify that the system has only a finite size and, therefore, cyclic boundary conditions must, as in the case of free electrons, be satisfied. The following is valid for a one-dimensional crystal made up of $\mathcal{N}_c$ cells:

$$\psi(x) = \psi(x + \mathcal{N}_c a) \quad . \tag{4.110}$$

Waves

$$\psi(x) \propto \exp(ikx)$$

with wavenumbers

$$k = i\frac{2\pi}{\mathcal{N}_c a} \quad , \quad i \text{ is an integer} \quad , \tag{4.111}$$

satisfy this condition. The same is true for a sum of waves of the type

$$\psi(x) = \sum_k c_k \exp(ikx) \quad (k \text{ from Eq. (4.111)}) \quad . \tag{4.112}$$

Inserting Eqs. (4.112) and (4.109) into the Schrödinger equation, Eq. (4.107), gives a product $u\psi$, which can be written as

$$\sum_{G_h,\, k'} u_{G_h} c_{k'} \exp[i(G_h + k')x] = \sum_{G_h,\, k} u_{G_h} c_{k-G_h} \exp(ikx) \quad . \tag{4.113}$$

Differentiating twice gives

$$\sum_k \left[\frac{\hbar^2 k^2}{2m_e} c_k + \sum_{G_h} u_{G_h} c_{k-G_h}\right] \exp(ikx) = \epsilon \sum_k c_k \exp(ikx) \quad . \tag{4.114}$$

The sums run here over discrete k (Eq. (4.111)) and G_h values (Eq. (4.106)). This equation can only be solved if it is fulfilled for each individual k separately, i.e., the following must apply

$$\frac{\hbar^2 k^2}{2m_e} c_k + \sum_{G_h} u_{G_h} c_{k-G_h} = \epsilon c_k \quad . \tag{4.115}$$

The effect of a periodic lattice potential can now be recognised. While all waves $\exp(ikx)$ represent independent solutions in the case of a homogeneous potential, i.e., for vanishing coefficients u_{G_h}, a periodic lattice potential leads to their coupling and mixing. It is to be noted that not all waves are coupled with each other, but rather only those that differ from each other by the wavenumber of a point of the reciprocal lattice, G_h.

Equation (4.115) can be used to determine the effect on the eigenstates and eigenenergies of switching on a weak periodic lattice potential. It is well known that an intensive mixing of states together with associated large energy changes occurs if the states in the unperturbed state are degenerate or almost degenerate. For the states which are to be coupled, this occurs exactly at positions where a splitting occurs in Fig. 4.8, i.e., for pairs

$$k_1 = -\frac{G_h}{2}, \quad k_2 = \frac{G_h}{2} \quad .$$

Both states have the same energy and differ from each other by G_h. For closeby states with

$$k_1 = -\frac{G_h}{2} + \Delta k, \quad k_2 = \frac{G_h}{2} + \Delta k \quad \left(\frac{\Delta k}{G_1} \ll 1\right)$$

it is always the case that

$$(k - G_h)^2 \approx k^2 \quad . \tag{4.116}$$

The states are almost degenerate, and thus marked energy changes and an especially strong mixing is to be expected. In such a situation, an approximate treatment of the infinite series of equations in Eq. (4.115) is to consider only the two most important ones:

$$\frac{\hbar^2 k^2}{2m_e} c_k + u_{G_h} c_{k-G_h} = \epsilon c_k \tag{4.117}$$

$$\frac{\hbar^2 (k - G_h)^2}{2m_e} c_{k-G_h} + u_{G_h} c_k = \epsilon c_{k-G_h} \quad . \tag{4.118}$$

The equations only depend on the amplitudes c_k and c_{k-G_h} of the two waves, which are strongly coupled together. The non-trivial solution of the set of homogeneous linear equations is obtained by setting the secular determinant to zero,

$$\begin{vmatrix} \epsilon_{0,k} - \epsilon & u_{G_h} \\ u_{G_h} & \epsilon_{0,k-G_h} - \epsilon \end{vmatrix} = 0 \quad , \tag{4.119}$$

where

$$\epsilon_{0,k} = \frac{\hbar^2 k^2}{2m_e} \quad \text{and} \quad \epsilon_{0,k-G_h} = \frac{\hbar^2 (k - G_h)^2}{2m_e} \tag{4.120}$$

are the energies in the unperturbed state. It thus follows that the eigenenergies after switching on the perturbation are given by

$$\epsilon = \frac{1}{2} [\epsilon_{0,k} + \epsilon_{0,k-G_h}] \pm \sqrt{\frac{1}{4} [\epsilon_{0,k} - \epsilon_{0,k-G_h}]^2 + u_{G_h}^2} \quad . \tag{4.121}$$

The energy change is largest at the boundaries of the Brillouin zone

$$-k_1 = k_2 = \frac{G_h}{2} \quad , \tag{4.122}$$

where the unperturbed energy is

$$\epsilon_{0,k} = \epsilon_{0,k-G_h} = \frac{\hbar^2}{2m_e} \frac{G_h^2}{4} = \epsilon_{0,G_h/2} \quad . \tag{4.123}$$

The degenerate energy levels are split here as

$$\epsilon = \epsilon_{0,G_h/2} \pm u_{G_h} \quad . \tag{4.124}$$

Upon proceeding a further Δk away from the two Brillouin zone boundaries, as indicated in Fig. 4.8, the energies after switching on the perturbation become

$$\epsilon = \epsilon_{0,G_h/2} + \frac{\hbar^2 \Delta k^2}{2m_e} \pm \sqrt{\frac{\hbar^2 \Delta k^2}{m_e} 2\epsilon_{0,G_h/2} + u_{G_h}^2} \ . \tag{4.125}$$

The following approximate result is obtained for small distances Δk:

$$\epsilon \approx \epsilon_{o,G_h/2} \pm u_{G_h} + \frac{\hbar^2 \Delta k^2}{2m_e} \left(1 \pm 2 \frac{\epsilon_{0,G_h/2}}{u_{G_h}} \right) \ . \tag{4.126}$$

This expression means that the curves $\epsilon(k)$ run horizontally at the Brillouin zone boundaries, as is shown in the figure. Equations (4.117) and (4.124) further show that both waves are, at the Brillouin zone boundaries, superimposed with the wavevectors being of the same magnitude but running in opposite directions, i.e., they form standing waves. The periodic lattice potential, thus, changes the character of the wavefunctions from that of a propagating to a standing wave. The two standing waves found at the Brillouin zone boundaries are of the form

$$\exp\left(i\frac{G_h}{2}x \right) + \exp\left(-i\frac{G_h}{2}x \right) \propto \cos\left(\frac{G_h}{2}x \right) \tag{4.127}$$

and

$$\exp\left(i\frac{G_h}{2}x \right) - \exp\left(-i\frac{G_h}{2}x \right) \propto \sin\left(\frac{G_h}{2}x \right) \ . \tag{4.128}$$

They are spatially shifted with respect to each other and have, since the interaction with the periodic lattice potential is then changed, different eigenenergies which determine the size of the energy gap.

In this perturbation theory treatment of the problem only two dominant wavefunctions have been combined, and the contributions of all others have been neglected. In fact, a general statement can be made about the form of the exact eigenfunctions, this being referred to as the **Bloch theorem**. Equation (4.115) demonstrated that the periodic lattice potential does not mix all free electron waves, but rather only those that differ from each other by a wavenumber G_h of the reciprocal lattice. Inversely, this means that the exact solution must be of the form

$$\psi(x) \propto \sum_{G_h} c_{k-G_h} \exp[i(k - G_h)x] \tag{4.129}$$

$$= \exp(ikx) \sum_{G_h} c_{k-G_h} \exp(-iG_h x) \ . \tag{4.130}$$

The sum

$$\sum_{G_h} c_{k-G_h} \exp(-iG_h x)$$

has the same form as the Fourier expansion, Eq. (4.109), of the periodic lattice potential and, thus, likewise represents a periodic lattice function. This means that the electron wavefunctions in a crystal are generally of the form

$$\psi(x) \propto \exp(ikx)\phi_k(x), \quad \text{with} \quad \phi_k(x) = \phi_k(x+la) \ , \tag{4.131}$$

i.e., they can be written as a product of a wave and a periodic lattice function. The free electron waves thus experience, in the crystal lattice, a modulation of the amplitude which oscillates regularly with period a. These are generally referred to as **Bloch waves**. The parameter k of the Bloch waves, in contrast to the case of free electrons, no longer corresponds to the momentum. Momentum is no longer conserved since the potential u is no longer homogeneous.

The k value of a Bloch wave is not unambiguously determined. This is apparent from the following rearrangement

$$\exp(ikx)\phi_k(x) = \exp\left[i(k+G_h)x\right]\exp\left(-iG_h x\right)\phi_k(x) = \exp\left[i(k+G_h)x\right]\phi'(x) \tag{4.132}$$

It can be recognised that the shift of k by a vector G_h of the reciprocal lattice again leads to a form which is in agreement with the Bloch theorem. There is a way by which k can be chosen in an unambiguous fashion: Let k be fixed such that it is always in the first Brillouin zone, i.e.,

$$-\frac{\pi}{a} < k \leq \frac{\pi}{a} \ . \tag{4.133}$$

This leads to an altered representation of the $\epsilon(k)$ dependence, with the diagram in Fig. 4.8 becoming that in Fig. 4.9. The lowest band stays unchanged, while the part between $G_1/2$ and $G_2/2$ is shifted by G_1 to the left and the part on the other side is shifted by the same amount to the right. All energybands in the new representation, which is referred to as the **reduced scheme**, are in the first Brillouin zone.

$\hbar k$ is, for reasons which will be given later, referred to as the **quasi momentum** of the Bloch wave. Within the first Brillouin zone, there are, according to Eq. (4.111), $\mathcal{N}_c$ possible values for the quasi momentum. This means that there are, in each individual band, $2\mathcal{N}_c$ states, which can be occupied by the electrons, whose spin has two possible orientations. The eigenstates can be classified by means of two quantum numbers, namely the wavenumber, k, which determines the quasi momentum and the number of the band, j. The corresponding wavefunction is

$$\psi_{kj} \propto \exp(ikx)\phi_{kj}(x) \ , \tag{4.134}$$

where ϕ_{kj} depends on both quantum numbers.

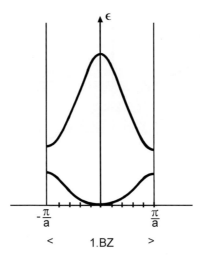

Fig. 4.9. The electron states in a one-dimensional lattice. The representation corresponding to the restriction of k values to the first Brillouin zone (the reduced scheme).

Real crystals are three dimensional. The corresponding theory is formally more complex, but the fundamental phenomena do not change. The discussion here is limited to a few important generalisations. For a three-dimensional crystal, the switching on of a lattice periodic potential again leads to a mixing of the free electron wavefunctions, where again only wavefunctions which differ from each other by a vector $\boldsymbol{G}_{hkl}$ of the reciprocal lattice are combined. Strong mixing effects, together with marked energy changes, are also only seen if the two wavefunctions in the potential-free case are degenerate or almost degenerate. The condition for this in the one-dimensional case was given in Eq. (4.116) and becomes, in the three-dimensional case

$$|\boldsymbol{k} - \boldsymbol{G}_{hkl}|^2 \approx |\boldsymbol{k}|^2 \quad . \tag{4.135}$$

The condition is exactly fulfilled, in the one-dimensional case, at the Brillouin zone boundaries. In the three-dimensional case, the boundaries become surfaces in $\boldsymbol{k}$ space; their geometric meaning becomes clear when the equation is written as

$$\boldsymbol{k} \cdot \frac{\boldsymbol{G}_{hkl}}{|\boldsymbol{G}_{hkl}|} = \frac{|\boldsymbol{G}_{hkl}|}{2} \quad . \tag{4.136}$$

This equation assigns a plane in $\boldsymbol{k}$ space to each vector $\boldsymbol{G}_{hkl}$ of the reciprocal lattice, such that the plane, which is perpendicular to $\boldsymbol{G}_{hkl}$, bisects the vector at its midpoint. The so-obtained Brillouin zone boundaries which are nearest to the origin enclose the first Brillouin zone of a three-dimensional crystal. Figure 4.10 illustrates this for the case of a two-dimensional crystal. It can be straightforwardly appreciated that the area of the first Brillouin zone of

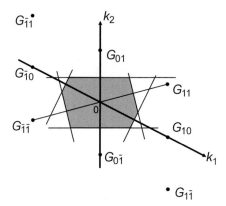

Fig. 4.10. The first Brillouin zone (*filled*) in the reciprocal space (k_1, k_2 space) of a two-dimensional crystal.

a two-dimensional crystal and correspondingly the volume of the first Brillouin zone of a three-dimensional crystal correspond to the area and the volume, respectively, of the unit cell of the reciprocal lattice.

It is evident that the Bloch theorem also applies in three dimensions, i.e., the exact electron wavefunctions in a three-dimensional periodic potential are of the form

$$\psi(\boldsymbol{r}) \propto \exp(\mathrm{i}\boldsymbol{k}\boldsymbol{r})\phi_{\boldsymbol{k}}(\boldsymbol{x}) \quad , \tag{4.137}$$

where $\phi_{\boldsymbol{k}}$ is also lattice periodic:

$$\phi_{\boldsymbol{k}}(\boldsymbol{r}) = \phi_{\boldsymbol{k}}(\boldsymbol{r} + \boldsymbol{R}_{uvw}) \quad . \tag{4.138}$$

The fact that the partitioning into the exponential function and the lattice periodic function $\phi_{\boldsymbol{k}}$ is not unambiguous also still applies, since $\boldsymbol{k}$ can always be shifted by a vector of the reciprocal lattice $\boldsymbol{G}_{hkl}$ such that the Bloch form is retained. The notation becomes unambiguous when $\boldsymbol{k}$ is restricted to the first Brillouin zone, with the result that the energy dependence for all bands as a function of $\boldsymbol{k}$ can be presented in the first Brillouin zone. The electron eigenstates

$$\psi_{\boldsymbol{k}j} \propto \exp(\mathrm{i}\boldsymbol{k}\boldsymbol{r})\phi_{\boldsymbol{k}j}(\boldsymbol{r}) \tag{4.139}$$

are now determined by two quantum numbers, the wavevector $\boldsymbol{k}$, which determines the quasi momentum $\hbar\boldsymbol{k}$, and the number of the band j.

A restriction to only finite extensions for a crystal, as in the one-dimensional case, can be formally achieved by imposing cyclic boundary conditions:

$$\psi(\boldsymbol{r}) = \psi(\boldsymbol{r} + \mathcal{N}_1 \boldsymbol{a}_1) = \psi(\boldsymbol{r} + \mathcal{N}_2 \boldsymbol{a}_2) = \psi(\boldsymbol{r} + \mathcal{N}_3 \boldsymbol{a}_3) \quad . \tag{4.140}$$

It is clear that the solution is the same as for a Fermi gas and is given by Eq. (4.34). There are, within the first Brillouin zone, exactly as many possible values $\boldsymbol{k}$ as there are cells in the crystal ($\mathcal{N}_c$), such that there are, considering the electron spin, $2\mathcal{N}_c$ states which can be occupied in each band.

The point has now been reached where the question raised at the start of the section about the reason for the existence of conductors, semiconductors, and insulators can be answered. These systems differ radically in their conductivity or the resistance which acts against a current. The best conductor has a resistance σ_{el}^{-1} on the order of $10^{-8}\,\Omega\mathrm{m}$, while semiconductors and insulators cover the enormous ranges 10^{-4} to $10^7\,\Omega\mathrm{m}$ and 10^{12} to $10^{20}\,\Omega\mathrm{m}$, respectively. The origin of these differences is indicated in Fig. 4.11. In the next section, it will be explained how the flow of current in a crystal with lattice periodic potential can be described. Even without this knowledge, it is clear, from symmetry, that no current, which has a definite direction, can flow when a band is completely full of electrons. A flow of current is only possible if there are energy bands in the crystal, which are only partially occupied by electrons. It is in exactly this respect that metals, semiconductors and insulators differ. Figure 4.11 shows the occupation of bands for the three cases. Only the uppermost – it is filled with diagonal hatching – of the completely occupied bands is shown (the vertical direction corresponds to the energy, while the horizontal direction has no meaning). For metals, the band above it is still occupied, but not completely. The highest occupied level of the electronic ground state, which is again and, in fact, generally referred to as the Fermi energy ϵ_F, is, for metals, within this band. The fact that there are, contrary to the assumptions of the Fermi gas model, energy gaps in a metal is not important since, as discussed above, only electrons in the vicinity of

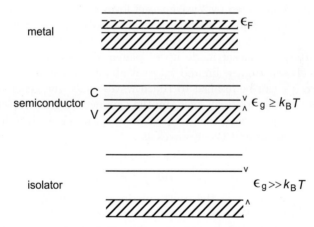

Fig. 4.11. The occupation of the highest energy bands for metals, semiconductors and insulators. For metals, there is a partially occupied band, in which the Fermi energy ϵ_F lies. For semiconductors, the conduction band (C) can only be occupied by the thermal excitation of electrons from the valence band (V). Insulators have a band gap ϵ_g, which is large compared to the thermal energy $k_B T$.

the Fermi surface can be excited and so contribute to the flow of current. The electronic ground state of a semiconductor, as it would be at absolute zero, is shown in the middle. The occupation ends with the completely filled **valence band** (V). The difference in energy ϵ_g to the initially empty **conduction band** (C) is such that it can be overcome by thermal excitation since ϵ_g is of the same order of magnitude or at least not too far above $k_B T$. The conduction of current is then possible, with its strength being determined by the number of thermally excited electrons. Finally, for insulators, the energy gaps from the last occupied to the first free band are so large that they can practically no longer be overcome thermally; in the absence of mobile electrons, there can be no flow of current.

4.2.2 Electron- and Hole-Conduction

Following the discussion of the flow of current in metals using the Fermi gas model, the question now arises as to how the currents in semiconductors can be described. For the treatment of metallic electrons, it was simply possible to use the classical equation of motion for the momentum $\hbar\mathbf{k}$. For the Bloch waves associated with semiconductors, $\hbar\mathbf{k}$ no longer corresponds to a momentum with a sharp value. However, it can be established that this observable behaves as a **quasi momentum** in that it likewise experiences a constant acceleration upon applying an electric field in the absence of any perturbations.

A similar approach to that for a Fermi gas can be followed in order to prove this. The motion of an individual localised electron is again considered, initially for a one-dimensional system. In the same way that a single electron in a metal was represented by a wavepacket formed from the overlap of planar waves, it is also possible to describe a single electron in a semiconductor as a wavepacket, which is generated this time from the superposition of Bloch waves. For the wavepacket made up of planar waves due to free particles, waves were chosen from a limited interval about an average wavevector k. The same recipe can be applied to Bloch waves, since the generation of the wave packet is brought about solely by the exponential term $\exp(ikx)$. Wave packets move in space with the group velocity v_g, which is given in the same way for free electrons and Bloch waves as

$$v_g(k) = \frac{\mathrm{d}\omega}{\mathrm{d}k} = \frac{1}{\hbar}\frac{\mathrm{d}\epsilon}{\mathrm{d}k} \quad . \tag{4.141}$$

k is here the wavevector in the centre of the k interval. Work is performed if a field acts on an electron in a semiconductor, such that the energy ϵ is correspondingly increased. In this way, the following applies, as in classical physics:

$$\mathrm{d}\epsilon = -eEv_g\mathrm{d}t \quad . \tag{4.142}$$

Since it can also be written that

$$\mathrm{d}\epsilon = \frac{\mathrm{d}\epsilon}{\mathrm{d}k}\mathrm{d}k = \hbar v_g \mathrm{d}k \quad , \tag{4.143}$$

it follows that

$$\hbar\frac{\mathrm{d}k}{\mathrm{d}t} = -eE \quad . \tag{4.144}$$

The equation states that the quasi-momentum $\hbar k$ of an electron in a semi-conductor experiences a uniform acceleration under the action of an electric field.

An individual mobile electron, which is represented by a wavepacket, makes a contribution

$$j_\eta \propto -ev_g(k) \tag{4.145}$$

to the electrical current density. The meaning of this is shown in Fig. 4.12. The electron, which is denoted by a point, is initially at the lowest point of the band at $k = 0$ and then moves in the negative k direction under the influence of the

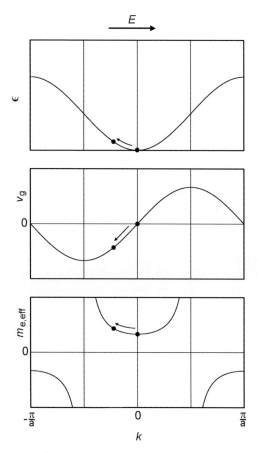

Fig. 4.12. The motion of an electron in the conduction band under the influence of an electric field (one-dimensional system): the energy, group velocity and effective mass as a function of the central k.

applied field E. The middle diagram represents the change in the group velocity – it is seen to adopt negative values. The negative group velocity together
with the negative elementary charge generate the expected positive current.
For an unperturbed motion of the electron, the quasi-momentum changes with
a constant rate according to Eq. (4.144). In contrast, a completely different
behaviour is observed for the group velocity. v_g can, as is apparent from the
figure, be negative and positive and correspondingly goes through phases of
acceleration and slowing down. In classical physics, the mass determines the
ratio between force and acceleration. This suggests, therefore, that an **effective mass** can be defined for the electrons in a semiconductor, such that the
ratio between the electrical force and the acceleration of the wavepacket is
correctly described. Using Eqs. (4.144) and (4.143), let us write

$$\frac{dv_g}{dt} = \frac{1}{\hbar} \frac{d^2\epsilon}{dk\,dt} = \frac{1}{\hbar} \frac{d^2\epsilon}{dk^2} \frac{dk}{dt} = -e \frac{1}{\hbar^2} \frac{d^2\epsilon}{dk^2} E \ . \tag{4.146}$$

It then follows that the effective mass $m_{e,eff}$ is given by

$$m_{e,eff} = \hbar^2 \left(\frac{d^2\epsilon}{dk^2} \right)^{-1} \ . \tag{4.147}$$

It can be seen that the effective mass of the electron is determined by the
curvature of the energy with respect to k in the band. The value of the effective
mass, thus, changes within a band, as is shown in the bottom part for the
example in Fig. 4.12. It is to be noted that $m_{e,eff}$ can adopt both positive and
negative values.

Although everything that has been stated above is theoretically correct, an
unperturbed motion of the electron is as rarely found in real semiconductors
as in metals. In semiconductors, the accelerating motion of electrons is also
interrupted again and again by collisions. It is to be emphasised that such
collisions only occur if the crystal lattice is perturbed. An ideal unperturbed
periodic potential has exactly the discussed effects on the velocity and mass
of the electrons, with the quasi-momentum maintaining its linear change with
time. Only distortions from ideality, such as imperfections in the crystal or
thermal motion, cause collisions and interrupt the constant motion of $\hbar k$. In
the real case, an electron, which is initially at $k = 0$, would not move far
from its starting point because of these collisions. A well-defined effective
mass can then be assigned to it. The dispersion relation $\epsilon(k)$ at $k = 0$ can be
derived from Eq. (4.126). From this, the following expression is obtained for
the effective mass of the electron:

$$m_{e,eff} = m_e \left(1 + 2 \frac{\epsilon_{0,G_h/2}}{u_{G_h}} \right)^{-1} \ . \tag{4.148}$$

In a semiconductor, the electrons move from the valence band to the conduction band by thermal excitation. In this way, they leave behind an unoccupied level in the valence band. The valence band is now no longer fully

occupied and can itself make a contribution to the flow of current. Figure 4.13 shows how this happens. Consider in the first instance the individual picture in the upper left. The open circle at the upper part of the band represents the unoccupied state which was created by the excitation. Applying a field E in the shown direction causes all electrons to move in the negative k direction and the unoccupied state simply moves with them. The valence band in this displaced state also now carries a current. All electrons with a single exception are present as pairs with quasi-momentum $\hbar k$ and $-\hbar k$ and thus cannot make a contribution to the flow of current. There remains only the one left-over uncompensated electron with a positive k value. It has, as can be seen, a negative group velocity which leads to a positive flow of current.

An alternative equivalent picture describes these facts in a simpler way, as is shown on the right-hand side of the figure. The given situation appears from

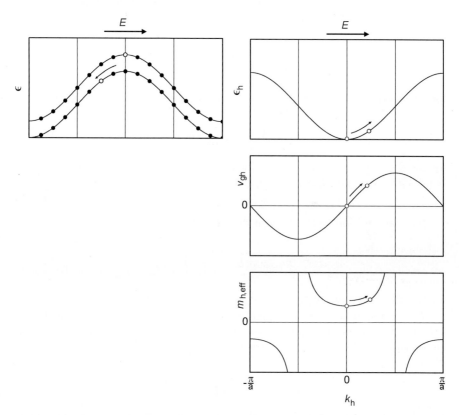

Fig. 4.13. The motion of a single unoccupied state in the valence band of a one-dimensional crystal under the influence of an electric field (*left*). The equivalent description as the motion of a positively charged quasi-particle referred to as a 'hole' (*right*): energy ϵ_h, group velocity v_{gh} and effective mass $m_{h,eff}$ as a function of the quasi-momentum k_h of the holes.

outside exactly as if there was a particle with a positive charge $+e$, which, as shown at the top, is the only particle moving in a mirror-image generated energy band. This quasi-particle is referred to as a **hole**. The hole has a quasi-momentum $\hbar k_h$, an energy ϵ_h, a group velocity v_{gh}, and an effective mass $m_{h,eff}$. All variables depend on k_h as is shown in the figure. The hole gains, under the influence of the field, a positive group velocity, and is accelerated with a corresponding positive effective mass, and thus gives rise to a positive current.

The description of the generated current as the reaction of all electrons in an almost full band or that of a single hole in an almost empty band leads to the same result, but the second form is obviously easier. By this approach, the conduction band and the valence band can formally be treated in the same way. Thermal excitation creates an electron-hole pair and both the electron and the hole move as individual particles in their respective bands, with both being characterised by the associated group velocity and effective mass. For the sake of completeness, the expressions are explicitly given for the hole. They are

$$\hbar\frac{dk_h}{dt} = +eE \tag{4.149}$$

for the change in the quasi-momentum of the hole in an external field,

$$v_g(k_h) = \frac{1}{\hbar}\frac{d\epsilon_h}{dk}(k_h) \tag{4.150}$$

for the dependence of the group velocity on the derivative of the energy in the band with respect to k,

$$j_\eta \propto +ev_g(k_h) \tag{4.151}$$

for the contribution to the current density, and

$$m_{h,eff} = \hbar^2\left(\frac{d^2\epsilon_h}{dk_h^2}\right)^{-1} \tag{4.152}$$

for the effective mass of the hole.

The transformation of the equations for the electrons and holes from the one- to the three-dimensional case is formally straightforward. The group velocity of the electrons is then given by

$$\boldsymbol{v}_g = \frac{1}{\hbar}\frac{d}{d\boldsymbol{k}}\epsilon \quad. \tag{4.153}$$

$\boldsymbol{v}_g$ is thus proportional to the gradient of the energy, i.e., it is perpendicular to the surfaces with constant energy. The equation of motion of the quasi-momentum is

$$\hbar\frac{d\boldsymbol{k}}{dt} = -e\boldsymbol{E} \quad. \tag{4.154}$$

The effective mass becomes, in three dimensions, a tensor property. Writing

$$\frac{d\boldsymbol{v}_{\mathrm{g}}}{dt} = \frac{1}{\hbar}\frac{d}{dt}\frac{d}{d\boldsymbol{k}}\epsilon = -e\frac{1}{\hbar^2}\frac{d}{d\boldsymbol{k}}\frac{d}{d\boldsymbol{k}}\epsilon\cdot\boldsymbol{E} \tag{4.155}$$

leads to the following expression for the ij components of the tensor of the reciprocal of the effective mass

$$\left(\frac{1}{m_{\mathrm{e,eff}}}\right)_{ij} = \frac{1}{\hbar^2}\frac{d^2\epsilon}{dk_i dk_j} \quad . \tag{4.156}$$

The nature of the corresponding expressions for the holes is obvious. It is necessary to use the energy ϵ_{h} and the quasi-momentum $\hbar k_{\mathrm{h}}$ together with the charge $+e$.

Intrinsic Conductivity and Doping Effects. What is the effect of all this on the electrical conductivity of semiconductors? The conductivity depends on temperature because of the thermal generation of electron-hole pairs and can in addition be influenced by the important technical procedure of **doping**, i.e., foreign atoms are introduced into the crystal lattice such that additional electrons or holes become available. In the first instance, it can be established that the contributions of electrons and holes to the conductivity are combined in an additive fashion. σ_{el} can be written as

$$\sigma_{\mathrm{el}} = e(\rho_{\mathrm{e}}\nu_{\mathrm{el,e}} + \rho_{\mathrm{h}}\nu_{\mathrm{el,h}}) \quad . \tag{4.157}$$

The expression contains the densities ρ_{e} and ρ_{h} of the electrons and holes, as well as for both charge carriers an additional term,

$$\nu_{\mathrm{el}} = \frac{|v_{\mathrm{g}}|}{E} \quad , \tag{4.158}$$

which is referred to as the **electric mobility**. The definition of ν_{el} in this way leads to the following statement: Applying an external field E leads, for both charge carriers, to a particle velocity v_{g}, which is proportional to E. The reason for this is the same as that for metallic electrons. In semiconductors, the applied field causes, as described in the previous section, a uniform acceleration of the charge carriers, which is however also in a semiconductor interrupted again and again by collisions. There is, as in the case of metallic electrons, a constant average velocity; the faster the collisions follow each other, the smaller it is (Eq. (4.70)). It is found, as is to be expected, that the electric mobilities $\nu_{\mathrm{el,e}}$ and $\nu_{\mathrm{el,h}}$ of both charge carriers decrease with increasing temperature. In contrast to metals, this, however, is not the property which determines the temperature dependence of the conductivity. The conductivity does not decrease, but rather it increases with increasing temperature because the densities of the electrons and holes increase.

We calculate first ρ_{e}, considering the level scheme on the left-hand side of Fig. 4.14. The occupation of the energy levels in the conduction band (C) is determined by the Fermi statistics, i.e., by Eq. (4.45). As will be shown later,

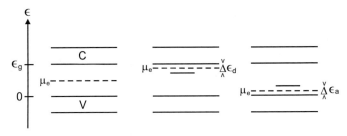

Fig. 4.14. Energy level schemes for the valence (V) and conduction (C) bands of differently treated semiconductors: without doping (*left*), n-doped (*middle*) and p-doped (*right*).

the chemical potential of the electrons μ_e, which together with T determines the distribution, lies approximately in the middle of the energy gap. For the electronic ground state, which is adopted at absolute zero, there is, as there must be, a completely occupied valence band and an empty conduction band. In semiconductors, the gap energies are normally, while not unovercomable, clearly larger than the thermal energy:

$$\frac{\epsilon_g}{k_B T} > 1 \quad . \tag{4.159}$$

The Fermi statistics distribution function can be approximated to

$$w \approx \exp{-\frac{\epsilon - \mu_e}{k_B T}} \quad . \tag{4.160}$$

The dependence $\epsilon(k)$ at the bottom of the conduction band can, as is indicated in Fig. 4.12 and also stated in Eq. (4.126), be described approximately by the parabolic form:

$$\epsilon - \epsilon_g \propto k^2 \quad . \tag{4.161}$$

In this way, the same situation as for a Fermi gas (Eq. (4.29)) is found, and, therefore, the previously derived expression for the density of levels $\mathcal{D}$ (Eq. (4.38)) can be applied here again:

$$\mathcal{D}(\epsilon) = \frac{V}{2\pi^2} \left(\frac{2m_{e,\text{eff}}}{\hbar^2}\right)^{3/2} (\epsilon - \epsilon_g)^{1/2} \quad . \tag{4.162}$$

Using this, the electron density in the conduction band ρ_e can be calculated as

$$\rho_e = \int\limits_{\epsilon_g}^{\infty} \frac{\mathcal{D}}{V}(\epsilon) w_e(\epsilon) d\epsilon = \frac{1}{2\pi^2} \left(\frac{2m_{e,\text{eff}}}{\hbar^2}\right)^{3/2} \int\limits_{\epsilon_g}^{\infty} (\epsilon - \epsilon_g)^{1/2} \exp{-\frac{\epsilon - \mu_e}{k_B T}} d\epsilon \quad . \tag{4.163}$$

Making the substitution

$$x = \frac{\epsilon - \epsilon_g}{k_B T} \tag{4.164}$$

and using the fact that

$$\int_0^\infty x^{1/2} \exp(-x) \, dx = \frac{\sqrt{\pi}}{2} \tag{4.165}$$

leads to the result

$$\rho_e = 2 \left(\frac{m_{e,\text{eff}} k_B T}{2\pi \hbar^2} \right)^{3/2} \exp - \frac{\epsilon_g - \mu_e}{k_B T} \quad . \tag{4.166}$$

An analogous approach yields the density of the holes ρ_h. Generally, the probability that an electron state is not occupied or, expressed in a different way, that this electron state is occupied with a hole, is given by

$$w_h = 1 - w = 1 - \frac{1}{\exp \frac{\epsilon - \mu_e}{k_B T} + 1} \approx \exp \frac{\epsilon - \mu_e}{k_B T} \quad . \tag{4.167}$$

Following the approach described by Fig. 4.13, the electron energy is replaced by the hole energy $\epsilon_h = -\epsilon$, while a parabolic approximation is likewise used for $\epsilon_h(k_h)$. The density of levels is thus given as:

$$\mathcal{D}_h = \frac{V}{2\pi^2} \left(\frac{2m_{h,\text{eff}}}{\hbar^2} \right)^{3/2} \epsilon_h^{1/2} \quad , \tag{4.168}$$

while

$$w_h \approx \exp \frac{-\epsilon_h - \mu_e}{k_B T} \quad . \tag{4.169}$$

The density ρ_h is then

$$\rho_h = \int_0^\infty \frac{\mathcal{D}_h(\epsilon_h)}{V} w_h(\epsilon_h) d\epsilon_h = 2 \left(\frac{m_{h,\text{eff}} k_B T}{2\pi \hbar^2} \right)^{3/2} \exp - \frac{\mu_e}{k_B T} \quad . \tag{4.170}$$

The expressions here for a three-dimensional crystal only use a single value for the effective mass of the electrons and holes. In fact, the effective masses are, in the general case, direction dependent, and the equations for the densities thus contain average values over all directions.

An important and interesting result is obtained upon forming the product of the electron density Eq. (4.166) and the hole density Eq. (4.170). The resulting expression

$$\rho_e \rho_h = 4 \left(\frac{k_B T}{2\pi \hbar^2} \right)^3 (m_{e,\text{eff}} m_{h,\text{eff}})^{3/2} \exp - \frac{\epsilon_g}{k_B T} \tag{4.171}$$

no longer contains the chemical potential. Since, in a pure semiconductor, which is being discussed here, electrons and holes are always formed as pairs, it follows that the densities must be the same:

$$\rho_e = \rho_h \quad , \tag{4.172}$$

i.e.,

$$\rho_e = \rho_h = 2 \left(\frac{k_B T}{2\pi\hbar^2} \right)^{3/2} (m_{e,\text{eff}} m_{h,\text{eff}})^{3/4} \exp{-\frac{\epsilon_g}{2 k_B T}} \quad . \tag{4.173}$$

A comparison with Eq. (4.166) yields an expression for the chemical potential:

$$\mu_e = \frac{\epsilon_g}{2} + \frac{3}{4} k_B T \ln \frac{m_{h,\text{eff}}}{m_{e,\text{eff}}} \quad . \tag{4.174}$$

It is to be recognised for the case where the effective masses of the electrons and holes are approximately the same,

$$m_{h,\text{eff}} \approx m_{e,\text{eff}} \quad , \tag{4.175}$$

that the chemical potential of the electron lies, as indicated in Fig. 4.14, roughly in the centre of the band gap.

It is now possible, since the densities are known, to determine the conductivity. Equations (4.157) and (4.173) give the result

$$\sigma_{el} = 2e \left(\frac{k_B T}{2\pi\hbar^2} \right)^{3/2} (m_{e,\text{eff}} m_{h,\text{eff}})^{3/4} \exp{-\frac{\epsilon_g}{2 k_B T}} \cdot (\nu_{el,e} + \nu_{el,h}) \quad . \tag{4.176}$$

The as-derived equation applies for the case of a pure semiconductor and describes its **intrinsic conductivity**. Commercially used materials are normally not pure at all, but rather foreign atoms are deliberately introduced in a controlled way. Such a doping allows the charge carrier concentration to be directly influenced, and this is a method which is of great practical importance. The most important material for the production of commercially used semiconductors is tetravalent silicon. If an arsenic or phosphorous atom – both are pentavalent elements – is deliberately introduced into the silicon lattice, the extra electron, which is not required for the four bonds in the lattice, is available for the conduction of current. At low temperatures, the electron remains, at first, still bonded to the arsenic or phosphorous atoms, because of the Coulomb force, in a hydrogen-like orbital. If the hydrogen scheme is assumed, the energy levels of this electron are given as

$$\epsilon = \epsilon_g - \frac{e^2}{4\pi\varepsilon_0\varepsilon r_1} \frac{1}{2n^2} \qquad n = 1, 2, 3 \ldots \quad , \tag{4.177}$$

where

$$r_1 = \frac{4\pi\varepsilon_0\varepsilon\hbar^2}{e^2 m_{\mathrm{e}}} \qquad (4.178)$$

denotes the radius of the orbit in the ground state $n = 1$. A complete separation from the mother atom, i.e., a dissociation, is possible for the extra electron. The associated energy is

$$\Delta\epsilon_{\mathrm{d}} = \frac{e^4 m_{\mathrm{e}}}{(4\pi\varepsilon_0\varepsilon)^2 2\hbar^2} \quad . \qquad (4.179)$$

At room temperature, this is easily achieved thermally. The energetic situation which results from introducing such **electron donating** atoms into the silicon crystal is represented in the middle of Fig. 4.14. The additional line corresponds to the level of the additional electron in its lowest bound state. The dissociation energy $\Delta\epsilon_{\mathrm{d}}$ determines how far this level lies below the conduction band. Introducing an electron donor atom into the semiconductor crystal shifts the chemical potential upwards. The additional electron, in the ground state of the electron system at absolute zero, remains bound and occupies this localised state. Therefore, μ_{e} must necessarily lie above this energy.

Additional holes are created if trivalent atoms such as aluminium or boron are introduced, instead of pentavalent atoms, into the silicon lattice. These **acceptor** atoms take an electron from the valence band in order to make the fourth covalent bond. Therefore, the inverse situation results. In the ground state at absolute zero, a hole is bound to the negatively charged mother atom. Thermal energy can release it, with an energy $\Delta\epsilon_{\mathrm{a}}$ being required. At typical operational temperatures, this thermal energy is available, and the holes are freely mobile. The representation of the associated electron level scheme is given on the right-hand side of Fig. 4.14; it shows that μ_{e} is now shifted downwards to a position near to the upper edge of the valence band.

After doping, in contrast to the case of a pure semiconductor, in which the number of electrons and holes are the same, there is a preponderance of one type of charge carrier. The terms n-type (negative charge carriers) and p-type (positive charge carriers) materials are used. The dominance of one type of charge carrier is further amplified because, according to Eq. (4.171), the product $\rho_{\mathrm{e}}\rho_{\mathrm{h}}$ remains constant upon doping, and thus, for example, an increase in the electron density automatically leads to a lowering in the hole density. Although Eq. (4.171) was derived in the discussion of pure semiconductors, it is generally applicable, i.e., for all materials, be they pure or doped, because μ_{e} is not included.

p-n Junctions and Schottky Barriers. The controlling properties of semiconductor components are in most cases a consequence of the physical properties of the interfaces between regions of different conductivity. As well as metal-semiconductor contacts, junctions between differently doped regions of a semiconductor are especially used. A basic element of this type is the

p-n junction at the interface between an acceptor- and donor-rich part of a silicon semiconductor. It is instructive to consider its properties in more detail.

If the surfaces of a p- and an n-doped semiconductor crystal are brought into contact with each other, the band structure at the junction, and also that of the nearby region, are changed, as is represented schematically in Fig. 4.15. Upon being put together, electrons from the n-region, where they exist with a higher concentration, diffuse into the p-region, and, in reverse, there is diffusion of holes in the opposite direction. In this way, an electrical double layer forms with a charge distribution η which is shown in the lower part of the figure. In the p-region, there is a negative charge because of the diffusion away of holes and the diffusion in of electrons, while in the n-region there is correspondingly a positive charge, with the two accumulations of charge being localised in the vicinity of the boundary. The double layer continues to build up until the system has reached a new equilibrium.

This is achieved when the chemical potential of the electrons, which is initially at a different level in the p- and n-parts, becomes constant everywhere. A contribution to this newly established chemical potential, which is

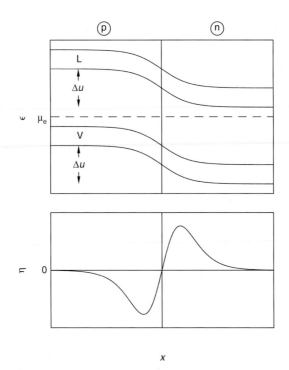

Fig. 4.15. p-n junction in the current-free equilibrium state: The position-dependent change in the energy ϵ of the electrons in the conduction and valence bands (*top*) and the charge distribution $\eta(x)$ (*bottom*).

often explicitly referred to as the **electrochemical potential**, is made by the electrostatic potential due to the double layer. The band scheme in the figure shows the energy of the electrons in the valence and conduction bands together with the changes in the transition region. The electrostatic part of the energy

$$u(x) = -eV(x) \qquad (4.180)$$

shows exactly the variation which is required to maintain the electrochemical potential at a constant level. The electrostatic potential $V(x)$ in the region of the double layer is, in this way, linked to the charge distribution $\eta(x)$ by the Poisson equation

$$\frac{\mathrm{d}^2 V}{\mathrm{d}x^2} = -\frac{\eta(x)}{\varepsilon_0} \qquad . \qquad (4.181)$$

The height Δu of the step in the overall electrostatic potential energy of the electrons is shown in the figure. A step of the same height is also found for the holes on account of their potential energy in the field of the electric double layer:

$$u_{\mathrm{h}}(x) = eV(x) \qquad . \qquad (4.182)$$

The equilibrium which establishes is of a dynamic nature. There are currents in both directions for both the electrons and the holes and the equilibrium is characterised by an exact compensation. The (particle-) current of electrons in the positive x direction, which is referred to as the **generation current**, arises spontaneously, since, for a decreasing potential energy, there is no barrier at all to be overcome. It depends solely on the density of electrons in the p-region:

$$j_{\mathrm{eg}}(0) \propto \rho_{\mathrm{e}}^{\mathrm{p}} \qquad . \qquad (4.183)$$

The opposing current in the opposite direction comes from a region with a high electron density $\rho_{\mathrm{e}}^{\mathrm{n}}$, but it has to climb up the potential step. There is only a certain probability, which is given by Boltzmann statistics, that the step in the potential energy will be realized. Correspondingly, the following is written for the (negative) current from n to p:

$$j_{\mathrm{er}}(0) \propto -\rho_{\mathrm{e}}^{\mathrm{n}} \cdot \exp -\frac{\Delta u}{k_{\mathrm{B}} T} \qquad . \qquad (4.184)$$

This part of the current is referred to in the literature as the **recombination current**. At equilibrium, generation and recombination currents compensate each other:

$$j_{\mathrm{eg}}(0) + j_{\mathrm{er}}(0) = 0 \qquad . \qquad (4.185)$$

Control elements react with specific current changes upon applying an external voltage. The p-n junction has rectification properties, i.e., its conductivity depends on the direction of the applied voltage. It is easy to understand why this is so. If a voltage with a drop off in the $+x$ direction is applied using metal electrodes to the p-n element of Fig. 4.15, the step in the potential energy of the electrons at the transition is reduced, let us assume, by the amount $e\delta V$. As a consequence, the recombination current is increased and is now given by

$$j_{\mathrm{er}}(\delta V) = j_{\mathrm{er}}(0) \exp \frac{e\delta V}{k_{\mathrm{B}}T} \quad . \tag{4.186}$$

On the other hand, the generation current, for which the step is of no importance, remains unchanged

$$j_{\mathrm{eg}}(\delta V) = j_{\mathrm{eg}}(0) \quad . \tag{4.187}$$

The sum of both currents, j_{e}, is then

$$j_{\mathrm{e}} = j_{\mathrm{er}} + j_{\mathrm{eg}} = j_{\mathrm{er}}(0) \exp \frac{e\delta V}{k_{\mathrm{B}}T} + j_{\mathrm{eg}}(0) = j_{\mathrm{er}}(0) \left(\exp \frac{e\delta V}{k_{\mathrm{B}}T} - 1 \right) \quad . \tag{4.188}$$

Thus, the current density increases exponentially as a function of δV for a voltage drop in the positive x direction. Equation (4.188) can also be used for the opposite voltage direction, i.e., a negative δV. In this case, it can be seen that the current rises to a low limiting value, $j_{\mathrm{eg}}(0)$, and then remains constant.

Analogous equations can be formulated for the holes. The generation current of the holes, which runs from the n- to the p-region in the negative direction, is determined by the number of holes in the n-region $\rho_{\mathrm{h}}^{\mathrm{n}}$:

$$j_{\mathrm{hg}}(0) \propto -\rho_{\mathrm{h}}^{\mathrm{n}} \quad , \tag{4.189}$$

while the recombination current in the positive direction is influenced, as for the electrons, by the height of the step in the potential energy Δu, and is given by

$$j_{\mathrm{hr}}(0) \propto \rho_{\mathrm{h}}^{\mathrm{p}} \cdot \exp -\frac{\Delta u}{k_{\mathrm{B}}T} \tag{4.190}$$

($\rho_{\mathrm{h}}^{\mathrm{p}}$ is the density of holes in the p-region). At equilibrium, the two currents compensate each other. If a voltage with a fall-off in the positive x direction is applied, the recombination current increases according to

$$j_{\mathrm{hr}}(\delta V) = j_{\mathrm{hr}}(0) \exp \frac{e\delta V}{k_{\mathrm{B}}T} \quad , \tag{4.191}$$

while the generation current of the holes remains unchanged, i.e.,

$$j_{\mathrm{hg}}(\delta V) = j_{\mathrm{hg}}(0) \quad , \tag{4.192}$$

such that the total particle current of the holes is

$$j_{\mathrm{h}} = j_{\mathrm{hr}} + j_{\mathrm{hg}} = j_{\mathrm{hr}}(0) \left(\exp \frac{e\delta V}{k_{\mathrm{B}}T} - 1 \right) \quad . \tag{4.193}$$

The charge current density j_η corresponds to the sum of the contributions of the electrons and the holes and is given by

$$j_\eta = ej_h + (-e)j_e = ej(0)\left(\exp\frac{e\delta V}{k_B T} - 1\right) ,$$
(4.194)

where

$$j(0) = j_{hr}(0) - j_{er}(0) .$$
(4.195)

It is, thus, also found for the sum of electrons and holes that the p-n junction acts like a rectifier, with a high transmittance in one direction, which becomes ever bigger with increasing voltage, and a vanishingly small transmittance in the other direction.

In the same way as at a p-n junction an equilibrium establishes itself also quite generally when two different conducting materials are brought into contact. There is always the build-up of a charge double layer, whose potential step brings the electrochemical potential to a constant value. The often encountered contacts between metals and semiconductors are another important case. Figure 4.16 considers, as an example, a metal in contact with a n-doped

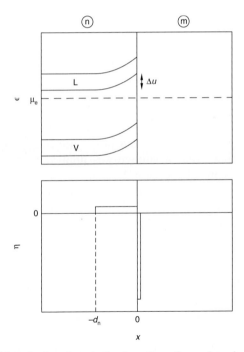

Fig. 4.16. The Schottky barrier at the junction of a n-doped semiconductor and a metal. The course of the conduction and valence bands in the current-free equilibrium state (*top*) and the charge density distribution in the Schottky approximation (*bottom*).

semiconductor with a higher chemical potential. The equilibrium state is again represented, with the course of the valence and the conduction bands in the semiconductor part, together with the electrochemical potential which carries over at the same constant value into the metallic part, at the top, and the charge density distribution at the bottom. The double layer now consists of a positively charged layer of thickness d_n in the semiconductor part and, by comparison, a very thin negatively charged layer on the metal side.

The box profile represents only an approximation. This was suggested by Schottky and allows an estimation of the width d_n of the transition region. It can be assumed, after making contact with the metal, that all charge carriers generated by n-doping flow from a layer of thickness d_n into the metal and are held there at the interface. For a concentration ρ_d of donor atoms and thus also electrons in the conduction band, it follows, from a two-fold integration of the Poisson equation (4.181), that the dependence of the potential for $-d_n < x < 0$ is given by

$$V(x) = -\frac{e\rho_d}{2\epsilon_0}(d_n + x)^2 \quad . \tag{4.196}$$

Up to the interface, a voltage drop

$$\Delta V = \frac{e\rho_d d_n^2}{2\epsilon_0} \tag{4.197}$$

and a corresponding step in the potential energy $\Delta u = e\Delta V$ establishes itself. This corresponds to the total value since the negative part of the double layer in the metal makes, on account of its small size, only a very small contribution, which can be neglected. Estimates of the thickness of the depleted zone which has been stripped of its electrons can be made using experimental values for the **Schottky barrier** ΔV, and values in the μm range are found.

It was stated above that a voltage was applied to a p-n element using double-sided metal contacts. Figure 4.17 shows the course of the potential for such a **heterostructure** in the equilibrium state without an external voltage. There are potential steps at all three interfaces. It is apparent that it is not only the rectification properties of the p-n junction which appear upon applying a voltage, with there also being the overlapping effects of the two Schottky barriers.

It is necessary, when comparing the chemical potential of the electrons in different materials, to choose a common zero point. The state of an electron without kinetic energy outside the sample body offers itself as a reference point. Chemical potentials thus assume negative values and describe the binding energy which is associated with the current carrying electrons in the solid, or, the work which is required for the electrons to be released. The latter can be determined using the photoelectric effect, i.e., the observation, upon irradiation by electromagnetic waves of variable frequency, that electrons are only released if the photon energy $\hbar\omega$ exceeds the binding energy. As was established above, the difference in the chemical potentials, upon the joining

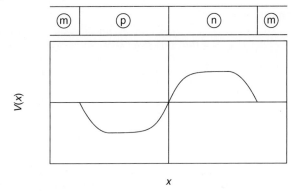

Fig. 4.17. The course of the potential in a metal:p-semiconductor:n-semi-conductor:metal heterostructure without an external voltage.

together of two conducting materials, is externally reflected by the contact potential ΔV. The associated step $e\Delta V$ in the electrostatic energy of the electrons thus reveals how large the difference in the work functions is. The bringing into contact of different metals leads, therefore, to the **Volta series of contact potentials**, whereby metals are arranged into a sequence by reference to their work functions.

4.3 Magnetic Field Effects

Magnetic fields affect mobile charges via the Lorentz force. Since it is directed perpendicular to the velocity, there is – classically – no change in kinetic energy. Quantum mechanics can introduce variations. The presence of a magnetic field modifies the electronic eigenstates, and sometimes in a way which changes the total energy of an electron system. In this section, observations for metals and semiconductors in magnetic fields will be discussed. As they are indicative of the properties of the electrons and holes in the vicinity of the Fermi surface, they can provide a view of its form in $\boldsymbol{k}$ space.

4.3.1 Cyclotron Resonance

To begin, let us consider how a magnetic field changes the motion in a classical electron gas. On account of the Lorentz force, the motion between two collisions no longer follows a straight line, but it rather follows a curved trajectory in a plane perpendicular to the direction of the magnetic field. The Lorentz force

$$\mathbf{f} = -e(\boldsymbol{v} \times \boldsymbol{B}) \tag{4.198}$$

acts as a centripetal force

$$m\omega_{\mathrm{c}}^2 r = e\omega_{\mathrm{c}} r B \quad .\tag{4.199}$$

It can be seen that the frequency with which orbits are undergone,

$$\omega_{\mathrm{c}} = \frac{eB}{m_{\mathrm{e}}} \quad ,\tag{4.200}$$

which is known as the **cyclotron frequency**, is always independent of the radius.

Which changes result upon changing over from the classical situation to a Fermi gas or to electrons and holes in the band model? As was emphasised repeatedly, it is only the charges in the vicinity of the Fermi surface which are mobile. The magnetic field acts on these and changes the trajectory from a straight line to a curved form. This can be described using the semi-classical approach adopted for the discussion of electrical conductivity. A single electron is represented as a wave packet, which for metals and semiconductors is formed from planar waves due to free particles and Bloch waves, respectively. The associated velocity is given by the group velocity

$$\boldsymbol{v}_{\mathrm{g}} = \frac{1}{\hbar}\frac{\mathrm{d}}{\mathrm{d}\boldsymbol{k}}\epsilon \quad .\tag{4.201}$$

The equation of motion for the (quasi-)momentum

$$\boldsymbol{p} = \hbar\boldsymbol{k}\tag{4.202}$$

of the electrons in the presence of a magnetic field $\boldsymbol{B}$ is now given, according to the classical expression, as

$$\hbar\frac{\mathrm{d}\boldsymbol{k}}{\mathrm{d}t} = -\frac{e}{\hbar}\frac{\mathrm{d}}{\mathrm{d}\boldsymbol{k}}\epsilon \times \boldsymbol{B} \quad .\tag{4.203}$$

This expression represents an equation of motion in $\boldsymbol{k}$ space. Since energy is conserved, it follows that the motion is at the Fermi surface and has a trajectory whose plane is perpendicular to $\boldsymbol{B}$. Figure 4.18 shows two possible orbits for the example of the Fermi surface of copper. These two are chosen, since the orbits enclose a maximum or a minimum surface in $\boldsymbol{k}$ space, and thus belong to the group of **extremal orbits**. The change in position in $\boldsymbol{k}$ space with respect to time also defines the time dependence of the group velocity, and thus determines an orbit in direct space. The orbital period τ can be determined for each orbit from the equation of motion. This yields, as is immediately apparent, the expression

$$\tau = \frac{\hbar^2}{eB}\oint \frac{|\mathrm{d}\boldsymbol{k}|}{\left|\frac{\mathrm{d}}{\mathrm{d}\boldsymbol{k}}\epsilon \times \frac{\boldsymbol{B}}{B}\right|} \quad ,\tag{4.204}$$

whereby the path integral is to be evaluated over the closed trajectory.

Fig. 4.18. The Fermi surface in the first Brillouin zone of copper. At some places, the surface reaches the zone limits. Two trajectories, which are followed by electrons upon applying a magnetic field, are shown. Both represent 'extremal orbits'.

On the basis of this equation, it can be easily ascertained, for an ideal Fermi gas with a spherical Fermi surface, that the orbital period is that of the corresponding cyclotron frequency Eq. (4.200) for all trajectories. Completely different periods can however result for different orbits for the case of real systems and, in particular, semiconductors. These periods can be determined using resonance experiments. As with electron spin resonance, it is necessary to irradiate with electromagnetic waves perpendicular to the stationary magnetic field, such that the electric field acts to accelerate the electrons, which can be achieved for a polarization perpendicular to $\boldsymbol{B}$. The take-up of energy, i.e., the absorption of radiation, occurs when the frequency of the radiation is the same as the orbital frequency of the electrons in the material. For usual magnetic fields in the Tesla region, it is found that these frequencies are, as in the case of electron spin resonance, in the microwave region (it should be noted that the Larmor frequency, which describes the effect of a magnetic field on electrons bound to an atom, is half the size of the cyclotron frequency of a free electron). In spite of varying orbital periods, a clear structure is usually recognisable in a cyclotron resonance experiment. This is proof for a non-uniform distribution with characteristic maxima. Figure 4.19 shows an example obtained for aluminium. The maxima are generally associated with the already mentioned extremal orbits. Upon going over to neighbouring orbits in parallely shifted planes, the orbital periods change only slowly in the region of these extremal orbits, and they, thus, have a large statistical weight. The experiment also shows that a resonance is achieved if the irradiated frequency is a multiple of the orbital frequency.

The discussion has so far only considered electrons. Holes react, of course, in an analogous fashion upon the application of a magnetic field. They likewise follow closed trajectories at the Fermi surface, but they differ from those of the electrons in the direction of orbit. Cyclotron resonance experiments can,

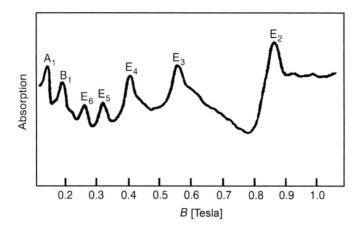

Fig. 4.19. A cyclotron resonance curve obtained for aluminium. The maxima can be assigned to three different extremal orbits (A, B, E). Higher order resonances are also observed for the extremal orbit E (from Moore and Spong [30]).

thus, by making use of circularly polarised electromagnetic waves, distinguish between electrons and holes.

4.3.2 The Hall Effect

If an electric field $\boldsymbol{E}$ is applied at the same time as a magnetic field $\boldsymbol{B}$ and the geometry is correctly chosen, it is possible to ensure that the trajectories of the charge carriers remain straight. The magnetic field then leads to a measurable separation of charge, which is referred to as the **Hall effect**. The to be chosen arrangement is shown in Figure 4.20. The magnetic field is aligned along z,

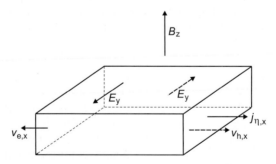

Fig. 4.20. Fields and currents associated with a measurement of the Hall effect: the charge current density j_η, the velocity of the electrons $(v_{e,x})$ and the holes $(v_{h,x})$ as well as the electric transverse field E_y arising if the conduction is due to electrons and holes (*dashed*) respectively.

while an electric field is applied by means of a voltage in the x direction. If the electric field leads to the flow of current, an electric field in the y direction would be observed to arise. The reason for this is the displacement of charges by the magnetic field and the resulting build-up of surface charges at the two surfaces perpendicular to the y axis. The surface charges continue to increase until the resulting field in the y direction has compensated the Lorentz force. The force on an electron resulting from the collective effect of an electric field $\boldsymbol{E}$ and a magnetic field $\boldsymbol{B}$ is given generally by

$$\mathbf{f} = -e(\boldsymbol{E} + \boldsymbol{v} \times \boldsymbol{B}) \ . \tag{4.205}$$

The stationary state is attained when the force in the y direction disappears, i.e.,

$$f_y = 0 \ , \tag{4.206}$$

which is the case for

$$E_y = v_x B_z \ . \tag{4.207}$$

Different applications of the Hall effect are possible. In one application, the density of the current in the x direction, $j_{\eta,x}$, and the induced electric field in the y direction which result upon applying a magnetic field and a voltage can be measured. The determination of the two parameters allows the calculation of the **Hall coefficient**

$$R_{\mathrm{H}} = \frac{E_y}{j_{\eta,x} B_z} \ . \tag{4.208}$$

Its meaning follows from

$$j_{\eta,x} = -e\rho_e v_x \tag{4.209}$$

and Eq. (4.207), which show that

$$R_{\mathrm{H}} = -\frac{1}{e\rho_e} \ . \tag{4.210}$$

If the charge carriers are holes, the Hall coefficient is given as

$$R_{\mathrm{H}} = \frac{1}{e\rho_{\mathrm{h}}} \ . \tag{4.211}$$

It is thus possible from the sign of a measured Hall coefficient to determine which of the two charge carriers dominates.

Alternately, the Hall effect can be used to determine the strength of a magnetic field. If the Hall coefficient is known, it follows that B can be determined by measuring the perpendicular field strength E_y for an applied fixed current.

4.3.3 Magnetisation Oscillations

In the last two sections, it was discussed how a magnetic field acts on the orbits of electrons. It can also be asked: How does the magnetic field change

the eigenstates of the electrons for the field-free case and does it influence the
density of levels? This is considered here for a metallic Fermi gas.

In the framework of classical mechanics, the discussion of a free electron
in a magnetic field begins with the Hamiltonian function

$$H = \frac{1}{2m_e}(\boldsymbol{p} - (-e)\boldsymbol{A})^2 \; . \tag{4.212}$$

It depends on, in addition to the canonical momentum $\boldsymbol{p}$, the vector potential $\boldsymbol{A}$. For a magnetic field in the z direction, a vector potential with

$$A_y = Bx \tag{4.213}$$

as the only non-vanishing component can be chosen. The Hamiltonian function
is then given as

$$H = \frac{1}{2m_e}\left[p_x^2 + (p_y + eBx)^2 + p_z^2\right] \; , \tag{4.214}$$

which becomes, upon introducing the cyclotron frequency ω_c from Eq. (4.200),

$$H = \frac{1}{2m_e}p_x^2 + \frac{m_e\omega_c^2}{2}\left(x + \frac{p_y}{m_e\omega_c}\right)^2 + \frac{p_z^2}{2m_e} \; . \tag{4.215}$$

Introducing the momentum operator

$$\mathbf{p} = -i\hbar\frac{\partial}{\partial \boldsymbol{r}} \tag{4.216}$$

allows the transition to the corresponding Schrödinger equation

$$\mathbf{H}\psi = \epsilon\psi \; . \tag{4.217}$$

This can be solved upon assuming the following product ansatz for the wavefunction:

$$\psi \propto \exp(ik_z z)\exp(ik_y y)\phi(x) \; . \tag{4.218}$$

The Hamiltonian operator acts on this wavefunction according to

$$\mathbf{H}\psi = \frac{\hbar^2}{2m_e}k_z^2\psi + \left(-\frac{\hbar^2}{2m_e}\frac{\partial^2}{\partial x^2} + \frac{m_e\omega_c^2}{2}(x - x_0)^2\right)\psi \; , \tag{4.219}$$

where the position parameter x_0, which depends on the wavevector component
k_y, is introduced according to the definition

$$x_0 = -\frac{\hbar k_y}{m_e\omega_c} \; . \tag{4.220}$$

An inspection of Eq. (4.219) reveals the form of the function $\phi(x)$ which is to
be chosen: The x-dependent part of the Hamiltonian operator, i.e., the part

in brackets, is the same as the Hamiltonian operator for a harmonic oscillator with frequency ω_c. The polynomial eigenfunctions of the harmonic oscillator can thus be chosen for $\phi(x)$. The application of the oscillator equivalent part of the Hamiltonian on $\phi(x)$ yields the energy eigenvalues

$$\epsilon_L = \left(n_L + \frac{1}{2} \right) \hbar \omega_c \;, \tag{4.221}$$

i.e., there are equidistant levels with a separation $\hbar \omega_c$. They are referred to as the **Landau levels**. The associated eigenstates, ϕ_{n_L}, are characterised by the quantum number n_L.

The wavefunctions of an electron in a magnetic field have thus been determined to be of the form

$$\psi_{k_z, k_y, n_L} \propto \exp(ik_z z) \exp(ik_y y) \phi_{n_L}(x - x_0(k_y)) \;. \tag{4.222}$$

Applying the total Hamiltonian operator to these wavefunctions gives the eigenenergies

$$\epsilon(k_z, n_L) = \frac{\hbar^2}{2m_e} k_z^2 + \left(n_L + \frac{1}{2} \right) \hbar \omega_c \;. \tag{4.223}$$

Switching on the magnetic field has thus changed the eigenstates. A new quantum number in the form of n_L has appeared, while the quantum number k_x has disappeared. It is important to note that the energy eigenvalues only depend on two of the three quantum numbers; k_y is not present. All levels, corresponding to different possible k_y values, are degenerate.

It is not surprising that the eigenfunctions have changed when it is considered that there has been a transition from an isotropic system to one with uniaxial symmetry, which is determined by the direction of $\boldsymbol{B}$. An interesting question concerns whether this also leads to changes in the density of levels. From classical physics, no energy changes are to be expected. However, as we will see, quantum mechanics causes characteristic changes. These can be recognised upon determining the density of levels for a chosen fixed k_z and then comparing the situation in the field-free case with that for an applied field.

Firstly, the density of levels is given for a Fermi gas considering all states k_x and k_y for a fixed k_z. This is referred to as $\mathcal{D}'(\epsilon)$ and can be determined, by an analogous derivation to that for the three-dimensional density of levels in Eq. (4.37), to be

$$\mathcal{D}' = 2 \frac{\mathcal{N}_1 \mathcal{N}_2 a^2}{(2\pi)^2} 2\pi k \frac{1}{(\hbar^2 k / m_e)} = \frac{\mathcal{N}_1 \mathcal{N}_2 a^2 m_e}{\pi \hbar^2} \;. \tag{4.224}$$

This result shows that the density of levels for a two-dimensional Fermi gas assumes a constant value.

For the electrons in the magnetic field, there are states corresponding to Eq. (4.221), i.e., there is a discrete distribution with energies at a separation

$$\Delta\epsilon = \hbar\omega_c$$

between the Landau levels. Figure 4.21 shows a comparison of the two distributions for three magnetic field strengths. To evaluate the differences between the three cases, it is necessary to take account of the degeneracy of the Landau levels. There are many different k_y values for each n_L; the energy does not depend on k_y, and the number of these k_y values is to be determined. The possible values of k_y can be determined using cyclic boundary conditions, as always for samples with a finite volume. If there are $\mathcal{N}_2$ cells in the y direction of the crystal, the selection rule follows as

$$k_y = i\frac{2\pi}{\mathcal{N}_2 a} \quad , \quad i \quad \text{is an integer} \ . \tag{4.225}$$

The separation between neighbouring k_y values is thus

$$\Delta k_y = \frac{2\pi}{\mathcal{N}_2 a} \ . \tag{4.226}$$

It, therefore, follows that there is a change in the position of the centre of the oscillator wavefunction by

$$\Delta x_0 = \frac{\hbar 2\pi}{m_e \omega_c \mathcal{N}_2 a} \ . \tag{4.227}$$

x_0 must lie within the sample, which means, when there are $\mathcal{N}_1$ cells in the x direction, that

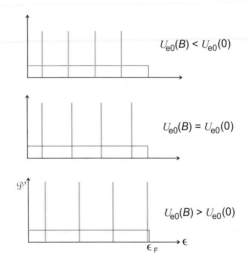

Fig. 4.21. The density of levels of a Fermi gas with $k_z = 0$. The field-free case (homogeneous distribution) is compared with the situations resulting from three magnetic fields of different strengths (discrete distributions).

$$-\frac{\mathcal{N}_1 a}{2} < x_0 < \frac{\mathcal{N}_1 a}{2} \ . \tag{4.228}$$

For the degree of degeneracy of the Landau levels a constant value is thus obtained:

$$2\frac{\mathcal{N}_1 a}{\Delta x_0} = 2\frac{\mathcal{N}_1 \mathcal{N}_2 a^2 m_e \omega_c}{2\pi\hbar} \ , \tag{4.229}$$

where the factor of 2 takes into account the fact that there are always two spin orientations. The knowledge of this degeneracy allows the average value of the density of levels in the two-dimensional electron gas in a magnetic field to be determined, as

$$\overline{\mathcal{D}'} = \frac{1}{\hbar\omega_c} 2\frac{\mathcal{N}_1 \mathcal{N}_2 a^2 m_e \omega_c}{2\pi\hbar} = \frac{\mathcal{N}_1 \mathcal{N}_2 a^2 m_e}{\pi\hbar^2} \ . \tag{4.230}$$

A comparison with Eq. (4.224) shows that this average density of levels is the same as that of the free electron gas in the absence of a field. The field, thus, only leads to the quantisation of the density of levels, with the average value being maintained.

Experiments, where the result only depends on the average density of levels, would not be able to verify the presence of a magnetic field. There are, however, observations, by which the quantisation is externally visible. An example is the oscillation of the magnetisation, which can be observed upon switching on and subsequently increasing a magnetic field at low temperature. This is known as the **de Haas–van Alphen effect**. Figure 4.22 gives an example of this in the form of the fluctuations in the magnetic susceptibility, which are visible at low temperature in a measurement on gold. To explain it, Eq. (4.223) is considered again. Upon introducing the Bohr magneton (Eq. (2.150))

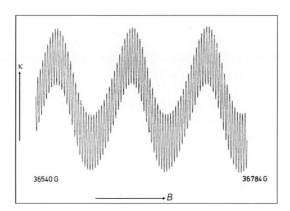

Fig. 4.22. The de Haas–van Alphen effect, as measured on gold at 2 K. A two-fold periodic fluctuation in the magnetic susceptibility is observed (from Ibach and Lüth [31]).

$$\hbar\omega_c = 2\mu_B B \quad , \tag{4.231}$$

it can be recognised that there is, in addition to the kinetic energy associated with free motion in the z direction, a second contribution to the eigenenergy of the state (k_z, n_L),

$$(2n_L + 1)\mu_B B \quad ,$$

which corresponds to the interaction energy of a magnetic dipole moment

$$m_z = -(2n_L + 1)\mu_B \tag{4.232}$$

with the external field B. Considering the contribution of all electrons, the energy of the electronic ground state, $U_{e,o}$, is correspondingly given as

$$U_{e,o} = U_{e,\mathrm{kin}} - \mathcal{V}M_z B \quad , \tag{4.233}$$

where $\mathcal{V}M_z$ represents the sum of the magnetic dipole moments of all electrons. It thus follows that fluctuations in the magnetisation upon switching on an external magnetic field, as they appear in Fig. 4.22, correspond to fluctuations in the electronic energy. The fact that the quantisation of the density of levels generated by a magnetic field can actually cause fluctuations in the energy content of the electrons is made clear by the three cases shown in Fig. 4.21. All three cases refer to the total number of electrons, which have no quasi-momentum in the z direction ($k_z = 0$). In the absence of a field, the density of levels $\mathcal{D}'$ ends here at the Fermi energy ϵ_F. All the shown situations can exist upon switching on a magnetic field. In the upper picture, the last occupied Landau level is almost $\hbar\omega_c$ away from the Fermi limit, while in the lower picture it is only just below the Fermi limit. As is immediately apparent, there is, upon comparing with the field-free situation, a decrease and an increase in the total energy in the first and second cases, respectively. Only for the field which leads to the arrangement of the Landau levels shown in the middle, does the energy remain unchanged. With increasing field strength B, the system adopts these different states again and again, and the magnetisation fluctuates correspondingly. For the magnetic susceptibility, a constant change between positive and negative values, and thus a quasi change between para- and diamagnetic behaviour, is observed.

Since the change is different for different k_z values, the question may, in the first instance, be asked as to why fluctuations are at all observable. The reason for this is again the dominant role of extremal orbits, as was already found for the case of cyclotron resonance. They also impose themselves here on the total behaviour. The observation of two superimposed oscillations with different periods in the experiment of Fig. 4.22 is due to there being two different extremal orbits. They differ from each other in a way similar to the extremal orbits in Fig. 4.18.

The de Haas–van Alphen effect is not only observed for metals but also for semiconductors, where it can be described in a qualitatively similar way, even though the exact solution of the Schrödinger equation for the determination of the Landau levels is, in the general case, then no longer possible.

4.4 Superconductivity

In 1911, Kamerlingh Onnes observed in a measurement of the electrical resistance of mercury at very low temperatures a completely unexpected and spectacular phenomenon, which is illustrated in Fig. 4.23: On going below a temperature of 4.2 K, the resistance falls suddenly to an unmeasurably small value. In this way, the **superconducting state** was discovered.

In the subsequent years, it was established that many conductors, both pure metals and also alloys, make the transition into this state at low temperature, and that this state possesses a wide range of further unusual properties. The chapter begins with an overview of the most important observations before moving onto a presentation of the current understanding, whereby the discussion makes use of qualitative pictures and is limited to simple treatments.

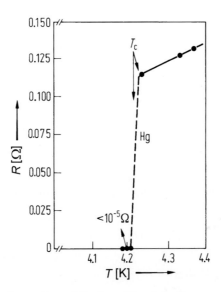

Fig. 4.23. The first observation of the transition into the superconducting state by Kamerlingh Onnes on mercury [32].

4.4.1 Meissner Effect and Energy Gap

For the superconducting state, it would be expected that a strong diamagnetism would be observed if a loss-free current is induced upon applying a magnetic field, with the current weakening the field according to Lenz's rule. In fact, the observations go beyond this. In Figure 4.24, the **Meissner effect** is illustrated. In the normal conducting state, which is usually diamagnetic, the field penetrates the sample without being much weakened. If the magnetic

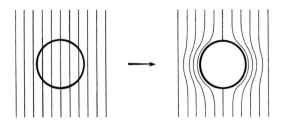

Fig. 4.24. The Meissner effect: On cooling a sample in a magnetic field, it is found, upon the transition into the superconducting state, that the field is completely displaced from the inside of the sample.

field is already applied to this normal conducting state and the sample is then cooled, it is found, that the field is completely driven out upon the transition into the superconducting state. From this it is evident that the cooling process alone leads to a generation of surface currents which have exactly the strength required to bring the external field inside the sample to zero. Within the sample there is

$$\boldsymbol{B} = 0 = \mu_0(\boldsymbol{H} + \boldsymbol{M}) \ , \tag{4.234}$$

which implies that the sample magnetisation $\boldsymbol{M}$ and the applied field $\boldsymbol{H}$ have the same magnitude with opposite signs:

$$\boldsymbol{M} = -\boldsymbol{H} \ . \tag{4.235}$$

The occurrence of this peculiar form of a perfect diamagnetism is, however, limited. It is only maintained up to a limiting value of the strength of the external magnetic field, namely the critical magnetic field H_c. The superconductor returns to a normal conducting state if H_c is exceeded. A general plot of the magnetisation curve $M(H)$ of a superconductor thus has the form shown in Fig. 4.25. With the help of a sufficiently strong magnetic field, it

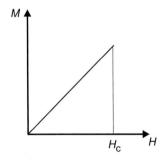

Fig. 4.25. The magnetisation curve of a superconductor with a critical magnetic field H_c.

is thus possible, even below the **critical temperature** T_c of the superconductor, to return into the normal conducting state. Thereby it is to be noted that this critical magnetic field strength is temperature dependent. The dependence is shown in Fig. 4.26 for different metals, where it can be seen that H_c increases continually, from an initial value of zero at the transition point, upon further cooling. The following linear dependence is valid over a wider temperature range

$$H_c(T) \propto T_c - T \quad . \tag{4.236}$$

Even if the transition into the superconducting state is not accompanied by any change in the crystal lattice, it is still a phase transition in the sense of thermodynamics because of the change in the electronic properties. It could at first be thought, on account of the jump to zero electric resistance, that it is a first-order phase transition. This, however, is not correct. Proof for this comes not only from the jump-free critical magnetic field, which sets in with a zero value, but also from measurements of the temperature dependence of the heat capacity. Figure 4.27 shows, as a typical example, experimental results for gallium. Upon the transition into the superconducting state, there is a jump in the heat capacity c_a; a latent heat which would be expected for a first-order phase transition is not observed. The figure also makes the comparison to the electronic heat capacity of a Fermi gas of normal electrons; this is proportional to T (Eq. (4.56)) and can be measured experimentally by applying a magnetic field $H > H_c$. In this way, the different nature of the charge carriers in the superconducting state is demonstrated. A characteristic

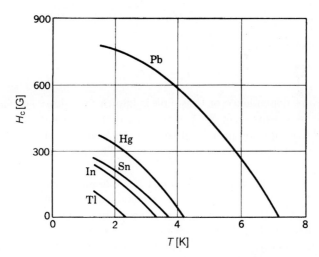

Fig. 4.26. The critical magnetic field H_c at the transition from a superconducting to a normal conducting state. Temperature dependence of H_c for different metals.

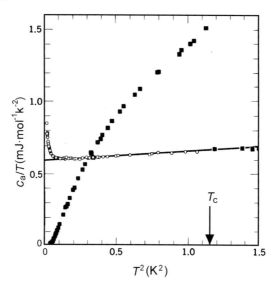

Fig. 4.27. The temperature dependence of the molar heat capacity of gallium in the normal conducting state ($H > H_c$, *open circles*) and the superconducting state ($H = 0$, *filled squares*) (from Phillips [33]).

of superconductors is illustrated by the part of the plot corresponding to the lowest temperatures, $T \to 0$, where the heat capacity can be described by

$$c_a \propto \exp -\frac{\Delta\epsilon_s}{k_B T} \quad , \tag{4.237}$$

with, to a first approximation,

$$\Delta\epsilon_s \simeq k_B T_c \quad . \tag{4.238}$$

A temperature dependence of this type is known for 2-level systems, for example molecules, which can exist in two different conformational states. The change in the occupation probabilities with temperature, which is determined by Boltzmann statistics, leads, for an energy difference $\Delta\epsilon$, to such a $c_a(T)$ dependence.

The two sets of experimental data in Figs. 4.28 and 4.29 demonstrate that the existence of an **energy gap** $\Delta\epsilon_s$ is a central feature of the superconducting state; indeed it is this which distinguishes it from the normal conducting state. The curves in Fig. 4.28 correspond to measurements, for three different metals, of the reflectivity of electromagnetic radiation in the infra-red region. An increase relative to the normal state is observed for the superconducting state, which, however, disappears at a characteristic frequency. The results in Fig. 4.29 come from **electron tunnelling experiments**, whereby a normal conducting metal was brought into contact with a superconductor via an

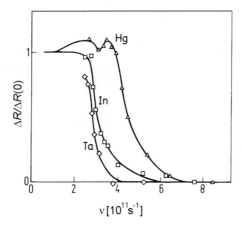

Fig. 4.28. The frequency dependence of the infra-red reflectivity of different super-conductors, plotted as the normalised difference ΔR from that of the corresponding normal conducting state (from Richards and Tinkham [34]).

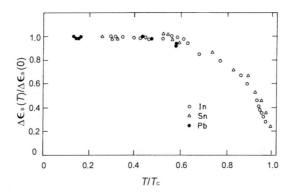

Fig. 4.29. The lowering in energy $\Delta\epsilon_s$ of the superconducting state as a function of temperature, as derived from electron tunnelling experiments on three different metals (from Giaver and Megerle [35]).

insulating layer with a thickness in the Å range. In principle, electrons can move back and forth between the two materials by means of the tunnel effect. However, it was experimentally found that current only flows if a characteristic minimum voltage $\Delta V_{\min}$ is exceeded. The points in the figure represent the values $e\Delta V_{\min}$ obtained for different metals at different temperatures below T_c. Why $e\Delta V_{\min}$ is equivalent to the excitation energy $\Delta\epsilon_s$ will be discussed later. Here, it will only be stated that $\Delta\epsilon_s = e\Delta V_{\min}$ is in the frequency range of the steps in the reflectivity, if $\Delta\epsilon_s = \hbar\omega$ is used, and that both sets of results are in good agreement with the calorimetric equations Eqs. (4.237) and (4.238). It is also important to note that the excitation energy $\Delta\epsilon_s$ is not constant, but rather, just as with H_c, it sets in at zero at the critical temperature (this is actually only hinted at by the experimental results) and then rises continually to its final value. All these results indicate that the transition from a normal to a superconducting state is a second-order phase transition with T_c as the critical temperature.

4.4.2 Cooper Pairs. The Ginzburg–Landau Wavefunction

The description of the electronic properties of metals and semiconductors in the Fermi gas and band model, respectively, is based on the assumption that the electrons can be treated as if they were independent particles. There are of course Coulombic interactions between the electrons, but it was possible to consider these commonly, as in the treatment of the electrons of an atom under the Hartree approximation, by means of a stationary average field. The eigenstates could then be calculated for individual electrons in an average field, and subsequently the Fermi–Dirac distribution function (Eq. (4.45)) could be used. The justification for this approach is provided by the results of the theory, by which the properties of electrons in metals and semiconductors can be well described. A different situation is found, now, in the superconducting state, where interactions appear which cannot be described using an average field. The observation of a characteristic energy gap $\Delta\epsilon_s$ indicates that the electrons here adopt bound states, and the simplest view for a theory is that of a pairwise coupling. In 1956, Cooper presented a theoretical analysis which demonstrated that such a formation of pairs is possible for electrons in the vicinity of the Fermi surface and, as required, a lowering of the energy then results. These **Cooper pairs** are the charge carriers in a superconductor. They possess a charge $-2e$, a mass $2m_e$ and a vanishing spin – the latter turns out to be of particular decisive importance. Their concentration and the pair binding energies are temperature dependent. Increasing the temperature inside the superconductor leads to an increasing dissociation of the pairs, up to the critical temperature T_c, where they completely disappear. Cooper pairs are carriers of a loss-free current. In a current loop, they lead, even at the lowest concentrations, to a collapse of the voltage, and this explains why a parameter such as the electric resistance exhibits a jump-like change, even for the given continuity of the transition.

How does a persistent current come into existence in a superconductor, and how can it be described? The property of Cooper pairs that they have no spin leads to a complete change in the behaviour as compared to that of individual electrons. Elementary particles with a vanishing or whole number spin are bosons, and these, in contrast to fermions, can occupy quantum-mechanical states in arbitrary numbers. Remember, for example, that unlimited numbers of photons can occupy the vibrational states in a radiation cavity such that a macroscopic process, namely an electromagnetic oscillation with an electric field strength amplitude E_0, arises. The dependence of E_0 on the density ρ_{ph} of the photons is given as

$$\frac{\varepsilon_0}{2} E_0^2 = \rho_{\mathrm{ph}} \hbar \omega \quad . \tag{4.239}$$

The charge carriers in a superconductor, the Cooper pairs, behave like bosons, which means that they, like photons, are able to occupy, in large numbers, an individual quantum-mechanical state. Ginzburg and Landau had suggested already in 1950, before the postulation of the possible existence of Cooper pairs, that the stationary states, which are occupied by charge carriers in a superconductor, can be described by a complex wavefunction $\Psi(\mathbf{r})$. The wavefunction determines by means of

$$|\Psi|^2 = \rho_{\mathrm{s}} \quad , \tag{4.240}$$

i.e., by its modulus, the local density ρ_{s}. Ψ also contains, like the electromagnetic oscillation, a further important attribute, namely a phase factor:

$$\Psi(\mathbf{r}) = |\Psi(\mathbf{r})| \exp[\mathrm{i}\theta(\mathbf{r})] \quad . \tag{4.241}$$

Eigenstates in superconductors can be associated with a current of particles. Schrödinger wave mechanics provides an expression for the current which is in a general form also valid in the presence of a magnetic field, and it is used in the **Ginzburg–Landau wave mechanics** of superconductors:

$$\mathbf{j}_{\mathrm{s}} = \frac{1}{2} \left(\Psi^* \frac{-\mathrm{i}\hbar\nabla - (-2e)\mathbf{A}}{2m_{\mathrm{e}}} \Psi + \Psi[\frac{-\mathrm{i}\hbar\nabla - (-2e)\mathbf{A}}{2m_{\mathrm{e}}}\Psi]^* \right) \quad . \tag{4.242}$$

Here $\mathbf{j}_{\mathrm{s}}$ is the particle current density of the Cooper pairs (charge $-2e$, mass $2m_{\mathrm{e}}$) and $\mathbf{A}$ denotes the vector potential. A frequently encountered case is that of a superconductor, inside of which there are macroscopic regions with a constant density of charge carriers. In this case, only the phase θ of the wavefunction changes, and the electric current density associated with the Cooper pairs

$$\mathbf{j}_{\eta\mathrm{s}} = -2e\mathbf{j}_{\mathrm{s}} \tag{4.243}$$

is given as

$$\mathbf{j}_{\eta\mathrm{s}} = -2e \left(\frac{\hbar}{2m_{\mathrm{e}}} \nabla\theta(\mathbf{r}) - \frac{(-2e)}{2m_{\mathrm{e}}} \mathbf{A} \right) |\Psi|^2 \quad . \tag{4.244}$$

In particular, it follows, for a planar wave

$$\Psi(\mathbf{r}) = |\Psi| \exp(\mathrm{i}\mathbf{k}\mathbf{r}) \tag{4.245}$$

that the current density is

$$\mathbf{j}_{\eta\mathrm{s}} = -2e|\Psi|^2 \left(\frac{\hbar\mathbf{k}}{2m_{\mathrm{e}}} + \frac{2e}{2m_{\mathrm{e}}} \mathbf{A} \right) \quad . \tag{4.246}$$

In this way, a description has been obtained of the situation which exists in a circular closed superconductor where a persistent current flows. The flow of current is associated with eigenstates which possess a non-zero velocity. As is immediately apparent, the expression in brackets has this meaning. All of the Cooper pairs which occupy the state have this velocity in common.

The wave-mechanical approach of Ginzburg and Landau is thus able to describe the existence of states with a persistent current and, as will become clear in the next section, it can also explain the Meissner effect and further phenomena. The approach is, however, basically of empirical nature in that it presupposes the existence of quasi-bosonic charge carriers in the super-conducting state. If their existence is to be explained, this must happen on a microscopic basis. In particular, the question arises as to the nature of the binding force which creates the Cooper pairs. In physics, binding forces are often active because of the exchange of particles, and the binding energies then equal the energy of the exchange particles. The binding energies $\Delta\epsilon_s$, which appear as the excitation energies in the experimental measurements represented in Figs. 4.27, 4.28 and 4.29, correspond without exception to frequencies in the infra-red region. The lattice vibrations are in this region, and this may lead to the conclusion that it is phonons which generate, as exchange particles, the electron-electron binding forces, which are the origin of the superconducting state. More evidence for this comes from observations of the type shown in Fig. 4.30. This figure shows, for tin, the change in the transition temperature as a function of the atomic weight A – the latter can be changed by choosing different isotopes. The following dependence is observed:

$$T_c \propto A^{-\alpha} \quad \text{with} \quad \alpha \approx 1/2 \ , \tag{4.247}$$

which corresponds to the change in the lattice vibration frequencies. In a simplified and intuitive picture, the role of the crystal lattice in promoting binding

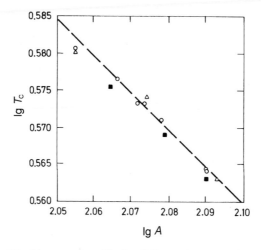

Fig. 4.30. The critical temperature T_c for different isotopes of tin. The dependence on the atomic weight (from Ibach and Lüth [36]).

lies in the generation of an electric field due to a local polarisation arising from the deforming effect of a first electron, which then exerts an attractive force on a second electron.

In 1957, Bardeen, Cooper and Schrieffer presented a theoretical explanation as to, first, why phonon-mediated forces of this type actually appear, and, second, why they, in spite of their low strength, are able to generate a bound state of two electrons. A full account of the **BCS theory** is not possible here, and only a brief sketch as to how the Cooper pairs are described can be presented. The associated 2-electron wavefunction $\psi_2(r_1, r_2)$ has the form

$$\psi_2(r_1, r_2) = \sum_k c_k \cos[k(r_1 - r_2)](\sigma_{\uparrow 1}\sigma_{\downarrow 2} - \sigma_{\uparrow 2}\sigma_{\downarrow 1}) \ . \tag{4.248}$$

It is composed of contributions with wavevectors k from a narrow region at the Fermi surface with

$$\epsilon_F - \Delta\epsilon_s/2 < \frac{\hbar^2 k^2}{2m_e} < \epsilon_F + \Delta\epsilon_s/2 \ . \tag{4.249}$$

The width $\Delta\epsilon_s$ corresponds to the interaction energy, which leads to a mixing of 2-electron states with amplitudes c_k. In this way, each individual 2-electron state consists of a symmetric position-dependent part, $\cos[k(r_1 - r_2)]$, and an anti-symmetric spin function, $\sigma_{\uparrow 1}\sigma_{\downarrow 2} - \sigma_{\uparrow 2}\sigma_{\downarrow 1}$, with a vanishing total spin.

Using the wavefunction $\psi_2(r_1, r_2)$ of the Cooper pair, it is possible to estimate its spatial extension. For this, only the reciprocal property of two functions which are linked by a Fourier transformation is used. The spatial extension δr_{12} of the wavefunction $\psi_2(r_1 - r_2)$ is inversely proportional to the width δk of the distribution of the amplitudes c_k in k space:

$$\delta r_{12} \simeq \frac{1}{\delta k} \ . \tag{4.250}$$

δk is given by Eq. (4.249), from which it follows that

$$\Delta\epsilon_s \simeq \Delta\frac{\hbar k^2}{2m_e} = \frac{\hbar^2 k_F}{m_e}\delta k \simeq \frac{\epsilon_F}{k_F}\delta k \ . \tag{4.251}$$

This leads to

$$\delta r_{12} \simeq \frac{\epsilon_F}{\Delta\epsilon_s k_F} \ . \tag{4.252}$$

Typical values are $\epsilon_F/\Delta\epsilon_s \simeq 10^3$, $k_F \simeq 10^{10}\,\mathrm{m}^{-1}$ and thus $\delta r_{12} \simeq 10^2\,\mathrm{nm}$. Hence, the average separation of the two electrons is large, indeed much larger than the lattice constant. A strong overlap of different Cooper pairs must thus be considered. Cooper pairs are, therefore, not bosons in the strict sense of the definition as applied to elementary particles. In spite of this, they can build-up, in a coherent, i.e., in-phase, manner, a collective multi-particle state, which is then the carrier of a macroscopic current.

London Penetration Depth, Flux Quantisation, and Electron Tunnelling. Using the wave-mechanical Eq. (4.244) as a starting point, it is possible to explain the Meissner effect and also a further characteristic phenomenon, the **flux quantisation**. The calculation of the curl on both sides of the equation yields

$$\nabla \times \boldsymbol{j}_{\eta s} = -\rho_s \frac{(-2e)^2}{2m_e} \boldsymbol{B} \quad . \tag{4.253}$$

This is referred to as the **London equation** and it describes a basic property of all superconducting materials. The London equation applies to all superconductors in the same way that Ohm's law $\boldsymbol{j}_\eta = \sigma \boldsymbol{E}$ applies to all conductors. Combining the London equation with one of the Maxwell equations,

$$\nabla \times \boldsymbol{B} = \mu_0 \boldsymbol{j}_{\eta s} \quad , \tag{4.254}$$

according to

$$\nabla \times (\nabla \times \boldsymbol{B}) = \nabla(\nabla \cdot \boldsymbol{B}) - \Delta \boldsymbol{B} = -\Delta \boldsymbol{B} = \mu_0 \nabla \times \boldsymbol{j}_{\eta s} \quad , \tag{4.255}$$

results in

$$\Delta \boldsymbol{B} = \mu_0 \frac{\rho_s(-2e)^2}{2m_e} \boldsymbol{B} \quad . \tag{4.256}$$

Using this differential equation, it is possible to describe the decay of a magnetic field in a superconductor, for the case where the field impinges the superconductor at its surface. As represented in Fig. 4.31, the fall-off is described by the solution of Eq. (4.256):

$$\boldsymbol{B} \propto \exp{-\frac{x}{\lambda_L}} \quad , \tag{4.257}$$

where λ_L is the **London penetration depth** given by

$$\lambda_L^2 = \frac{2m_e}{\mu_0 \rho_s(-2e)^2} \quad . \tag{4.258}$$

There is a current of superconducting charge carriers in the y direction associated with the decaying magnetic field, which is polarised in the z direction. It correspondingly only flows in a surface zone of thickness λ_L.

Figure 4.32 shows what happens if an initially present external magnetic field, which passes through a closed superconducting loop, is switched off. In the general case, it is observed that surface currents of superconducting charge carriers remain in existence, with this being apparent because of a magnetic field which remains. It is observed that the remaining flux, which flows through the surface enclosed by the loop, is quantised, and increases in discrete steps of

$$\Phi_0 = \frac{\pi\hbar}{e} \quad . \tag{4.259}$$

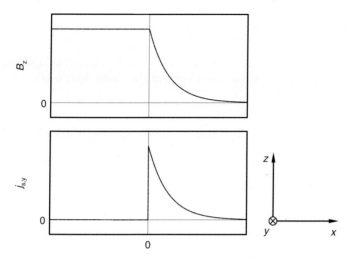

Fig. 4.31. The surface of a superconductor ($x > 0$) upon applying an external magnetic field: the exponential decay of the magnetic field strength B_z and the current $j_{s,y}$ of the Cooper pairs.

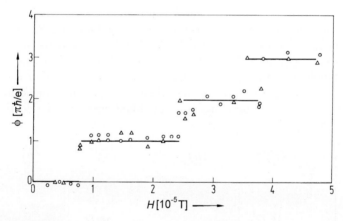

Fig. 4.32. The quantisation of the magnetic flux Φ which remains in a ring of superconducting tin after switching off an external field. The jump-like changes in the flux upon changing the strength of the field H applied before the transition into the superconducting phase (from Deaver and Fairbank [37]).

Equation (4.244) can also be used to describe this phenomenon of flux quantisation. Consider a closed path inside the loop-like conductor where no current flows. Since

$$j_{\eta,s} = 0 \ , \tag{4.260}$$

it is true for the integral along a circle that

$$\oint \hbar \nabla \theta \mathrm{d}\boldsymbol{r} = \oint (-2e)\boldsymbol{A}\mathrm{d}\boldsymbol{r} = (-2e)\int \boldsymbol{B}\mathrm{d}^2\boldsymbol{r} = (-2e)\varPhi \ , \qquad (4.261)$$

where $\mathrm{d}^2\boldsymbol{r}$ is an element of the surface enclosed by the circle. The change in the phase angle upon making one circuit of the circle is then given by

$$\hbar(\theta_2 - \theta_1) = (-2e)\varPhi \ . \qquad (4.262)$$

Since the wavefunction in a stationary state is unique, it must be true that

$$\theta_2 - \theta_1 = p2\pi \ , \qquad (4.263)$$

which leads for $p = 1$ to Eq. (4.259).

We now come back to the experiment in Fig. 4.29 for a short discussion. The left-hand side of Fig. 4.33 describes the energetic effect on the charge carriers of bringing together a normal conducting metal and a superconductor, with a thin insulating layer in between them. The Fermi surface in the normal conducting metal and the energy level of the charge carriers in the superconductor are also of the same height if the external potential is the same on both sides. The level scheme for the superconductor differs from that of the normal conductor in two ways. Firstly, all superconducting charge carriers collectively occupy the same level, while, secondly, this is separated from the next highest state by an energy gap of magnitude $\Delta\epsilon_s/2$ (the energies in the scheme are always, for both electrons and Cooper pairs, to be understood as the energy per individual electron, and thus the energy gap becomes $\Delta\epsilon_s/2$). The higher states can only be occupied if a Cooper pair dissociates. The individual electrons of the metal could also, in principal, move within the superconductor, where there are also individual electrons. To do this, they

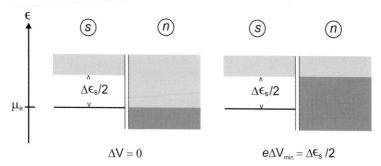

Fig. 4.33. The basis of the electron tunnelling experiment for the determination of $\Delta\epsilon_s$. 's' and 'n' denote the energy schemes for the superconductor and the normal conducting metal, respectively. The energy is given per individual electron in both cases, i.e., also for the superconductor. For the same potential, the energy level of the Cooper pairs and the Fermi energy of the metal are at the same height (*left*). Electron tunnelling sets in upon increasing the potential of the metal by $\Delta V = \Delta\epsilon_s/(2e)$ (*right*).

must, in the first instance, penetrate the insulating layer, something which is possible by means of the tunnel effect for microscopic layers. The electrons can only do this, though, if there exists on the side of the superconductor an unoccupied level with the same energy which is, in principle, accessible for individual electrons. In the initial state, this is evidently not the case, and this remains so until a voltage is applied which is sufficiently strong that the situation on the right-hand side of the figure is attained. This means that current only flows if the energy gap is overcome with the help of an applied voltage which exceeds the minimum value

$$e\Delta V_{\min} = \frac{\Delta\epsilon_{\mathrm{s}}}{2} \quad . \tag{4.264}$$

4.4.3 The Critical Magnetic Field. Type II Superconductors

Ginzburg and Landau also showed how a thermodynamic theory for describing the transition from the normal to the superconducting state can be developed within the framework of the description of superconductors by the wavefunction Ψ. The approach is the same as that which was used to treat other second-order phase transitions, such as the transition into the ferromagnetic or ferroelectric state: A suitable expansion of the Gibbs free energy density in terms of powers of the order parameter, which controls the phase transition, is chosen. This order parameter is now the wavefunction Ψ. Ψ exhibits the desired behaviour in that, upon cooling, it sets in at zero at the critical point T_{c} and then continually rises. In fact, the phase transition is correctly described by the following expression:

$$g(\Psi) = g(0) + b(T - T_{\mathrm{c}})|\Psi|^2 + c_4|\Psi|^4 \quad , \quad \text{where } b > 0 \text{ and } c_4 > 0 . \tag{4.265}$$

The expression is identical with those in Eqs. (3.15) and (3.31) for ferro-electrics and ferromagnets, and the conclusions derived there can be directly carried over. The order parameter, now Ψ, earlier P and M, is thermody-namically an inner variable. Equilibrium corresponds to the minimum in the Gibbs free energy. Plots of the Gibbs free energy density above and below T_{c} are shown in Fig. 3.5. In the superconducting state, the equilibrium value is given by analogy to Eq. (3.17) as

$$|\Psi|^2_{\mathrm{eq}} = \frac{b(T_{\mathrm{c}} - T)}{2c_4} \quad , \tag{4.266}$$

while, in the normal conducting state, it must be equal to zero. The result shows that the concentration of superconducting charge carriers, i.e., the Cooper pairs, increases linearly upon cooling below T_{c}:

$$\rho_{\mathrm{s}} = |\Psi|^2_{\mathrm{eq}} \propto T_{\mathrm{c}} - T \quad . \tag{4.267}$$

The lowering of the Gibbs free energy density as compared to the normal conducting state amounts to

$$g(0) - g(|\Psi|_{\mathrm{eq}}) = \frac{b^2}{2c_4}(T_c - T)^2 \quad . \tag{4.268}$$

In a second step, it is possible to introduce the effect of an applied magnetic field into the description and to determine the **critical field strength** H_c. In the general case, the change of g upon applying a magnetic field is given by Eq. (A.10) as

$$\mathrm{d}g = -B\mathrm{d}H \quad . \tag{4.269}$$

In the normal conducting state, which is only weakly diamagnetic ($B \approx \mu_0 H$), it follows that

$$g(H) = g(0) - \frac{\mu_0}{2}H^2 \quad . \tag{4.270}$$

However, in a superconducting state, since $B = 0$,

$$g(H) = g(0) \quad . \tag{4.271}$$

The two Gibbs free energies are the same at the critical field strength, i.e., we have

$$g(|\Psi|_{\mathrm{eq}}) - g(0) = -\frac{b^2}{2c_4}(T_c - T)^2 = -\frac{\mu_0}{2}H_c^2 \quad , \tag{4.272}$$

and thus

$$H_c = \frac{b}{(\mu_0 c_4)^{1/2}}(T_c - T) \quad , \tag{4.273}$$

which is in agreement with the experimental observations (Fig. 4.26, Eq. (4.236)). As has been made clear by the derivation, the expression $\mu_0 H_c^2/2$ denotes, for each temperature, the extent to which the Gibbs free energy is lowered in the superconducting state as compared to the normal conducting state.

There are superconducting materials, which, upon applying a magnetic field, do not react in the previously discussed way, i.e., corresponding to Fig. 4.25, but rather a magnetisation dependence like that in Fig. 4.34 is observed. Here, the ideal diamagnetic behaviour ends at a first critical field strength H_{c1} not with a jump-like transition into the normally conducting state, but rather a continuous decrease in the magnetisation is then observed. This ends at a second critical field strength H_{c2}, where the value of zero is reached. Such materials are referred to as **type II superconductors**. Figure 4.35 shows how these differ structurally from **type I superconductors**. The picture, which was obtained for a magnetic field strength in the intermediate region $H_{c1} < H < H_{c2}$, depicts a two-phase state, which consists of a superconducting matrix interpenetrated by normal conducting tubes. The tubes are regularly arranged in the form of a two-dimensional lattice.

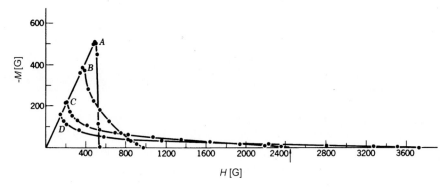

Fig. 4.34. Magnetisation curves for lead (A) and different lead-indium (2.8–20.4%) alloys (B, C, D). The alloys are type II superconductors (from Livingston in [23]).

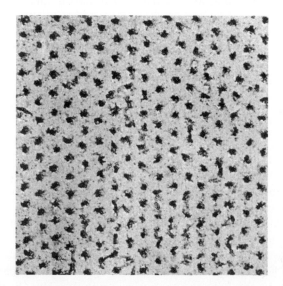

Fig. 4.35. Type II superconductors with regularly arranged normal conducting tubes, which were made visible on the surface of a sample with a cylinder form with the help of ferromagnetic particles (from Träuble and Essmann [23]).

Ginzburg and Landau, together with Abrikosov, provided an explanation for the appearance of such two-phase structures. Again starting with the wavefunction Ψ, an equation for the density of the Gibbs free energy was formulated which took into account the possibility of a spatial variation in the concentration of superconducting charge carriers as well as the presence of an external magnetic field H. Equation (4.265) is extended in the following way:

$$g(\Psi(r)) = g(0)+b(T-T_{\mathrm{c}})|\Psi|^2+c_4|\Psi|^4+\frac{1}{4m_{\mathrm{e}}}\left|(-i\hbar\nabla-(-2e)\boldsymbol{A})\,\Psi\right|^2-\int\limits_0^H B\mathrm{d}H'$$

$$(4.274)$$

Seen from the point of view of quantum mechanics, the fourth term has the meaning of a kinetic energy of the superconducting particles (charge $-2e$, mass $2m_{\mathrm{e}}$). Empirically, it is also the simplest term which can take into account a spatial variation of Ψ. The fifth term describes the lowering of the density of the Gibbs free energy upon the penetration of a magnetic field into the sample (Eq. (A.10)), something which happens for the two-phase states of type II superconductors in the normal conducting parts and a transition region with a thickness corresponding to the London penetration depth.

On the basis of the Ginzburg–Landau equation (4.274), two-phase structures can be handled and an explanation can also be provided as to under which circumstances they appear. It is not possible to explain this exactly here, and the discussion is limited to a few short explanations. In principle, in order to determine the equilibrium state, it is necessary to find the wavefunction $\Psi(r)$ which is associated, for the given boundary conditions, with the minimum in the total Gibbs free energy of the sample

$$\mathcal{G} = \int\limits_{\mathcal{V}} g(\Psi(r))\mathrm{d}^3 r \quad . \tag{4.275}$$

Standard variation methods can be used to solve this problem. Figure 4.36 shows a schematic representation of one of the solutions found in this way. It represents an individual normal conducting tube in a superconducting matrix. The upper and lower pictures depict the concentration distribution of the superconducting charge carriers $\rho_{\mathrm{s}}(r) = |\Psi|^2(r)$ and the spatial variation of the magnetic field B, respectively. For both parameters, there is not an abrupt change between the two phases. As was established above, an external magnetic field also penetrates into the superconducting part to a depth corresponding to the London penetration depth λ_{L}. The Ginzburg–Landau theory now yields a second characteristic length, which is referred to as the **coherence length** ξ_{G}; this characterises the width of the region within which the concentration of the superconducting charge carriers changes from a value of zero in the normal conductor to the end value $|\Psi|^2_{\mathrm{eq}}$ in the superconductor. The parameter ξ_{G} can be derived from the Ginzburg–Landau equation (4.274) by making a plausible argument. If all field contributions are neglected, g can be re-expressed as

$$g(\Psi(r)) - g(0) \propto -\frac{|\Psi|^2(r)}{|\Psi|^2_{\mathrm{eq}}} + \frac{|\psi|^4(r)}{2|\Psi|^4_{\mathrm{eq}}} + \xi_{\mathrm{G}}^2\frac{|\nabla\Psi|^2(r)}{|\Psi|^2_{\mathrm{eq}}} \quad , \tag{4.276}$$

where, according to Eq. (4.266), the end value of the superconducting charge carrier concentration is given by

$$|\Psi|^2_{\text{eq}} = \frac{b(T_c - T)}{2c_4}$$

and

$$\xi_G^2 = \frac{\hbar^2}{4m_e b(T_c - T)} \quad . \tag{4.277}$$

ξ_G is the only parameter in the expression, and it has the dimensions of length. In Fig. 4.36, the only length is the width of the transition region, and it must, therefore, be determined by ξ_G. Equation (4.277) states that ξ_G diverges upon approaching T_c and follows the power law

$$\xi_G \propto (T_c - T)^{-1/2} \quad . \tag{4.278}$$

In the figure, λ_L is chosen to be larger than ξ_G. In fact, this is a fundamental prerequisite for the appearance of a two-phase structure, i.e., a type II superconductivity. The reason for this is qualitatively easy to recognise. A normal conducting tube only spontaneously forms in a superconducting matrix if a lowering of the Gibbs free energy of the system results in the presence of a magnetic field H. Here, two terms, which act against each other, come into consideration: Firstly, the increase in the Gibbs free energy because of the higher value in the normal conducting phase corresponding to Eq. (4.272), and, secondly, the lowering in the Gibbs free energy due to the penetration of the magnetic field, as described by Eq. (A.10). Even if the core region of the normal conducting tube should be vanishingly small, it has, because the

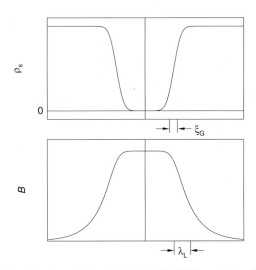

Fig. 4.36. The distribution of the magnetic field B and the density of Cooper pairs ρ_s in the region of a normal conducting tube. The London penetration depth λ_L and the coherence length ξ_G.

coherence length ξ_G is independent of this, a cylindrical region with a cross sectional area $\simeq \pi\xi_G^2$ with an increased Gibbs free energy. There is, thus, an increase per unit length of the tube of

$$\Delta\mathcal{G}_1 \simeq \pi\xi_G^2 \frac{\mu_0}{2} H_c^2 \quad . \tag{4.279}$$

The penetrating magnetic field isn't restricted to the central part of the tube, but rather it encompasses, according to the penetration depth λ_L, a larger cylinder, with a cross sectional area of approximately $\pi\lambda_L^2$. The resulting lowering of the Gibbs free energy, again considered per unit length, amounts for an applied external field H to

$$\Delta\mathcal{G}_2 \simeq -\pi\lambda_L^2 \int_0^H B(H')\mathrm{d}H' \quad . \tag{4.280}$$

Using

$$B \approx \mu_0 H \quad , \tag{4.281}$$

leads to

$$\Delta\mathcal{G}_2 \simeq -\pi\lambda_L^2 \frac{\mu_0}{2} H^2 \quad . \tag{4.282}$$

The sum is, thus, given as

$$\Delta\mathcal{G} = \Delta\mathcal{G}_1 + \Delta\mathcal{G}_2 \tag{4.283}$$

$$= \pi\xi_G^2 \frac{\mu_0}{2} H_c^2 - \pi\lambda_L^2 \frac{\mu_0}{2} H^2 \quad . \tag{4.284}$$

The result shows that a tube can form even before the critical field strength H_c is reached, but only if

$$\lambda_L > \xi_G \quad . \tag{4.285}$$

The argumentation is, in fact, not yet completely correct, since the quantisation of the magnetic flux has been neglected, and this also applies for the flux due to the normal conducting tubes. Equation (4.281), which describes a continuous increase in B, which is proportional to the external field H, must be corrected. In reality, B changes in a jump-like fashion. The field can only penetrate if a quantum of flux Φ_0 can be created, i.e., the following applies:

$$\pi\lambda_L^2 \mu_0 H \geq \Phi_0 \quad . \tag{4.286}$$

The gain in Gibbs free energy, which is initially zero, amounts then, in the first instance, to

$$\Delta\mathcal{G}_2 = \frac{\Phi_0 H}{2} \quad . \tag{4.287}$$

The field strength H_{c1}, at which normal conducting tubes first appear in a type II superconductor, can now be estimated. The value follows from

$$0 = \Delta\mathcal{G} = \pi\xi_G^2 \frac{\mu_0}{2} H_c^2 - \frac{\Phi_0 H_{c1}}{2} \tag{4.288}$$

as

$$H_{c1} = \frac{\pi\xi_G^2 \mu_0 H_c^2}{\Phi_0} \quad . \tag{4.289}$$

4.5 Ionic Conduction in Electrolyte Systems

All discussions in this chapter so far have considered electric currents in crystals, which are carried by electrons, be it in the conduction or valence band or in the bound state of a Cooper pair in a superconductor. Completely different conditions exist in liquids. Currents are not only of an electronic nature, but, on account of the high, spatially unrestricted motion of atoms or molecules in the liquid state, they can also be due to the motion of positive and negatively charged ions.

Ions are found in particularly high concentrations in **electrolyte systems**. Currents can thus be generated which, although not as strong as those in metallic conductors, have strengths which far exceed those of semiconductors. Figure 4.37 depicts a typical arrangement by which the conduction of current in an electrolyte system can be investigated. Two electrodes are immersed into a trough which is filled with an ion containing electrolyte liquid. Electrolytes are often materials which are already composed of ions in the solid state, i.e., they exist as ionic crystals. The ionic character is then maintained in the liquid state, be it either in solution or as a melt. **Strong electrolytes** remain almost completely dissociated in the liquid state, while, for **weak electrolytes**, the degree of dissociation is limited and thus temperature dependent. As will be shown in the following, chemical reactions take place at the electrodes, and a chemically inert material is thus chosen for them, e.g., platinum. The

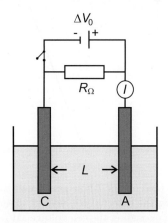

Fig. 4.37. An electrolyte cell with the anode (A) and cathode (C).

electrode which is connected with the positive pole of the d.c. voltage supply is termed the **anode**, while that which is connected with the negative pole is referred to as the **cathode**.

What is observed upon applying a voltage? Figure 4.38 shows a schematic current-voltage curve corresponding to the measurement of the current I upon increasing the external voltage ΔV_0. The plot shows that, at low voltages, only a very weak current flows, until the situation changes upon exceeding a critical value, namely the **cell voltage** ΔV_z. For values above ΔV_z, the current strength increases in a linear fashion:

$$I \propto \Delta V_0 - \Delta V_z \ . \tag{4.290}$$

At the same time, chemical reactions set in at the electrodes. For an aqueous solution of hydrochloric acid (HCl), which is chosen as an example here, hydrogen and chlorine gas is released at the cathode and anode, respectively. For a constant voltage, these chemical reactions proceed initially at a constant rate and only slow down later as a consequence of the falling electrolyte concentration in the solution. The process which is happening here, namely the controlled carrying out of a chemical reaction through the supply of charge at a minimum voltage is referred to as **electrolysis**.

If the voltage is suddenly switched off by opening the switch, after the chemical reaction has proceeded for a certain time, the current does not cease to flow, but rather it flows for a limited time in the opposite direction using the connection via the resistance R_Ω. The voltage drop at the resistance is identical with the potential difference between the anode and the cathode, and is equal to ΔV_z. This observation means that the electrolyte cell has become a voltage source with the cathode and anode as the minus and plus poles, respectively. It is found, for the case of HCl solution, that the flow of current

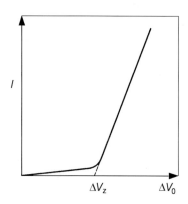

Fig. 4.38. The change in the current I flowing through an electrolyte cell observed upon continually increasing the external d.c. voltage ΔV_0. A noticeable flow of current is only observed upon exceeding the cell voltage ΔV_z. The current is associated with chemical reactions at the electrodes.

is associated with the return into solution of hydrogen and chlorine gas, and correspondingly this only carries on as long as there is hydrogen and chlorine available at the cathode and the anode, respectively. This set-up is referred to as a **galvanic element**; the electrolyte cell has become a 'galvanic cell'.

Figure 4.39 shows the dependence of the electrical conductivity σ_{el} on the molar electrolyte concentration $\tilde{c}$ for some different aqueous solutions. σ_{el} depends on the immersion area A and the separation L of the electrodes according to

$$\sigma_{el} = \frac{IL}{A(\Delta V_0 - \Delta V_z)} \quad . \tag{4.291}$$

The curves show that the conductivity certainly doesn't always continue to rise upon increasing the ion concentration. As well as the observation of a maximum, it is to be noted that there is no proportional increase at low concentration. Kohlrausch found that the concentration dependence of the **molar conductivity**

$$\tilde{\sigma}_{el} = \frac{\sigma_{el}}{\tilde{c}} \tag{4.292}$$

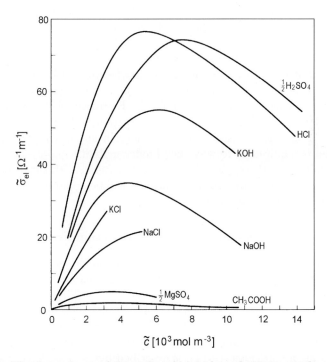

Fig. 4.39. The dependence of the molar electrical conductivity on the molar concentration of the electrolyte for different aqueous electrolyte solutions (from Wedler [38]).

is described empirically, for the case of diluted electrolyte solutions, by

$$\tilde{\sigma}_{el}(\tilde{c}) = \tilde{\sigma}_{el}(0) - \text{const } \tilde{c}^{1/2} \quad . \tag{4.293}$$

What does this all mean? Some conclusions can be drawn immediately. The current in the connecting wires and in the electrolyte solution is carried at a uniformly constant strength by electrons and ions, respectively; the question then arises as to how the change in the charge carrier type is achieved. The answer is that it is exactly this which is achieved by the chemical reactions at the anode and cathode. These are, therefore, not secondary phenomena, but rather essential and important parts of the process. A constraint is, however, immediately imposed: The reactions can only proceed if electrical work is supplied, and thus current does not flow in the mixed electron-ion circuit for an arbitrarily small voltage, but only if this work can be performed. The corresponding criterion is given by thermodynamics and will now be formulated.

The dependence in Eq. (4.291) indicates how the applied voltage ΔV_0 is distributed in the circuit. Since there is, first, effectively no voltage drop in the connecting wires and, second, a flow of current, with a resulting voltage drop, in the electrolyte only upon exceeding ΔV_z, it follows that steps in the potential must appear at the interface of the platinum electrodes and the electrolyte liquid. Up to an applied voltage ΔV_z, they carry all the voltage, while, afterwards, the step heights remain as ΔV_z in the sum. It is necessary to explain how these steps in the potential arise.

If the conductivity does not increase proportionally with the charge carrier concentration in the solution, it follows that interactions exist between the charge carriers which reduce their mobility. **Kohlrausch's law** shows that this effect increases with increasing concentration. While it is obvious that Coulombic interactions exist between the ions in an electrolyte solution, an explanation is required as to why they hinder mobility.

In the following, these questions will be addressed, with the discussion being limited to the most important features. It will be apparent that previously discussed facts can be used in part.

4.5.1 The Cell Voltage and the Gibbs Reaction Free Energy

Chemical reactions, be they in a laboratory or on an industrial scale, normally proceed spontaneously, in a thermodynamic sense as an irreversible process associated with a decrease in the Gibbs free energy. They can, in principle, be reversed, but the decrease in Gibbs energy must then be compensated by correspondingly performing work. Exactly this happens in electrolysis, and it there occurs in a controllable way. The electrolysis reaction, for the example introduced in the previous section, is described by

$$\text{H}^+ + \text{Cl}^- \text{ (in solution)} \xrightarrow{\Delta g_r > 0} \frac{1}{2}\text{H}_2 + \frac{1}{2}\text{Cl}_2 \text{ (gas under normal conditions)} \quad .$$
$$\tag{4.294}$$

The work per mole to be provided, ΔW, corresponds, under the existing conditions of constant temperature and pressure, to the Gibbs reaction free energy:

$$\Delta W = \Delta \mathcal{G}_r \quad . \tag{4.295}$$

The fact that the reaction begins at the cell voltage ΔV_z indicates that it is exactly then that the necessary work is provided as electrical power, i.e.,

$$\Delta W_{el} = e N_A \Delta V_z = \Delta \mathcal{G}_r \quad . \tag{4.296}$$

In the reaction formula the aggregation states were noted. This is necessary since they influence the Gibbs reaction free energy. It is to be noted that the molar Gibbs free energy of a dissolved compound depends on the concentration of the solution. The chemical potentials of the two components are given, for the case in the example of a univalent electrolyte, by

$$\tilde{\mu}_+ = \tilde{\mu}_+^0 + \tilde{R}T \ln \phi_+ + \tilde{R}T \ln \gamma_+ \tag{4.297}$$

$$\tilde{\mu}_- = \tilde{\mu}_-^0 + \tilde{R}T \ln \phi_- + \tilde{R}T \ln \gamma_- \quad . \tag{4.298}$$

In this way, the second term, which depends on the volume fraction $\phi_{+/-}$, corresponds to the increasing entropy upon increasing the dilution (see Eq. (3.69)), while the third term expresses, by means of the **activity coefficient** $\gamma_{+/-}$, all deviations from the behaviour of an ideal solution, in which the dissolved atoms would be distributed statistically without order ($\gamma_{+/-}(\text{ideal}) = 1$). These deviations arise from order phenomena, which on account of the Coulombic forces are always to be expected in electrolyte solutions. Since the first terms, $\mu_{+/-}^0$, are by definition independent of concentration, it follows that the product of the activity coefficient $\gamma_+ \gamma_-$ can be derived from the cell voltage ΔV_z according to

$$eN_A \Delta V_z = \Delta \mathcal{G}_r$$
$$= \Delta \mathcal{G}_r^0 - \tilde{R}T \ln \phi_+ \phi_- - \tilde{R}T \ln \gamma_+ \gamma_- \quad . \tag{4.299}$$

In electrochemistry, the corresponding equation valid for an electrolyte of arbitrary valency is referred to as the **Nernst equation**; the fundamental concentration-independent part $\Delta \mathcal{G}_r^0$ is termed the **standard Gibbs reaction free energy**.

If the supply of work is interrupted, i.e., by opening the switch in Fig. 4.37, the chemical reaction is immediately reversed into the direction which is driven by the decrease in Gibbs free energy, namely:

$$\frac{1}{2}H_2 + \frac{1}{2}Cl_2 \; (gas) \xrightarrow{-\Delta \mathcal{G}_r} H^+ + Cl^- \; (in \; solution) \quad . \tag{4.300}$$

For the given cell arrangement, the reaction enthalpy is not wasted, i.e., turned into disordered thermal motion, but rather it is available for the controlled release of electrical work. In the example here, it is simply necessary that hydrogen and chlorine are brought into contact with the cathode and the anode, respectively. Work can then be performed as long as both gases are available.

For an external voltage ΔV_0 which is greater than the cell voltage, it is found that the part of the voltage exceeding ΔV_z falls off in the electrolyte solution, such that a constant current is maintained. The associated electrical power $I(\Delta V_0 - \Delta V_z)$ is, exactly like in a solid Ohmian resistor, transformed into thermal motion. The conversion happens by means of frictional forces, which are experienced by the ions in their motion in the fluid environment. It is to be noted that there is, as in the case of electrons in metals and semiconductors, a proportionality between the strength of the electric field E to which the ions are exposed and the resulting constant velocity. The concept of the electrical mobility ν_{el} which was introduced above (Eq. (4.158)) can thus be used, with the velocities v_+ and v_- of the ions hence being given as

$$v_+ = \nu_{el,+} E \quad \text{und} \quad v_- = \nu_{el,-} E \ . \tag{4.301}$$

The electrical conductivity is thus, using the expression in Eq. (4.157) for the electrons and holes in semiconductors correspondingly, given by

$$\sigma_{el} = e(\rho_+ \nu_{el,+} + \rho_- \nu_{el,-}) \ , \tag{4.302}$$

where the ion densities are

$$\rho_+ = \rho_- = N_A \tilde{c} \ . \tag{4.303}$$

As has been established above, there must be a jump-like voltage drop between the inside of the electrodes and the part of the electrolyte solution in the immediate vicinity of the electrodes, with a total step height of ΔV_z for $\Delta V_0 > \Delta V_z$ and ΔV_0 for $\Delta V_0 < \Delta V_z$. These jumps arise because of the formation of charge double layers – the reason for their formation is exactly the same as in the case of the charge double layers at p-n junctions and metal-metal or metal-semiconductor contacts. If the same charge carriers are present with different chemical potentials in two different materials, it is found that diffusion processes across the interface set in upon contact. An electrical double layer comprising an enriched and a depleted area continues to form until the diffusion currents in the two directions are brought to the same value. The condition for achieving thermodynamic equilibrium was already identified when discussing the nature of the equilibrium state at a p-n junction (Fig. 4.15): The double layer forms such that the electrochemical potentials of the charge carriers in the two materials, which are making contact, are at the same level. This also applies here, i.e., for the solid-liquid, solid-gas, or liquid-gas interfaces in electrolyte cells. The charge carriers which cross the interfaces can now be ions or electrons.

In order to describe the current-voltage curve above the cell voltage ΔV_z, it is possible, on account of the above discussion, to use Eqs. (4.183) and (4.188), which were formulated for the case of the flow of current at a p-n junction. Consider first the equilibrium state at

$$\Delta V_0 = \Delta V_z \quad ,$$

i.e., at the transition between the galvanic and electrolytic operation of the cell. Electron currents, which are transported by Cl^- ions, flow in both directions at the anode: There are electron currents of density $j_{e,ox}$ due to the conversion from a species in solution to a gas as well as one of density $j_{e,red}$ associated with the reverse process involving the ionisation of a chlorine atom by the take-up of an electron from the anode. The indices express the fact that the current acts in an oxidising, i.e., electrons are removed, or a reducing, i.e., electrons are gained, fashion. Since only $j_{e,red}$ must overcome the potential step $e\Delta V_A$ at the anode, it follows that

$$j_{e,ox}(0) \propto \rho^E \quad , \tag{4.304}$$

$$j_{e,red}(0) \propto -\rho^A \cdot \exp -\frac{N_A e \Delta V_A}{k_B T} \quad , \tag{4.305}$$

where ρ^E and ρ^A denote the densities of the Cl^- ions associated with the electrons in the electrolyte and the gas. At equilibrium, it follows, in analogy to Eq. (4.185), that

$$j_{e,red}(0) + j_{e,ox}(0) = 0 \quad . \tag{4.306}$$

If the potential step is increased, by applying an external voltage, to

$$\Delta V_A + \delta V_A \quad , \tag{4.307}$$

the reducing current correspondingly drops down to

$$j_{e,red}(\delta V_A) = j_{e,red}(0) \exp\left(-\frac{N_A e \delta V_A}{\tilde{R} T}\right) \quad . \tag{4.308}$$

The overall current density of the electrons is, thus,

$$j_e(\delta V_A) = j_{e,ox}(\delta V_A) + j_{e,red}(\delta V_A) \tag{4.309}$$

$$= j_{e,ox}(0)\left(1 - \exp -\frac{N_A e \delta V_A}{\tilde{R} T}\right) \quad , \tag{4.310}$$

and the observed linear dependence for small **overvoltages** δV_A

$$j_e \propto \delta V_A \tag{4.311}$$

follows from a series expansion. The same applies for the cathode. The overvoltage $\Delta V_0 - \Delta V_z$ applied to the cell is divided between the anode and the cathode, such that there is the same current strength.

Deviations from linearity are frequently observed for higher overvoltages, and there is a transition to an exponential dependence

$$j_e \propto \exp(\mathrm{const}\; \delta V_A) \; . \tag{4.312}$$

This indicates that the oxidation current is, in contrast to the generation current of a p-n junction, not a constant quantity, but rather it increases exponentially for increasing overvoltages. This is expressed in the **Butler–Volmer equation**

$$j_{e,\mathrm{ox}} = j_0 \exp\left[-\frac{\tilde{g}^* - \alpha N_A e(\Delta V_A + \delta V_A)}{\tilde{R}T}\right] = j_{e,\mathrm{ox}}(0) \exp\frac{\alpha N_A e \delta V_A}{\tilde{R}T} \; ,$$
$$\tag{4.313}$$

which is based on the assumption that the oxidation current upon entering the anode from solution must overcome an activation barrier, the height of which is influenced by the voltage step at the anode. The barrier height has, without a voltage step, the value $\tilde{g}^*$, which is given by the molar Gibbs free energy of a certain activated state which the Cl^- ion must overcome during the transfer. Upon applying a voltage, the barrier height decreases as described in the equation. When replacing $j_{e,\mathrm{ox}}(0)$ in Eq. (4.310) by Eq. (4.313) with the empirical parameter α, the observed exponential increase in the current for high overvoltages – known for over 100 years as the **Tafel law** – is correctly described.

The details of the transfer process have been neglected in the formal description of the currents which flow between the electrolyte solution and the electrode – these details, which are often difficult to explain, determine the values of $\tilde{g}^*$ and α.

4.5.2 Debye–Hückel Theory and Ion Mobility

In a landmark paper which was of major importance for the development of electrochemistry, Debye and Hückel, in 1923, presented a model by which the Coulombic interaction forces in weak electrolyte solutions could be quantitatively described. The approach in a generalised form was later extended to problems outside electrochemistry, and it thus had a far reaching effect. The aim of the Debye–Hückel theory is the calculation of the change in the Gibbs free energy of the electrolyte solution which results, starting from an ideal solution, from the Coulombic forces between the ions. The starting point for this is the equation

$$\Delta \mathcal{G} = \tilde{R}T \ln \gamma_+ \gamma_- = \frac{1}{2} N_A e \bar{V}(0) \; . \tag{4.314}$$

It states that the lowering of the Gibbs free energy – expressed, according to Eqs. (4.297) and (4.298), by the values of the activity coefficients – is determined by the electric potential which is generated by all other ions at the location of a particular positive ion. Since this potential exhibits time-dependent fluctuations because of the motion of the ions, the time-averaged potential must be considered – this is denoted $\bar{V}(0)$. The division by two is necessary to avoid counting all pair interactions twice.

$\bar{V}(0)$ is determined by the charge distribution around a particular ion. For a negative and a positive ion, an accumulation of positive and negative ions, respectively, is to be expected. Ionic crystals demonstrate that both possibilities can exist at the same time while maintaining global charge neutrality. The pair distribution function $g_2(r)$ was introduced in Sect. 1.2.2 in the discussion of liquid structures. This function, in a somewhat generalised form, is again employed here. Our interest focuses on the two pair distribution functions

$$g_2^{+-}(r) = g_2^{-+}(r) \tag{4.315}$$

$$g_2^{++}(r) = g_2^{--}(r) \ , \tag{4.316}$$

which describe the average local structure around an ion. $g_2^{+-}(r)$ denotes the density of negative ions at a distance r from a selected positive ion, while $g_2^{-+}(r), g_2^{--}(r)$ and $g_2^{++}(r)$ analogously describe the other pairings. From these definitions, it follows that

$$g_2^{+-}(r) + g_2^{++}(r) = 1 \ . \tag{4.317}$$

$\bar{V}(0)$ can be evaluated if the average charge distribution around the ion is known. The latter is referred to as $\bar{\eta}(r)$. For the chosen example of a univalent electrolyte, it is given, in terms of the pair distribution functions, as

$$\bar{\eta}(r) = -eg_2^{+-}(r) + eg_2^{++}(r) \ . \tag{4.318}$$

On account of electroneutrality it must be true that

$$\int \bar{\eta}(r)\mathrm{d}^3 r = -e \qquad \text{i.e.,} \qquad \int \left[g_2^{+-}(r) - g_2^{++}(r)\right]\mathrm{d}^3 r = 1 \ . \tag{4.319}$$

$\bar{V}(0)$ can be determined, for a given charge distribution $\bar{\eta}(r)$, by solving the Poisson equation

$$\nabla^2 \bar{V}(\boldsymbol{r}) = -\frac{\bar{\eta}(\boldsymbol{r})}{\varepsilon_0\varepsilon} \tag{4.320}$$

in the form which applies for the case of an isotropic system

$$\frac{1}{r^2}\frac{\mathrm{d}}{\mathrm{d}r}(r^2\frac{\mathrm{d}}{\mathrm{d}r})\bar{V}(r) = -\frac{\bar{\eta}(r)}{\varepsilon_0\varepsilon} \ . \tag{4.321}$$

The inclusion of the dielectric constant ε allows the polarisation properties of the solvent to be taken into account.

How can $\bar{\eta}(r)$ be obtained? First of all, it must be stated once again that the determination of the pair distribution function – a central problem in the theory of liquids – cannot be achieved exactly. Debye and Hückel chose an approach, which while being approximate in nature, proved to be very successful. The account begins with a consideration of the behaviour of a system composed of non-interacting, but charged, particles which is exposed to an electric field with a potential $V(\boldsymbol{r})$. A particle density distribution as determined by Boltzmann statistics is established. The local densities $\rho_+(\boldsymbol{r})$ and $\rho_-(\boldsymbol{r})$ are given, for a 1:1 mixture of univalent positive and negative ions as in the electrolyte solution example, by

$$\rho_+(\boldsymbol{r}) = \bar{\rho}_+ \exp -\frac{eV(\boldsymbol{r})}{k_B T} \tag{4.322}$$

$$\rho_-(\boldsymbol{r}) = \bar{\rho}_- \exp \frac{eV(\boldsymbol{r})}{k_B T} \quad , \tag{4.323}$$

where it is assumed that for a vanishing potential the densities

$$\bar{\rho}_+ = \bar{\rho}_-$$

exist. The charge density distribution $\eta(\boldsymbol{r})$ follows as

$$\eta(\boldsymbol{r}) = e\bar{\rho}_+ \exp -\frac{eV(\boldsymbol{r})}{k_B T} - e\bar{\rho}_- \exp \frac{eV(\boldsymbol{r})}{k_B T} \quad . \tag{4.324}$$

If the kinetic energy is large compared to the potential energy, i.e.,

$$\frac{eV}{k_B T} \ll 1 \quad , \tag{4.325}$$

a linear approximation can be made such that

$$\eta(\boldsymbol{r}) \approx -2e\bar{\rho}_+ \frac{eV(\boldsymbol{r})}{k_B T} \quad . \tag{4.326}$$

The suggestion of Debye and Hückel was simply to use this equation for the charge distribution and the variation of the potential in the environment of an ion, i.e., to relate $\bar{\eta}(r)$ and $\bar{V}(r)$ as

$$\bar{\eta}(r) = -2e\bar{\rho}_+ \frac{e\bar{V}(r)}{k_B T} \quad . \tag{4.327}$$

How can this be justified? Can a system of interacting particles be treated as a system of individual particles in a collective potential field? This approach has actually been chosen previously, namely again and again in Chap. 3: It was assumed that the forces arising from all other particles which act upon a particular particle generate a **molecular field**. In this molecular field the particles then move independently from each other. Representative examples of molecular fields were the polarisation in ferroelectrics, the magnetisation in ferromagnetics or the nematic potential in liquid crystals. $\bar{V}(r)$ represents here again such a molecular field, in which the ions move now as individual particles. Under this condition the charge distribution $\bar{\eta}(r)$ is in fact given by Eq. (4.327).

There is always a self-consistency condition associated with the **molecular field approximation** in so far as the strength of the molecular field is always proportional to an associated **order parameter**, and, vice versa, in that the order parameter establishes itself in the field. This also applies here where there are two different equations, Eqs. (4.321) and (4.327), linking the **order parameter** $\bar{\eta}(r)$ and the molecular field $\bar{V}(r)$. The solution can be found by eliminating $\bar{\eta}(r)$, which leads to the differential equation

$$\frac{1}{r^2}\frac{\mathrm{d}}{\mathrm{d}r}(r^2\frac{\mathrm{d}}{\mathrm{d}r})\bar{V}(r) = \frac{1}{\xi_{\mathrm{D}}^2}\bar{V}(r) \ . \tag{4.328}$$

It is known as the **Poisson–Boltzmann equation**. The only coefficient in the equation is the **Debye length** ξ_{D}, given by

$$\xi_{\mathrm{D}} = \left(\frac{\varepsilon_0\varepsilon k_{\mathrm{B}}T}{2\bar{\rho}_+e^2}\right)^{1/2} \ . \tag{4.329}$$

Performing the same derivation for the case of an electrolyte of arbitrary composition, with ion densities $\bar{\rho}_i$ and charges z_ie, gives

$$\xi_{\mathrm{D}} = \left(\frac{\varepsilon_0\varepsilon k_{\mathrm{B}}T}{2I_{\mathrm{s}}N_{\mathrm{A}}}\right)^{1/2} \ , \tag{4.330}$$

where

$$I_{\mathrm{s}} = \frac{1}{2N_{\mathrm{A}}}\Sigma_i\bar{\rho}_iz_i^2e^2 \tag{4.331}$$

is referred to as the **ionic strength**.

The Poisson–Boltzmann equation can be solved exactly. The general solution is

$$\bar{V}(r) = \frac{c_1}{r}\exp-\frac{r}{\xi_{\mathrm{D}}} + \frac{c_2}{r}\exp\frac{r}{\xi_{\mathrm{D}}} \ . \tag{4.332}$$

This can be verified by inserting Eq. (4.332) back into Eq. (4.328). The obvious requirement $\bar{V}(r \to \infty) = 0$ implies that the second term must be zero, and thus

$$\bar{V}(r) = \frac{c_1}{r}\exp-\frac{r}{\xi_{\mathrm{D}}} \ . \tag{4.333}$$

The physical meaning of the result can be immediately recognised: The accumulation of oppositely charged ions in the vicinity of an ion leads to a screening effect such that the Coulombic field of ions at separations which are large compared to the Debye length completely vanishes. There are thus no long-range interactions in an electrolyte. The screening sets in with increasing ion strengths at ever shorter distances. It can be recognised from Eq. (4.327) that ξ_{D} also characterises the extension of the ion cloud.

The Coulombic potential is still completely active in the immediate vicinity of the ions, i.e., for $r \ll \xi_{\mathrm{D}}$. This allows the determination of the integration constant c_1, such that the final solution is

$$\bar{V}(r) = \frac{e}{4\pi\varepsilon_0\varepsilon r} \exp{-\frac{r}{\xi_D}} \tag{4.334}$$

for a simple positively charged ion as the probe ion.

At the start, we asked about the nature of the potential which is generated at the location of the probe ion by all the other ions. In Eq. (4.334), a solution has been obtained which includes the contribution of the central ion as well as that of the induced negatively charged cloud, which provides the screening. The potential generated solely by the charged cloud is obtained by subtracting the contribution of the central ion:

$$\bar{V}(r) = \frac{e^2}{4\pi\varepsilon_0\varepsilon r} \left(\exp(-\frac{r}{\xi_D}) - 1 \right) . \tag{4.335}$$

From this $\bar{V}(0)$ is obtained from this expression by a series expansion:

$$\bar{V}(0) = -\frac{e^2}{4\pi\varepsilon_0\varepsilon\xi_D} . \tag{4.336}$$

The ion-ion interaction leads, as expected, to a lowering of the energy and thus the chemical potential. The amount by which the energy is lowered increases together with the ion strength according to Eq. (4.330), and it is found that

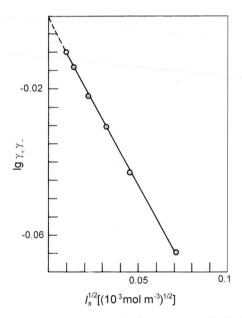

Fig. 4.40. The change in the product of the activity coefficients upon increasing the electrolyte concentration of a HCl solution. I_s denotes the ionic strength (from Bockris and Reddy [39]).

$$\tilde{R}T \ln \gamma_+ \gamma_- \propto -I_{\mathrm{s}}^{1/2} \ . \tag{4.337}$$

This prediction can be checked. Changes in the activity coefficients are apparent in measurements of the cell voltage, as is described by the Nernst equation (Eq. (4.299)). Many such investigations were carried out, and they showed that the Debye–Hückel theory is able to explain very well the behaviour observed for dilute electrolyte solutions. Figure 4.40 shows, as an example, the result of an investigation of a dilute hydrochloric acid solution – a perfect agreement between theory and experiment is observed.

The Motion of the Ion Clouds. A consideration of Eq. (4.302) and Kohlrausch's law (Eq. (4.293)) reveals that the ion mobility in an electrolyte solution decreases upon increasing the concentration according to the following square root law

$$\nu_{\mathrm{el},+}(\tilde{c}) + \nu_{\mathrm{el},-}(\tilde{c}) = \nu_{\mathrm{el},+}(0) + \nu_{\mathrm{el},-}(0) - \beta \tilde{c}^{1/2} \ . \tag{4.338}$$

This is a consequence of the structure within the solution, or expressed more precisely, it is, considering the Debye–Hückel result, a consequence of the formation of the ion ion-cloud local structure. The characteristic structural parameter is the Debye length. The fact that, first, ξ_{D} also has a square-root dependence on the concentration, i.e.,

$$\xi_{\mathrm{D}} \propto \tilde{c}^{-1/2} \ , \tag{4.339}$$

and, second, that it is the only structural parameter in the system suggest that the reduction in mobility is intimately linked with ξ_{D}. In this respect, two different effects can be envisaged. Firstly, the local structure in the electrolyte, i.e., individual ions being surrounded by oppositely charged ion clouds, will inevitably be disturbed upon applying an electric field because of the opposite forces acting on the two parts of this structure. A stationary structure establishes itself even here; it is, however, made up around the moving ions which have a constant average velocity in the direction of (or opposite to) the field. This local structure is no longer isotropic. Additional forces arise now, because of the opposite velocities of the ion and the surrounding cloud and also the loss of isotropy. Debye, Hückel and Onsager presented a statistical hydrodynamic theory by which these forces can be calculated. This cannot be described here, but rather a plausible explanation of the results will be presented.

A simple argument can be used in the case of the force originating from the opposite movement of the cloud. The ion cloud can be treated as if it were a colloid with a radius ξ_{D} and a charge $-e$. If a force $-eE$ is applied to this colloid in solution, it has a resulting velocity, which is given according to Stokes' law (Eq. (5.172)) by

$$v_{\mathrm{cl}} = \frac{-eE}{6\pi \xi_{\mathrm{D}} \eta_{\mathrm{w}}} \ , \tag{4.340}$$

where η_w denotes the viscosity of water. The central ion would simply be carried along, had it not a positive charge drawing it in the field direction. Being incorporated in the moving cloud, its velocity is reduced by the cloud velocity v_{cl}, which implies a change in the electric mobility of the ion by

$$\Delta \nu_{el,+} = -\frac{e}{6\pi\xi_D\eta_w} \quad . \tag{4.341}$$

The reduction is inversely proportional to ξ_D and is thus in agreement with Kohlrausch's law.

A second force arises from the asymmetry of the average charge distribution in the cloud. Charges are redistributed from the front of the moving region to the back; this results in a force which acts against the ion motion. The cloud is undergoing a constant reorganisation, with the asymmetry being determined by the reorganisation time; it will be ever more extended in the direction opposite to the motion, the longer this time is. The reorganisation time can be estimated. Consider first how long it would take for the cloud to disperse if the central ion was rapidly removed. For this to occur, it is necessary for the particles in the cloud to diffuse over distances on the order of ξ_D. It will be explained in Sect. 5.2.1 that a time on the order of

$$\tau \simeq \frac{\xi_D^2}{6D_s} \tag{4.342}$$

is required where D_s denotes the self-diffusion coefficient of the ionic particles. It is the distance Δx, which the central ion moves in this time, which is essential in determining the asymmetry. It is given, for a velocity v, by

$$\Delta x \simeq v\tau \tag{4.343}$$

$$= \frac{v\xi_D^2}{6D_s} \quad . \tag{4.344}$$

In a perturbation treatment values we chose for v and D_s the values found in a solution without Coulombic interactions, i.e., at a vanishing concentration. We write

$$v = \nu_{el,+}(0)E \quad , \tag{4.345}$$

while D_s is given, on the basis of the Einstein relation (Eq. (5.171)), by

$$D_s = k_B T \nu_-(0) = k_B T \frac{\nu_{el,-}(0)}{e} \quad . \tag{4.346}$$

Making the assumption

$$\nu_{el,+} \approx \nu_{el,-} \tag{4.347}$$

leads to the result

$$\Delta x \simeq \frac{e\xi_D^2 E}{6k_B T} \quad . \tag{4.348}$$

The force f_{asy} acting on the central ion increases with increasing asymmetry. In the absence of more precise knowledge about the nature of the deformed cloud, we consider $\Delta x/\xi_{\mathrm{D}}$ as representing an 'asymmetry coefficient' and expand the force as a power series in this coefficient. One obtains in linear approximation

$$f_{\mathrm{asy}} \simeq -f_0 \frac{\Delta x}{\xi_{\mathrm{D}}} = -f_0 \frac{e\xi_{\mathrm{D}}E}{6k_{\mathrm{B}}T} \quad . \tag{4.349}$$

The fact that ξ_{D} is the only structural parameter suggests, starting from Coulomb's law, the following form for the parameter f_0 which determines the strength of the force:

$$f_0 = \frac{e^2}{4\pi\varepsilon_0\varepsilon\xi_{\mathrm{D}}^2} \quad . \tag{4.350}$$

It then follows that

$$f_{\mathrm{asy}} \simeq -\frac{e^3}{24\pi\varepsilon_0\varepsilon k_{\mathrm{B}}T\xi_{\mathrm{D}}}E \quad . \tag{4.351}$$

f_{asy} causes the total force to be reduced to

$$f + f_{\mathrm{asy}} = eE\left(1 - \frac{e^2}{24\pi\varepsilon_0\varepsilon k_{\mathrm{B}}T\xi_{\mathrm{D}}}\right) \tag{4.352}$$

and leads to a corresponding reduction in the electrical mobility

$$\Delta\nu_{\mathrm{el},+} = -\frac{e^2}{24\pi\varepsilon_0\varepsilon\xi_{\mathrm{D}}k_{\mathrm{B}}T}\nu_{\mathrm{el},+}(0) \quad . \tag{4.353}$$

An inverse dependence on ξ_{D} is, thus, also found here. When taken together, the two effects, which are referred to in the literature as the **electrophoretic effect** and the **relaxation effect**, lead to the following overall result in the sum for both positive and negative ions:

$$\nu_{\mathrm{el},+} + \nu_{\mathrm{el},-} = \nu_{\mathrm{el},+}(0) + \nu_{\mathrm{el},-}(0) - \frac{\beta_1}{\xi_{\mathrm{D}}} - [\nu_{\mathrm{el},+}(0) + \nu_{\mathrm{el},-}(0)]\frac{\beta_2}{\xi_{\mathrm{D}}} \quad . \tag{4.354}$$

The Debye–Hückel–Onsager theory gives the same dependencies albeit with slightly changed prefactors. The predictions of the theory describe the experimental results very well. Figure 4.41 shows plots, corresponding to Kohlrausch's law, of the molar conductivity $\tilde{\sigma}_{\mathrm{el}}$ against $\tilde{c}^{1/2}$ for different electrolytes in aqueous solution. The slopes of the best-fit straight lines agree with the predictions of the theory to a precision in the percentage range.

Finally, a comment will be made about the zero concentration limit of the **electrical mobility** $\nu_{\mathrm{el},+/-}(0)$, which can be derived from measurements in aqueous solution of the molar **limiting conductivity**

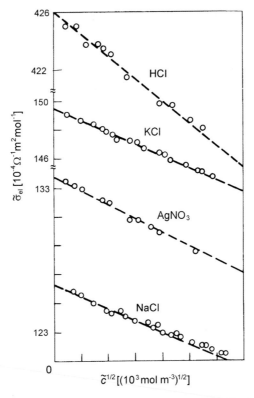

Fig. 4.41. The decrease in the molar conductivity upon increasing the electrolyte concentration for different aqueous electrolyte solutions. The experimental data follow Kohlrausch's law (Eq. (4.293)) (from Bockris and Reddy [40]).

$$\tilde{\sigma}_{el}(\tilde{c} \to 0) \propto \nu_{el,+}(0) + \nu_{el,-}(0) \ . \tag{4.355}$$

In the first instance, it would be expected, in the absence of ion-ion interactions, that the electrical mobility is, on the basis of Stokes' law, given by

$$\nu_{el,+/-}(0) \approx \frac{\pm e}{6\pi\eta_w a_I} \ , \tag{4.356}$$

i.e., it would be expected that it depends on the radius of the ion, a_I. A value for a_I can be derived from the packing density in ionic crystals. However, the experimental results differ significantly from such expected values. In particular, a direct contradiction is found with the zero concentration limits of the electrical mobility for the alkali metal ions Li^+, Na^+, K^+ and Rb^+. A decrease in the limiting mobility would be expected to be associated here with the increasing radius from Li^+ to Rb^+. However, the inverse is actually found, with Li^+ having the lowest electrical mobility. The reason for this, at first, surprising behaviour is the **hydration** of the ions. In aqueous solution, the ions are surrounded by a shell of associated water molecules because of the large dipole moment of water. The Coulombic force at and in the vicinity of the surface of the ions acts to orientate the dipole moment of the H_2O molecules such that a shell forms. At a constant charge e, the Coulombic force is found to be stronger the smaller the ionic radius is. Li^+ ions, thus, have the most extended hydrated shell. Since this shell moves with the ion on applying an electric field, it is the radius of the hydration shell and not the ionic radius which determines the limiting ionic mobility, which explains the experimental observations.

4.6 Exercises

1. The Fermi–Dirac distribution function is given by

$$w(\epsilon) = \cfrac{1}{\exp\left(\frac{\epsilon - \mu_e}{k_B T}\right) + 1}.$$

   Using a linear approximation for $\epsilon = \mu_e$, make an approximate plot of the function for $T \neq 0$, $k_B T \ll \mu_e$. How large is the region in which the distribution $w(\epsilon)$ for $T \neq 0$ deviates from that for $T = 0$?

2. The compression modulus K is given by the second derivative of the free energy $\mathcal{F}$ with respect to the volume: $K = \mathcal{V} \partial^2 \mathcal{F} / \partial \mathcal{V}^2$. Estimate the compression modulus for the alkali metals – assume that the Helmholtz free energy corresponds to the kinetic energy of the corresponding Fermi electron gas at $T = 0$. (Sodium: $\rho_m = 0.97\,\mathrm{g\,cm^{-3}}$, $A = 23\,\mathrm{g\,mol^{-1}}$).

3. (a) Calculate the density of levels for a two-dimensional gas of free electrons in a 'quantum film'. The boundary conditions for the electronic wavefunctions are $\psi(x, y, z) = 0$ for $|x| > a$, where a has atomic dimensions. The gas spreads itself out over a macroscopic area A in the y, z plane. A standing wave is obtained in the ground state in the x direction, while there are travelling waves in the y and z directions.

   (b) Calculate the density of levels for a one-dimensional gas of free electrons in a 'quantum wire'. The boundary conditions for the electronic wavefunctions are $\psi(x, y, z) = 0$ for $|x| > a$ and $|y| > b$, where a and b have atomic dimensions. The quantum wire spreads itself out over a macroscopic length L in the z direction. A standing wave is obtained in the ground state in the x and y directions, while there is a travelling wave in the z direction.

4. For a sufficiently thin metallic film, there is a dependence of the electrical conductivity σ_{el} on the thickness of the film d. The effect begins to become clear when the film thickness is of the same order of magnitude as the mean free path λ_f. Show that the following dependence is to be expected for $d < \lambda_f$:

$$\frac{\sigma_{\mathrm{el}}}{\sigma_{\mathrm{el},0}} = \frac{3d}{4\lambda_{\mathrm{f}}} + \frac{d}{2\lambda_{\mathrm{f}}} \ln\left(\frac{\lambda_{\mathrm{f}}}{d}\right)$$

Hint: Electrons far away from the surface freely move, on average, a distance λ_{f} between two collisions. Electrons, in the vicinity of the surface, can be scattered at the surface before reaching this distance. This means, considering the average for all electrons, that there is a reduction in the mean free path and thus the conductivity, since $\sigma_{\mathrm{el}} \propto \lambda_{\mathrm{f}}$. $\lambda_{\mathrm{f}}(d)$ can be calculated from simple statistical considerations.

5. (a) Calculate the plasma frequency ω_{pl} of sodium. At which wavelength does the metal become transparent to electromagnetic radiation ($\rho_{\mathrm{m}} = 0.97\,\mathrm{g\,cm^{-3}}, A = 23\,\mathrm{g\,mol^{-1}}$)?

   (b) Calculate the dielectric susceptibility χ, the dielectric constant ε and the refractive index $n = \sqrt{\varepsilon}$ of sodium for X-ray radiation of energy $10\,\mathrm{keV}$.

6. Solutions of the Maxwell equations with the following form ('surface plasmons') can exist at the surface of a plasma (plasma for $z > 0$, vacuum for $z < 0$):

$$z > 0: \quad V(x, z)_+ = V_0 \cos(kx)e^{-kz}$$
$$z < 0: \quad V(x, z)_- = V_0 \cos(kx)e^{kz} \ .$$

Calculate $\boldsymbol{E} = -\nabla V$. The tangential component of $\boldsymbol{E}$ and the normal component of $\boldsymbol{D}$ must be continuous. For which frequencies can these conditions be fulfilled? Use the expression for $\varepsilon(\omega)$ of metals for the case of $\omega\tau \gg 1$.

7. If an electron is found in a weak, periodic potential with a period a, changes are above all found, to a first approximation, for the wavefunctions of the electrons with wavevectors k at the edges of the Brillouin zones, $k = p\pi/a$ (p is an integer). A perturbation calculation shows that the modified wavefunctions can be described by standing waves. The wavefunctions at the edge of the first Brillouin zone have the following form:

$$\psi_+(x) = A\cos\frac{\pi}{a}x \qquad \psi_- = A\sin\frac{\pi}{a}x \ .$$

Normalise ψ_+ and ψ_- in the interval a, such that

$$\int_0^a |\psi_+(x)|^2\,\mathrm{d}x = \int_0^a |\psi_-(x)|^2\,\mathrm{d}x = 1 \ ,$$

and calculate the energy of the two solutions ψ_+ and ψ_-:

$$\epsilon = \int_0^a \psi_\pm^* \left(\frac{-\hbar^2}{2m_{\mathrm{e}}}\frac{\mathrm{d}^2}{\mathrm{d}x^2} + u(x)\right)\psi_\pm\,\mathrm{d}x \quad \text{where} \quad u(x) = -u_0\cos\frac{2\pi}{a}x.$$

Sketch the functions $u(x)$, $|\psi_+(x)|^2$ and $|\psi_-(x)|^2$ and use the sketches to discuss the result of the above calculation.

8. Electrons in a magnetic field move within closed orbits. The time τ for one circuit around a closed orbit on the Fermi surface is given, using the equation of motion $\hbar d\boldsymbol{k}/dt = \mathbf{f}$, as

$$\tau = \frac{\hbar^2}{e} \oint \frac{|d\boldsymbol{k}|}{\left|\frac{d}{d\boldsymbol{k}}\epsilon \times \boldsymbol{B}\right|} \quad .$$

Calculate τ for a Fermi electron gas and show that $\tau = 2\pi/\omega_c$, where ω_c is the cyclotron frequency.

5

Microscopic Dynamics

The discussion of the characteristics of solids, liquids, liquid crystals and polymers in the previous chapters has been largely restricted to their structural properties. In this final chapter, we move on to the subject of microscopic dynamics in the different states of aggregation. In fact it must be said that this separation is somewhat artificial, since structure and dynamics have a strong interdependence and cannot be separated from each other. To every structural state, there belongs a closely-linked characteristic form of motion; the one cannot be properly understood without the other. This chapter now catches up with the important second part and will deal with the characteristics of the microscopic motion in different states of condensed matter, discussing concepts and models which can be used to describe them. We will treat here the following state-determined motional forms:

Crystals are characterised by motional modes, which are, independent of the type of excitation, always described by propagating waves. Three important types of excitation will be considered: firstly, lattice vibrations, secondly, the spin waves found in ferromagnetics, and thirdly, the exciton waves associated with electronic excitations in crystals. Wave-like excitations are encountered in crystals for all length scales down to the Å range. Their amplitudes, however, are typically limited, this being, in particular, the case for lattice vibrations. Displacements of the lattice building blocks from their equilibrium positions by more than about 15% of the lattice constants do not occur in the crystalline state. If this were to happen a crystal would exceed the limit of its inner stability such that an equilibrium state could no longer exist.

Wave-like excitations are also found in the nematic phase of liquid-crystalline substances. Corresponding spatial variations in the director orientation lead to the build-up of elastic forces, which act in a restoring fashion. There results, however, no propagating vibrational motion because of the high inner viscosity. The return to the homogeneous equilibrium state occurs in the manner of a relaxator, i.e., via an exponential decay process.

Relaxatory eigenmodes are also found in polymer melts. The connectivity leads, for a stretching of a sequence of segments within the chain, to the generation of elastic restoring forces. The dynamics of the individual chains in the melt can be described as a superposition of collective modes, which are referred to as 'Rouse modes'.

Both the Rouse modes in polymers and the fluctuations in the director field of a liquid crystal are excitations where the associated length-scales are in the mesoscopic region, from 10 nm to μm. For atomic liquids, the dynamics in this range is of a complex nature. In contrast to crystals, where wave-like mechanical excitations even exist down into the microscopic region, in liquids, these waves are increasingly perturbed upon shortening the wavelength due to the growing structural inhomogeneities. Vibrational motions still exist, but they are local in nature and have limited lifetimes. Atoms and molecules vibrate in a cage which is formed by their neighbours. This process is interrupted again and again as the configuration of the cage changes. In particular, the cage can open for a short time and then a diffusive step becomes possible. Such diffusive steps are characteristic of liquids, with the diffusion resulting in a long-range transport of particles.

All these motional forms will be discussed in this chapter. At the end, there is an experimentally oriented section, which considers time-resolved scattering experiments. Such experiments are universally applicable and allow the different motional forms in all states of aggregation to be analysed.

5.1 Crystals: Propagating Waves

5.1.1 Lattice Vibrations

Thermal energy in a crystal expresses itself in the displacement of all atoms from their equilibrium positions. On account of the coupling between the neighbours in a crystal, it follows that these displacements are not separate for each atom but occur in a cooperative fashion. As was mentioned above, it is not possible, in the crystalline state, for these displacements to exceed a certain upper limit. To a good approximation, however, it is possible, within this range, to describe the interaction potential by a harmonic approach, i.e., the forces are treated as spring-like obeying a linear law. In this way, the crystal can be considered to be a system of masses with spring-like couplings. In a theoretical analysis, it is necessary to determine the eigenmodes of this macroscopically large vibrational system. The eigenvibrations possess, as mentioned above, a wave-like character and it will be shown that this is a consequence of the translational symmetry of the crystal. The dynamics of the crystal is theoretically explained if all the eigenmodes are known. It is, as ever for vibrational systems which are driven by linear forces, possible to describe every motional state as a superposition of eigenmodes, in this case wave-like lattice vibrations. For the description, it is necessary to classify the lattice vibrations. In particular, the question arises as to the relationship between the wavevector and the frequency of a vibration, i.e., the **dispersion relation** $\omega(\boldsymbol{k})$.

The Dynamics of Linear Two-Atom Chains. There is a simple system, which encompasses all the significant characteristics of the vibrational behaviour of crystals. It is a linear two-atom chain. The discussion of this will act as an introduction, straightforward generalisations of it will then allow the transition to a three-dimensional crystal. Figure 5.1 shows a schematic representation of the model system. The diagram contains all the parameters as well as the coordinates required in the discussion. The unit cell of a chain of length a consists of two atoms of mass m_A and m_B, while b represents the force constant of the spring, which couples two neighbouring atoms together. $z_A(l)$ and $z_B(l)$ describe the displacements from their equilibrium positions for the two atoms in cell l.

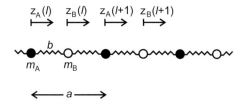

Fig. 5.1. A linear chain with two atoms per unit cell.

The equations of motion for this linear system are:

$$m_A \ddot{z}_A(l) = b[z_B(l) - z_A(l)] + b[z_B(l-1) - z_A(l)]$$
$$m_B \ddot{z}_B(l) = b[z_A(l+1) - z_B(l)] + b[z_A(l) - z_B(l)] \ . \tag{5.1}$$

Consider, as a trial solution, a wave with a frequency ω and a wavevector k, such that

$$z_A(l) = Z_A \exp[-i\omega t + ikla]$$
$$z_B(l) = Z_B \exp\left[-i\omega t + ik\left(l + \frac{1}{2}\right)a\right] \ . \tag{5.2}$$

Z_A and Z_B denote the amplitudes of the two partial waves associated with the atoms A and B. Substituting into the equations of motion gives

$$-m_A Z_A \omega^2 = Z_B b\left(\exp ik\frac{a}{2} + \exp -ik\frac{a}{2}\right) - 2Z_A b$$
$$-m_B Z_B \omega^2 = Z_A b\left(\exp ik\frac{a}{2} + \exp -ik\frac{a}{2}\right) - 2Z_B b \ . \tag{5.3}$$

The exponential terms on the right-hand side of the two expressions in Eq. (5.2) appear, after the substitution, on both sides in Eq. (5.3) and can be eliminated. This shows, in fact, that wave-like solutions exist. As can be easily recognised, it is a consequence of the periodicity; the result would not be obtained if distances, masses or force constants fluctuated.

Equation (5.3) represents a homogeneous linear system of equations in terms of the amplitudes Z_A and Z_B. Non-trivial solutions are found when

$$\begin{vmatrix} m_A\omega^2 - 2b & 2b\cos(ka/2) \\ 2b\cos(ka/2) & m_B\omega^2 - 2b \end{vmatrix} = 0 \ , \tag{5.4}$$

i.e.,

$$(m_A\omega^2 - 2b)(m_B\omega^2 - 2b) - 4b^2\cos^2 k\frac{a}{2} = 0 \ . \tag{5.5}$$

A quadratic expression for ω^2 has thus been obtained, the solutions of which are

$$\omega^2 = \frac{b}{m_r} \pm \left[\frac{b^2}{m_r^2} - \frac{4b^2}{m_A m_B} \sin^2 k\frac{a}{2} \right]^{1/2} \ , \tag{5.6}$$

where the reduced mass is given by

$$m_r = \frac{m_B m_A}{m_B + m_A} \ . \tag{5.7}$$

Equation (5.6) shows that the frequency changes periodically with the wavenumber. For every integer p we have

$$\omega^2(k) = \omega^2\left(k + p\frac{2\pi}{a} \right) \ . \tag{5.8}$$

The physical basis for this behaviour becomes clear upon comparing the displacement patterns of two waves whose wavevectors differ by a whole-number multiple of $2\pi/a$:

$$z_A(l) \propto \exp\left[-i\omega t + i\left(k + p\frac{2\pi}{a} \right) la \right] = \exp[-i\omega t + ikla] \ . \tag{5.9}$$

It is apparent that the two forms describe identical displacements and, therefore, the frequencies must also be the same. This, however, leads to an important conclusion: Since wavevectors, which differ from one another by whole-number multiples of $2\pi/a$, are completely equivalent, it is sensible to limit the consideration to the range

$$-\frac{\pi}{a} < k \le \frac{\pi}{a} \ . \tag{5.10}$$

In this way, all of the existing lattice vibrations are included. The region defined in Eq. (5.10) was encountered previously in the discussion of the electron properties in a one-dimensional lattice, where it was termed the **first Brillouin zone**. Its extension is determined by the period of the reciprocal lattice (see Sect. 1.5.3). For a one-dimensional crystal, the vectors G_h of the reciprocal lattice are given, according to Eq. (1.131), by

$$G_h = \frac{2\pi}{a}h \ , \tag{5.11}$$

and Eq. (5.10) can thus be re-expressed as

$$-G_1/2 < k \le G_1/2 \ . \tag{5.12}$$

Equation (5.9) therefore means that the same displacement patterns are found for waves, whose wavevectors differ by a vector of the reciprocal lattice G_h.

For each k, there are two solutions to Eq. (5.6). Figure 5.2 shows plots of ω against k for the two branches for all k values within the first Brillouin zone. It is not difficult to recognise the nature of the two solutions. For the lower branch, the frequency tends to zero for a vanishing wavenumber. This is precisely a basic property of acoustic waves, and this solution is of this nature. The gradient at the origin, which corresponds to the proportionality constant between the frequency and the wavenumber, should thus be equal to the velocity of sound. It is easy to verify that this is indeed the case. A power series expansion transforms Eq. (5.6) into

$$\omega^2 = \frac{2bm_r}{m_B m_A}\left(\frac{ka}{2}\right)^2 = \frac{ba/2}{(m_B + m_A)/a}k^2 \ . \tag{5.13}$$

The numerator corresponds to the modulus of elasticity of the linear chain

$$b\frac{a}{2} = E_t \tag{5.14}$$

(transforming Eq. (2.3) into that for the case of a linear chain leads to a force $f = E_t z/(a/2)$) and the denominator describes the linear mass density

$$\frac{m_B + m_A}{a} = \rho_m \ . \tag{5.15}$$

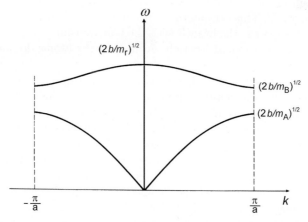

Fig. 5.2. Wave-like excitations in a linear two-atom chain: the dispersion relations $\omega(k)$ for the acoustic and the optical branches. The range of the representation corresponds to the first Brillouin zone.

It thus follows that

$$\omega = \sqrt{\frac{E_t}{\rho_m}} k \quad . \tag{5.16}$$

The proportionality factor corresponds to the well-known expression for the velocity of sound. Acoustic waves must, in the limit of infinitely large wavelengths, lead to identical displacements of both atoms, which is also the case:

$$\frac{Z_A}{Z_B} = \lim_{k,\omega \to 0} \frac{-2b\cos(ka/2)}{m_A\omega^2 - 2b} = 1 \quad . \tag{5.17}$$

The upper branch has a completely different character. A non-vanishing frequency, which is given by

$$\omega^2 = \frac{2b}{m_r} \quad , \tag{5.18}$$

is characteristically found here in the limiting case of $k \to 0$. An evaluation of the ratio of the amplitudes of the two atoms gives

$$\frac{Z_A}{Z_B} = \frac{-2b}{(m_A 2b/m_r) - 2b} = -\frac{m_B}{m_A} \quad . \tag{5.19}$$

The character of the vibration is thus clear: In each unit cell, it is found that the displacements of the two atoms A and B occur in opposite directions. A restoring force and thus a non-vanishing frequency establishes itself even if there is no phase difference between the unit cells ($k = 0$). If vibrations of this type occur in an ionic crystal, the result is the appearance of an oscillating dipole in every cell. This can interact with an electric field, i.e., it is 'optically active'. For this reason, the branch with the non-vanishing frequency at $k = 0$ is referred to as the **optical branch**, even when the atoms do not carry any charge.

In the above discussion, it has been assumed, without it ever being explicitly stated, that the chain is infinitely long; only in this way is it possible for waves to propagate in an unperturbed fashion. Real systems, in contrast, always have a finite extension. This has the consequence that the number of eigenvibrations remains finite. When modelling the properties of real systems, this limiting of the number of eigenmodes must always be taken into account, even if the number is very large; if this is not done, it is not possible to calculate thermal properties, such as the specific heat. Doing it, however, leads to a change: For a finite stretched chain, standing waves appear as eigenvibrations for the case of both free and fixed ends, rather than the propagating waves of Eq. (5.2). Fortunately, it is possible, to circumvent this problem, in the same manner as in the treatment of the electrons in the lattice: If the linear chain is made into a closed circle, the travelling waves continue to be eigenmodes. The forming into a circle can be introduced into the equations in the same easy way as before. A circular chain of $\mathcal{N}_c$ cells requires that

$$z_A(l + \mathcal{N}_c) = z_A(l) \tag{5.20}$$

and analogously for z_B. This leads to the following cyclic boundary condition:

$$\exp(ikla) = \exp[ik(l + \mathcal{N}_c)a] \quad . \tag{5.21}$$

This is fulfilled by

$$\exp(ik\mathcal{N}_c a) = 1 \quad . \tag{5.22}$$

In this way, a selection rule has been obtained: For a closed chain, it is found that only eigenvibrations with wavevectors

$$k = i\frac{2\pi}{\mathcal{N}_c a} \qquad (i \text{ is an integer}) \tag{5.23}$$

appear, with the limitation to the first Brillouin zone meaning that

$$-\frac{\mathcal{N}_c}{2} < i \le \frac{\mathcal{N}_c}{2} \quad . \tag{5.24}$$

Overall, there are thus $\mathcal{N}_c$ different values for the wavevector k. Since there are two branches, there are therefore $2\mathcal{N}_c$ different eigenvibrations. This corresponds, as required, to the number of degrees of freedom.

Eigenvibrations in Crystals. It is evident how the results obtained for a linear two-atom chain can be generalised for the case of a three-dimensional crystal. A planar wave in a three-dimensional lattice can be described by

$$z_j(\boldsymbol{R}_{uvw}) = z_j \exp(-i\omega t + i\boldsymbol{k}\boldsymbol{R}_{uvw}) \quad . \tag{5.25}$$

$\boldsymbol{k}$ is the wavevector and $\boldsymbol{R}_{uvw}$ denotes the position of the corner of the unit cell uvw. If the cell contains n atoms, $3n$ coordinates are required to describe the displacements in a cell; these coordinates are denoted by z_j, where

$$j = 1, \ldots, 3n \quad . \tag{5.26}$$

Corresponding to the number of degrees of freedom per cell, it is to be expected that there are $3n$ different branches, each with a different dispersion relation $\omega_j(\boldsymbol{k})$.

A displacement of a wavevector $\boldsymbol{k}$ by a vector of the reciprocal lattice, i.e., the transition from $\boldsymbol{k}$ to $\boldsymbol{k} + \boldsymbol{G}_{hkl}$, leaves the displacements of all atoms completely unchanged, because the following is true:

$$z_j(\boldsymbol{R}_{uvw}) \propto \exp[i(\boldsymbol{k} + \boldsymbol{G}_{hkl})\boldsymbol{R}_{uvw}] = \exp(i\boldsymbol{k}\boldsymbol{R}_{uvw}) \tag{5.27}$$

since

$$\boldsymbol{G}_{hkl}\boldsymbol{R}_{uvw} = p2\pi \qquad (p \text{ is an integer}) \; . \qquad (5.28)$$

As a consequence, it is also in the three-dimensional case appropriate to impose a restriction on the wavevectors. It was already discussed for the band model of electron states how such a choice can be made: All possibilities are included if the $\boldsymbol{k}$ vectors are limited to the first Brillouin zone.

In a last step of the generalisation, it is necessary to take account of the finite extension and the finite number of degrees of freedom in a three-dimensional crystal. This is achieved, as in the case of a Fermi electron gas, by imposing the Born–von Karman cyclic boundary conditions. For a crystal, consisting of $\mathcal{N}_1$, $\mathcal{N}_2$ and $\mathcal{N}_3$ unit cells in the direction of the three basis vectors of the Bravais lattice, such that the total number of unit cells is

$$\mathcal{N}_c = \mathcal{N}_1 \mathcal{N}_2 \mathcal{N}_3 \qquad (5.29)$$

the cyclic boundary conditions are

$$z_j(\boldsymbol{R}_{uvw}) = z_j(\boldsymbol{R}_{uvw} + \mathcal{N}_1\boldsymbol{a}_1) = z_j(\boldsymbol{R}_{uvw} + \mathcal{N}_2\boldsymbol{a}_2) = z_j(\boldsymbol{R}_{uvw} + \mathcal{N}_3\boldsymbol{a}_3) \; . \qquad (5.30)$$

These are fulfilled if the wavevectors $\boldsymbol{k}$ are chosen such that

$$\exp(\mathrm{i}\boldsymbol{k}\mathcal{N}_1\boldsymbol{a}_1) = \exp(\mathrm{i}\boldsymbol{k}\mathcal{N}_2\boldsymbol{a}_2) = \exp(\mathrm{i}\boldsymbol{k}\mathcal{N}_3\boldsymbol{a}_3) = 1 \; . \qquad (5.31)$$

This is the case if $\boldsymbol{k}$ is of the form

$$\boldsymbol{k}_{i_1 i_2 i_3} = i_1\frac{\widehat{\boldsymbol{a}}_1}{\mathcal{N}_1} + i_2\frac{\widehat{\boldsymbol{a}}_2}{\mathcal{N}_2} + i_3\frac{\widehat{\boldsymbol{a}}_3}{\mathcal{N}_3} \qquad \text{where} \quad i_1, i_2, i_3 \quad \text{are integers} \; . \qquad (5.32)$$

The set of numbers i_1, i_2, i_3 are to be chosen such that $\boldsymbol{k}$ is in the first Brillouin zone; this then leads to $\mathcal{N}_c$ different wavevectors. The great advantage of the Born–von Karman condition is that it ensures that the travelling waves in Eq. (5.25) are maintained as solutions.

A clear scheme for the classification of lattice vibrations in a three-dimensional crystal has thus been achieved. Each lattice vibration possesses one of $\mathcal{N}_c$ different wavevectors

$$\boldsymbol{k}_i \quad , \quad i = 1, \ldots, \mathcal{N}_c \; .$$

In addition, each lattice vibration belongs to one of $3n$ branches, where the branch determines the relationship between the frequency and the wavevector:

$$\omega_j(\boldsymbol{k}_i) \quad , \quad j = 1, \ldots, 3n \; .$$

In agreement with the number of spatial degrees of freedom of all atoms in the crystal, there are therefore altogether

$$3n\mathcal{N}_1\mathcal{N}_2\mathcal{N}_3 = 3n\mathcal{N}_c \tag{5.33}$$

different lattice vibrations. If the selected wavevectors are represented by points in reciprocal space, it follows that the point density is, as given in the previous chapter (Eq. (4.35)), equal to

$$\frac{\mathcal{N}_c}{\widehat{V}_c} = \frac{\mathcal{N}_c V_c}{(2\pi)^3} = \frac{\mathcal{V}}{(2\pi)^3} \quad, \tag{5.34}$$

where V_c, $\widehat{V}_c$, and $\mathcal{V}$ denote the volume of the unit cell, the reciprocal lattice cell, and the crystal, respectively.

It is clearly to be expected that there will again be two different types among the $3n$ different branches. Three of the branches correspond, for $\boldsymbol{k} \to 0$, to the three acoustic waves which appear in a crystal. These three acoustic waves differ in their polarisation direction, whereby it is found – always for cubic crystals or otherwise often along certain propagation directions – that one and two of the waves are longitudinally and transversely polarised, respectively. The remaining $3n$-3 branches have the character of optical vibrations and exhibit different non-zero frequencies for a vanishing k. Figure 5.3 shows, as an example, the different branches which were found for a crystal of silicon with the wavevectors oriented in the 100-direction. A longitudinal and a (two-fold degenerate) transverse acoustic branch (denoted LA and TA) as well as a longitudinal and a (two-fold degenerate) transverse optical branch

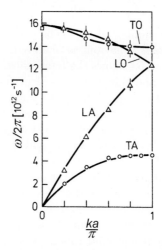

Fig. 5.3. The lattice vibration spectrum of silicon with two acoustic (LA, TA) and two optical (LO, TO) branches, as determined by inelastic neutron scattering ($\boldsymbol{k}$ in the [100] direction; from Dolling and Cowley in [41]).

(denoted LO and TO) are observed. The data were recorded using the method of inelastic neutron scattering, which will be described in the final section of this chapter.

Phonons – Debye's Theory of Specific Heats. We now consider the question: How much energy is taken up by the lattice vibrations at thermal equilibrium? In the context of classical physics, it is possible to give a direct answer. Since each lattice vibration in its own right represents a harmonic oscillator, even if it is carried by an enormous number of atoms, it follows that the take-up of energy at thermal equilibrium is given by

$$\langle u_{\text{pot}} \rangle = \langle u_{\text{kin}} \rangle = \frac{k_{\text{B}} T}{2} \quad , \tag{5.35}$$

with the sum of potential and kinetic energy thus being

$$\langle u_{\text{pot}} \rangle + \langle u_{\text{kin}} \rangle = k_{\text{B}} T \quad . \tag{5.36}$$

This would imply that the take-up of energy is always the same regardless of the wavevector or the frequency. The total vibrational energy $\mathcal{U}_{\text{vib}}$ of the crystal would then be given as

$$\mathcal{U}_{\text{vib}} = 3n\mathcal{N}_{\text{c}} k_{\text{B}} T \quad , \tag{5.37}$$

and the specific heat $c_{v,\text{vib}}$ would have the temperature-independent value

$$c_{v,\text{vib}} = \frac{1}{V} \frac{\partial \mathcal{U}_{\text{vib}}}{\partial T} = 3n\rho_{\text{c}} k_{\text{B}} \tag{5.38}$$

(ρ_{c} denotes, as ever, the number of unit cells per unit volume). Experimentally, a completely different behaviour is observed, in particular in the low-temperature region. Figure 5.4 shows the temperature dependence of the specific heats of germanium and silicon. It is evident that the specific heat is in no way constant, but rather, starting at low temperature from a vanishing value, it rises gradually until a final value is reached. Interestingly, this final value is here, and also for many other atomic crystals, in good agreement with the prediction of Eq. (5.38), an observation which is expressed generally by the **Dulong and Petit law**.

The reason for the complete failure of the classical prediction are quantum effects. Quantum mechanics stipulates that the energy of a harmonic oscillator does not change continuously but in quantum steps of magnitude $\hbar\omega$. This means that a certain minimum amount of thermal energy must already be available for the first excitation, and this is only achieved at the appropriate temperature. Consider a particular lattice vibration ij (wavevector $\boldsymbol{k}_i$, branch j): Using the quantum-mechanical result about the energy eigenvalues ϵ_{ij} of this harmonic oscillator, we have

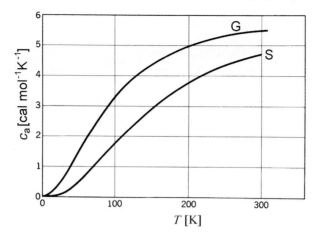

Fig. 5.4. The specific heats of germanium (G) and silicon (S) in the low-temperature region.

$$\epsilon_{ij} = \hbar\omega_{ij}\left(n_{ij} + \frac{1}{2}\right) \tag{5.39}$$

with the short notation

$$\omega_{ij} = \omega_j(\boldsymbol{k}_i) \ . \tag{5.40}$$

n_{ij} denotes the number of vibrational quanta occupying the state. The energy quanta of the lattice vibrations are called **phonons**. The total vibrational energy $\mathcal{U}_{\mathrm{vib}}$ of a crystal at a particular temperature corresponds to the sum of the mean energies of all the eigenvibrations and is therefore determined by the associated mean phonon occupation numbers:

$$\mathcal{U}_{\mathrm{vib}} = \sum_{ij}\langle\epsilon_{ij}\rangle = \sum_{ij}\hbar\omega_{ij}\left(\langle n_{ij}\rangle + \frac{1}{2}\right) \ . \tag{5.41}$$

The mean numbers $\langle n_{ij}\rangle$ can be calculated using Boltzmann statistics. The expression

$$w(n_{ij}) = \mathcal{Z}^{-1}\exp-\frac{\hbar\omega_{ij}(n_{ij} + \frac{1}{2})}{k_{\mathrm{B}}T} \tag{5.42}$$

gives the probability that a particular lattice vibration is occupied by n_{ij} phonons. The partition function $\mathcal{Z}$ of the lattice vibration ij follows from the normalisation condition

$$\sum_{n_{ij}=0}^{\infty} w(n_{ij}) = 1 \tag{5.43}$$

as

$$\mathcal{Z} = \sum_{n_{ij}=0}^{\infty}\exp-\frac{\hbar\omega_{ij}(n_{ij} + \frac{1}{2})}{k_{\mathrm{B}}T} = \frac{\exp-(\hbar\omega_{ij}/2k_{\mathrm{B}}T)}{1 - \exp-(\hbar\omega_{ij}/k_{\mathrm{B}}T)} \ . \tag{5.44}$$

The mean phonon occupation number can be calculated:

$$\langle n_{ij} \rangle = \frac{1 - \exp-(\hbar\omega_{ij}/k_{\mathrm{B}}T)}{\exp-(\hbar\omega_{ij}/2k_{\mathrm{B}}T)} \sum_{n_{ij}=0}^{\infty} n_{ij} \exp-\left(\frac{\hbar\omega_{ij}n_{ij}}{k_{\mathrm{B}}T} + \frac{\hbar\omega_{ij}}{2k_{\mathrm{B}}T} \right)$$

$$= \left(1 - \exp-\frac{\hbar\omega_{ij}}{k_{\mathrm{B}}T} \right) \frac{\mathrm{d}}{\mathrm{d}\left(-\frac{\hbar\omega_{ij}}{k_{\mathrm{B}}T}\right)} \sum_{n_{ij}=0}^{\infty} \exp-\frac{\hbar\omega_{ij}n_{ij}}{k_{\mathrm{B}}T}$$

$$= \left(1 - \exp-\frac{\hbar\omega_{ij}}{k_{\mathrm{B}}T} \right) \frac{\mathrm{d}}{\mathrm{d}\left(-\frac{\hbar\omega_{ij}}{k_{\mathrm{B}}T}\right)} \frac{1}{1 - \exp-\frac{\hbar\omega_{ij}}{k_{\mathrm{B}}T}}$$

$$= \frac{\exp-(\hbar\omega_{ij}/k_{\mathrm{B}}T)}{1 - \exp-(\hbar\omega_{ij}/k_{\mathrm{B}}T)} = \frac{1}{\exp(\hbar\omega_{ij}/k_{\mathrm{B}}T) - 1} \quad . \tag{5.45}$$

The last expression is also the fundamental law of Bose–Einstein statistics – the latter could have been used directly to determine $\langle n_{ij} \rangle$ on the basis of the fact that phonons are Bose particles. The average energy is thus given as

$$\langle \epsilon_{ij} \rangle = \left(\langle n_{ij} \rangle + \frac{1}{2} \right) \hbar\omega_{ij} = \frac{\hbar\omega_{ij}}{\exp(\hbar\omega_{ij}/k_{\mathrm{B}}T) - 1} + \frac{\hbar\omega_{ij}}{2} \quad . \tag{5.46}$$

Summing over all lattice vibrations then leads to the following expression for the total vibrational energy

$$U_{\mathrm{vib}} - U_{\mathrm{vib}}(T = 0) = \sum_{ij} \langle \epsilon_{ij} \rangle = \sum_{ij} \frac{\hbar\omega_{ij}}{\exp(\hbar\omega_{ij}/k_{\mathrm{B}}T) - 1} \quad . \tag{5.47}$$

Debye showed how this sum can be evaluated when the heat capacity at low temperatures is of interest. Equation (5.47) is, first, written in a non-discrete, continuous form, as

$$U_{\mathrm{vib}} - U_{\mathrm{vib}}(0) = \int_{\omega} \mathcal{D}(\omega) \frac{\hbar\omega}{\exp(\hbar\omega/k_{\mathrm{B}}T) - 1} \mathrm{d}\omega \quad . \tag{5.48}$$

Here,

$$\mathcal{D}(\omega)\mathrm{d}\omega$$

denotes the number of lattice vibrations with frequencies in the interval between ω and $\omega + \mathrm{d}\omega$. $\mathcal{D}$, correspondingly, has the meaning of a **spectral density**. It follows from Eq. (5.45) that high-frequency vibrations with

$$\hbar\omega_{ij} \gg k_{\mathrm{B}}T$$

cannot be excited since

$$\langle n_{ij} \rangle \approx \exp-\frac{\hbar\omega_{ij}}{k_{\mathrm{B}}T} \approx 0 \quad . \tag{5.49}$$

As a consequence, this means that only the acoustic vibrations, and in fact only those in the vicinity of $k = 0$, can take up energy at low temperatures. Here, the dispersion law for sound waves can be applied, namely

$$\omega_1(\boldsymbol{k}_i) = c_{\mathrm{sl}}|\boldsymbol{k}_i| \tag{5.50}$$

for longitudinally-polarised waves, and

$$\omega_2(\boldsymbol{k}_i) = \omega_3(\boldsymbol{k}_i) = c_{\mathrm{st}}|\boldsymbol{k}_i| \tag{5.51}$$

for transversally-polarised waves; c_{sl} and c_{st} denote the corresponding phase velocities. Debye suggested that only these sound waves should be considered in calculating the specific heats at low temperatures. To derive the associated spectral density, we first ask about the corresponding density in $\boldsymbol{k}$ space, which can be immediately specified. Using Equation (5.34), it follows that the number of lattice vibrations, for which the wavevector is between $k \to k + \mathrm{d}k$, is given by

$$D(k)\mathrm{d}k = 4\pi k^2 \frac{V}{(2\pi)^3}\mathrm{d}k \quad . \tag{5.52}$$

A sum of all three acoustic branches yields then for $D(\omega)$ the expression

$$D(\omega)\mathrm{d}\omega = \sum_{j\,=\,1,2,3} D(k)\frac{\mathrm{d}k}{\mathrm{d}\omega_j}\mathrm{d}\omega \tag{5.53}$$

and, thus, the result

$$D(\omega) = \frac{V}{2\pi^2}\left(\frac{\omega^2}{c_{\mathrm{sl}}^2}\frac{1}{c_{\mathrm{sl}}} + 2\frac{\omega^2}{c_{\mathrm{st}}^2}\frac{1}{c_{\mathrm{st}}}\right) \quad . \tag{5.54}$$

The Debye approximation for the spectral density correctly describes $D(\omega)$ at low frequencies, but then leads to deviations. This is shown in Fig. 5.5, which also illustrates a further requirement: Since, in a calculation of the specific heat, even if an approximation procedure is applied, the total number of degrees of freedom must be maintained at the correct value, it is necessary that the distribution function is truncated at a certain value of ω. The requirement is

$$\int_0^{\omega_{\mathrm{D}}} D(\omega)\mathrm{d}\omega = 3\mathcal{N}_{\mathrm{c}} \quad , \tag{5.55}$$

and provides an equation which determines the cut-off at the **Debye frequency** ω_{D}. Substituting in Eq. (5.54) yields the expression

$$\frac{V}{2\pi^2}\left(\frac{1}{3c_{\mathrm{sl}}^3} + \frac{2}{3c_{\mathrm{st}}^3}\right)\omega_{\mathrm{D}}^3 = 3\mathcal{N}_{\mathrm{c}} \quad . \tag{5.56}$$

The take-up of energy can now be calculated by using the Debye approximation in the general Eq. (5.48). This leads to

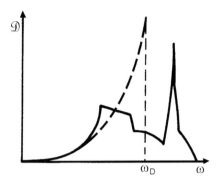

Fig. 5.5. A schematic representation of a typical spectral density of the acoustic vibrations together with the prediction of Debye theory. Peaks are observed when the dispersion relation of a branch possesses a horizontal tangent.

$$\frac{\mathcal{U}_{\mathrm{vib}} - \mathcal{U}_{\mathrm{vib}}(0)}{\mathcal{V}} = \frac{1}{2\pi^2}\left(\frac{1}{c_{\mathrm{sl}}^3} + \frac{2}{c_{\mathrm{st}}^3}\right)\int_0^{\omega_{\mathrm{D}}}\frac{\hbar\omega^3}{\exp(\hbar\omega/k_{\mathrm{B}}T) - 1}\mathrm{d}\omega$$

$$= \frac{1}{2\pi^2}\left(\frac{1}{c_{\mathrm{sl}}^3} + \frac{2}{c_{\mathrm{st}}^3}\right)\frac{(k_{\mathrm{B}}T)^4}{\hbar^3}\int_0^{x_{\mathrm{D}}}\frac{x^3}{\exp x - 1}\mathrm{d}x \quad , \qquad (5.57)$$

whereby the following substitutions were made in the second step:

$$x = \frac{\hbar\omega}{k_{\mathrm{B}}T} \tag{5.58}$$

$$x_{\mathrm{D}} = \frac{\hbar\omega_{\mathrm{D}}}{k_{\mathrm{B}}T} \quad . \tag{5.59}$$

The integral represents a well-defined function Φ, and Eq. (5.57) can thus be re-expressed as

$$\frac{\mathcal{U}_{\mathrm{vib}} - \mathcal{U}_{\mathrm{vib}}(0)}{\mathcal{V}} = 9\frac{\mathcal{N}_{\mathrm{c}}}{\mathcal{V}}\frac{(k_{\mathrm{B}}T)^4}{(\hbar\omega_{\mathrm{D}})^3}\Phi\left(\frac{\hbar\omega_{\mathrm{D}}}{k_{\mathrm{B}}T}\right)$$

$$= 9\rho_{\mathrm{c}}k_{\mathrm{B}}T_{\mathrm{D}}\left(\frac{T}{T_{\mathrm{D}}}\right)^4\Phi\left(\frac{T_{\mathrm{D}}}{T}\right) \quad . \tag{5.60}$$

The **Debye temperature** T_{D} is introduced in Eq. (5.60) according to

$$\hbar\omega_{\mathrm{D}} = k_{\mathrm{B}}T_{\mathrm{D}} \quad . \tag{5.61}$$

The limiting value of the function Φ

$$\lim_{T\to 0}\Phi = \frac{\pi^4}{15} \tag{5.62}$$

can be used for low temperatures. The following final result is thus obtained:

$$c_{v,\text{vib}} = \frac{1}{T_D}\frac{\partial(\mathcal{U}_{\text{vib}}/\mathcal{V})}{\partial(T/T_D)} = \frac{36\pi^4}{15}\rho_c k_B \left(\frac{T}{T_D}\right)^3 \; . \tag{5.63}$$

It is referred to as the **Debye T^3 law**. It describes very well the temperature dependence of the specific heat at low temperatures, with T_D being the only material-specific parameter. Values for the Debye temperature spread over a wide range. For example, T_D is 36, 105, 321, 370, 467, and 1460 K for caesium, lead, sodium chloride, germanium, iron, and beryllium, respectively.

5.1.2 Spin Waves

In ferromagnetics, there is, in addition to the displacement of the atoms from their equilibrium positions, a further degree of freedom, namely the reorientation of the spins away from the direction of the uniform alignment in the ground state adopted at very low temperatures near 0 K. In this section, the nature of this additional group of elementary excitations will be discussed in detail. They are thermally activated in a ferromagnet, and thus lead to a decrease in the magnetisation with increasing temperature. This decrease will be calculated.

For a crystal with its constant coupling forces, it is to be expected that the eigenmodes of the spin system again possess a wave-like character. Such a **spin wave** is shown in Fig. 5.6, for the simple case of a linear chain. All the spins precess about a common fixed direction, with there being a constant angle difference between neighbours. Each spin wave has a definite (precession) frequency ω and wavelength, and, in the crystal, a fixed propagation direction, i.e., a certain wavevector $\boldsymbol{k}$.

In order to investigate the relationship between ω and $\boldsymbol{k}$, i.e., the dispersion relation of the spin waves, it is necessary to construct the equations of motion and then search for wave-like solutions. Let us again, in the first instance, consider a one-dimensional system, i.e., the rotational displacements in a linear chain of spins, as shown in the figure. The expression for the interaction energy between two neighbouring spins in a ferromagnet was given in Sect. 3.2 (Eq. (3.30)): The interaction energy for the spins at positions l and $l+1$ is

Fig. 5.6. Spin waves in a linear atomic chain in the ferromagnetic state.

$$u(l, l+1) = -\beta_{\mathrm{ex}}\boldsymbol{S}(l) \cdot \boldsymbol{S}(l+1) \quad. \tag{5.64}$$

The resulting torque which acts between the two spins appears in the equation of motion. It is given by

$$\boldsymbol{T}(l, l+1) = \beta_{\mathrm{ex}}\boldsymbol{S}(l) \times \boldsymbol{S}(l+1) \quad. \tag{5.65}$$

This follows from remembering that the potential energy and the torque in the case of a magnetic moment $\boldsymbol{m}$ in an external magnetic field $\boldsymbol{B}$ are given by the similar equations:

$$u = -\boldsymbol{m} \cdot \boldsymbol{B} \tag{5.66}$$

and

$$\boldsymbol{T} = \boldsymbol{m} \times \boldsymbol{B} \quad. \tag{5.67}$$

Exactly as in this well-known case, there arises for a spin system with an interaction energy described by a scalar product a torque with the form of a vector product of the same variables. Torques due to both neighbours act on each spin $\boldsymbol{S}(l)$. The equation of motion is thus given as

$$\frac{\mathrm{d}}{\mathrm{d}t}\boldsymbol{S}(l) = \beta_{\mathrm{ex}}[\boldsymbol{S}(l) \times \boldsymbol{S}(l-1) + \boldsymbol{S}(l) \times \boldsymbol{S}(l+1)] \quad. \tag{5.68}$$

Elementary excitations in the form of spin waves with small amplitudes will be discussed here. If the precession axis is in the z direction, this means that

$$S_x(l), S_y(l) \ll S_z(l) \approx |\boldsymbol{S}| \quad. \tag{5.69}$$

The following expressions are then found for the lateral components of the spins, S_x and S_y:

$$\begin{aligned}
\frac{\mathrm{d}}{\mathrm{d}t}S_x(l) &= \beta_{\mathrm{ex}}[S_y(l)S_z(l-1) - S_z(l)S_y(l-1) \\
&\quad + S_y(l)S_z(l+1) - S_z(l)S_y(l+1)] \\
&\approx \beta_{\mathrm{ex}}|\boldsymbol{S}|[2S_y(l) - S_y(l-1) - S_y(l+1)] \\
\frac{\mathrm{d}}{\mathrm{d}t}S_y(l) &= -\beta_{\mathrm{ex}}|\boldsymbol{S}|[2S_x(l) - S_x(l-1) - S_x(l+1)] \quad.
\end{aligned} \tag{5.70}$$

The right-hand side of the differential equation for the longitudinal component S_z contains products of the small variables S_x and S_y. In a linear approximation, we may set

$$\frac{\mathrm{d}}{\mathrm{d}t}S_z(l) \approx 0 \quad, \tag{5.71}$$

and consequently replace S_z by the modulus of $\boldsymbol{S}$. As a starting point in the search for the properties of wave-like solutions, consider the ansatz

$$\begin{aligned}
S_x(l) &= S_{x0}\exp(-\mathrm{i}\omega t + \mathrm{i}kla) \\
S_y(l) &= S_{y0}\exp(-\mathrm{i}\omega t + \mathrm{i}kla) \\
S_z(l) &\approx \mathrm{const} \approx |\boldsymbol{S}| \quad.
\end{aligned} \tag{5.72}$$

A check confirms, as expected, that wave-like solutions result, with the amplitudes S_{x0} and S_{y0} being given by the following system of linear equations:

$$i\omega S_{x0} = 2\beta_{\mathrm{ex}}|S|(1 - \cos ka)S_{y0}$$
$$i\omega S_{y0} = -2\beta_{\mathrm{ex}}|S|(1 - \cos ka)S_{x0} \ . \tag{5.73}$$

Non-trivial solutions are found upon setting the secular determinant to zero, i.e.,

$$\begin{vmatrix} i\omega & -2\beta_{\mathrm{ex}}|S|(1 - \cos ka) \\ 2\beta_{\mathrm{ex}}|S|(1 - \cos ka) & i\omega \end{vmatrix} = 0 \ . \tag{5.74}$$

The dispersion relation is, thus, given as

$$\omega = \pm 2\beta_{\mathrm{ex}}|S|(1 - \cos ka) \ . \tag{5.75}$$

The different signs are associated with spin waves with opposite propagation directions.

It is to be noted that the dispersion relation has a periodic dependence on k, since the following is true:

$$w(k) = w\left(k + p\frac{2\pi}{a}\right) \qquad (p \text{ is an integer}) \ . \tag{5.76}$$

The reason for this behaviour is known from the above discussion of lattice vibrations: A change in the wavenumber by a whole-number multiple of the basis vector $G_1 = 2\pi/a$ of the reciprocal lattice does not change the displacement pattern of the spin wave. As in the case of the lattice vibrations, it, therefore, again makes sense to restrict the wavevectors k to the first Brillouin zone:

$$-\frac{G_1}{2} < k \leq \frac{G_1}{2} \ . \tag{5.77}$$

A qualitative difference is apparent upon comparing the dispersion relations for spin waves and lattice vibrations. While the frequency of the spin waves tends to zero for $k \rightarrow 0$ as in the case of acoustic lattice vibrations, it is found that the increase of the frequency as a function of k is no longer linear but rather quadratic, as can be shown by a power series expansion:

$$\omega \approx \beta_{\mathrm{ex}}|S|(ka)^2 \ . \tag{5.78}$$

This means that neither the phase velocity ω/k nor the group velocity $d\omega/dk$ are constant, and thus a wavepacket, i.e., a local excitation, quickly dissipates. The quadratic dependence is illustrated by the example in Fig. 5.7. The dispersion relation was obtained for a cobalt-iron alloy by inelastic neutron scattering along three equivalent directions of the (cubic) crystal.

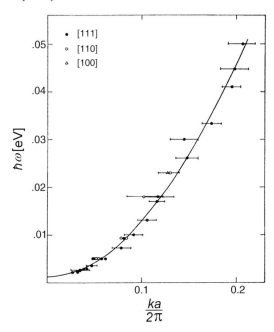

Fig. 5.7. The dispersion relation for spin waves in a cobalt-iron alloy, as measured by inelastic neutron scattering (from Sinclair and Brockhouse [42]).

The fact that the sketch in Fig. 5.6 does indeed correctly describe the spin wave becomes clear upon a consideration of the ratio of the amplitudes

$$\frac{S_{x0}}{S_{y0}} = \frac{2\beta_{\mathrm{ex}}|\boldsymbol{S}|(1 - \cos ka)}{-\mathrm{i}\omega} = \pm\mathrm{i} \quad . \tag{5.79}$$

Since S_{x0} and S_{y0} are of the same magnitude, the following can be written

$$S_{x0} = S_{y0} = S_{\perp} \tag{5.80}$$

and thus

$$\begin{aligned} S_x(l) &= S_{\perp} \exp(-\mathrm{i}\omega t + \mathrm{i}kla) \\ S_y(l) &= \pm\mathrm{i}S_{\perp} \exp(-\mathrm{i}\omega t + \mathrm{i}kla) \end{aligned} \tag{5.81}$$

or, in a real-valued notation,

$$\begin{aligned} S_x(l) &= S_{\perp} \cos(-\omega t + kla) \\ S_y(l) &= \pm S_{\perp} \sin(-\omega t + kla) \quad . \end{aligned} \tag{5.82}$$

The spin wave in Fig. 5.6 corresponds to one of these patterns.

The discussion so far has considered an infinitely long chain. If the spin wave is to be maintained as a solution while limiting the number of degrees of freedom, it is necessary, as in the case of lattice vibrations, to introduce cyclic boundary conditions. These are, for a chain of $\mathcal{N}_c$ spins,

$$S(l + \mathcal{N}_c) = S(l) \quad . \tag{5.83}$$

As in Eq. (5.23), there again follows a selection of discrete values of k

$$k = i\frac{2\pi}{\mathcal{N}_c a} \qquad (i \text{ is an integer}) \quad . \tag{5.84}$$

The restriction to the first Brillouin zone then means that

$$-\frac{\mathcal{N}_c}{2} < i \leq \frac{\mathcal{N}_c}{2} \quad . \tag{5.85}$$

Magnons – The Bloch $T^{3/2}$ Law. For spin waves, the question must also be asked as to whether is it necessary to consider quantum effects, i.e., can the amplitude $S_\perp$ be continuously increased from a starting point of zero? The answer is no, with the reason being the quantisation of the angular momentum. A fundamental requirement of quantum mechanics is that the angular momentum of a system in a particular direction can always only be changed by a whole number multiple of $\hbar$. As will be shown here, this leads to a quantisation of the take-up of energy. If energy can only be gained in quanta, it follows that there is a lower limit for the to be provided thermal energy. Therefore, as in the case of lattice vibrations, visible quantum effects are to be expected at low temperatures.

The interaction energy between two neighbouring spins in a spin wave depends on the difference φ in the precession angle according to

$$u(l, l + 1) = -\beta_{ex}S(l) \cdot S(l + 1) = -\beta_{ex}|S|^2 \cos\varphi \quad . \tag{5.86}$$

Since the value is the same for all pairs in the chain, it follows that the total interaction energy is

$$U = -\beta_{ex}\sum_l S(l) \cdot S(l + 1) = -\beta_{ex}\mathcal{N}_c|S|^2 \cos\varphi \approx -\beta_{ex}\mathcal{N}_c|S|^2 \left(1 - \frac{\varphi^2}{2}\right) \quad . \tag{5.87}$$

A geometric interrelation is illustrated in Fig. 5.8. The spins of two neighbours are shown in one and the same representation, such that it is clear that the following applies

$$|S|\varphi \approx S_\perp ka \quad . \tag{5.88}$$

The energy gain upon excitation of a spin wave with wavenumber k and amplitude $S_\perp$, beginning at an energy minimum corresponding to $\varphi = 0$, is then given by

$$U - U(\varphi = 0) = \frac{\beta_{ex}\mathcal{N}_c S_\perp^2 (ka)^2}{2} \quad . \tag{5.89}$$

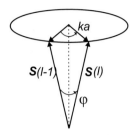

Fig. 5.8. An illustration of Eq. (5.88).

The excitations for the lowest energy appear for small wavenumbers k. The dispersion relation in the quadratic approximation of Eq. (5.78) applies here, and, therefore,

$$U - U(0) = \frac{\mathcal{N}_{\mathrm{c}} S_\perp^2}{2|\boldsymbol{S}|} \omega \ . \tag{5.90}$$

There is necessarily a change in the spin components in the z direction upon the excitation of a spin wave. The change is very small and could be neglected in the derivation of the dispersion relation. It, however, comes into play for the analysis of quantum effects. Making an approximation, it can be written that

$$S_z = \left(|\boldsymbol{S}|^2 - S_\perp^2\right)^{1/2} \approx |\boldsymbol{S}| - \frac{S_\perp^2}{2|\boldsymbol{S}|} \ . \tag{5.91}$$

The total angular momentum in the z direction, i.e., that due to all spins in the chain, is given by

$$S_{z,\mathrm{tot}} = \mathcal{N}_{\mathrm{c}} S_z = \mathcal{N}_{\mathrm{c}} |\boldsymbol{S}| - \mathcal{N}_{\mathrm{c}} \frac{S_\perp^2}{2|\boldsymbol{S}|} \ . \tag{5.92}$$

For the total spin, there is, however, the requirement that it can only change in steps of $\hbar$. This means that the amplitude of the spin wave $S_\perp$ can only adopt values, for which

$$\Delta \left(\mathcal{N}_{\mathrm{c}} \frac{S_\perp^2}{2|\boldsymbol{S}|} \right) = n\hbar \qquad , \qquad n = 1, 2, \ldots \ . \tag{5.93}$$

This has a surprisingly simple effect upon the energy gain which is associated with a change of $S_{z,\mathrm{tot}}$ by $\hbar$: When considered together, Eqs. (5.93) and (5.90) mean that the energy of a spin wave is increased in steps, which are always equal, i.e., the eigenenergies U_i of a spin wave with frequency ω_i are given by

$$U_i - U(0) = n\hbar\omega_i \ . \tag{5.94}$$

The level scheme is, thus, that of an harmonic oscillator. The basis for this, though, is of a completely different nature; for spin waves, it is caused, in contrast to the case of lattice vibrations, by the quantisation of the total angular momentum. The quanta with angular momentum $\hbar$ and energy $\hbar\omega$, which are possessed by a spin wave and determine its amplitude and energy, are termed **magnons**.

It can now be calculated how the magnetisation in a ferromagnet at low temperature begins to be reduced from the maximum of the ground state value because of the thermal excitation of spin waves. The average number of magnons, which a particular spin wave possesses at a given temperature and which determines the amplitude of the spin wave, can be evaluated using Boltzmann statistics. Since only the energy eigenvalues of the states enter into the expression, it follows that the harmonic oscillator equations are valid. Equation (5.45) can, thus, again be used, and the average number of magnons in a spin wave i with frequency ω_i is

$$\langle n_i \rangle = \frac{1}{\exp(\hbar\omega_i / k_B T) - 1} \; . \tag{5.95}$$

For a change in the total spin $S_{z,\text{tot}}$ by $\hbar$, there is an associated change in the magnetic moment by $2\mu_B$ (see Eq. (2.155)). The overall loss of magnetisation at a temperature T is thus given from the contributions of all spin waves by

$$M(0) - M(T) = \Delta M(T) = 2\mu_B \sum_{k_i} \langle n_i \rangle(T) \; . \tag{5.96}$$

So far, only linear chains have been considered. The transition to a three-dimensional crystal can be carried out immediately in that it simply means that spin waves are to be considered which run in all directions through the crystal. For the possible wavevectors $\boldsymbol{k}_i$ exactly the same as for the lattice vibrations applies:

- All possible spin waves are included if only the wavevectors from the first Brillouin zone (now that in three-dimensional reciprocal space) are allowed.
- For the case of spin waves, the finite extension of a crystal, i.e., the finite number of degrees of freedom, now given by the number of rotating spins, can again be taken into account by imposing cyclic boundary conditions.

The sum in Eq. (5.96) is now over all spin waves which can be excited in a 3D-ferromagnet of finite extension. In order to calculate it, the sum is transformed into an integral:

$$\sum_{k_i} \langle n_i \rangle = \int \frac{\mathcal{D}(\omega)}{\exp(\hbar\omega / k_B T) - 1} d\omega \; . \tag{5.97}$$

The integral again contains a density of levels in frequency space, which can be calculated as follows. In the spirit of the Debye approximation used in the calculation of the contribution of the lattice vibrations to the specific heat, the dispersion relation for small wavevectors

$$\omega = \beta k^2 \tag{5.98}$$

is employed. The density of levels in k space is again given by Eq. (5.34), here leading to

$$\mathcal{D}(\omega)\mathrm{d}\omega = \frac{\mathcal{V}}{2\pi^2}k^2\mathrm{d}k \quad . \tag{5.99}$$

From the dispersion relation, it follows that

$$\mathrm{d}\omega = 2\beta k \mathrm{d}k \tag{5.100}$$

and, thus,

$$k^2\mathrm{d}k = \frac{\omega^{1/2}}{2\beta^{3/2}}\mathrm{d}\omega \quad . \tag{5.101}$$

As in the Debye treatment, it is again necessary to introduce a cut-off frequency in the approximate expression for the density of levels in frequency space, so as to ensure that there is the correct number of independent lattice vibrations. For a crystal comprised of $\mathcal{N}_\mathrm{c}$ cells, the cut-off frequency, ω_{max}, is determined by

$$\int_0^{\omega_{max}} \mathcal{D}(\omega)\mathrm{d}\omega = \mathcal{N}_\mathrm{c} \quad . \tag{5.102}$$

Bringing all the equations together leads to the following expression for the total number of all magnons:

$$\sum_{k_i}\langle n_i \rangle = \frac{\mathcal{V}}{4\pi^2\beta^{3/2}} \int_0^{\omega_{max}} \frac{\omega^{\frac{1}{2}}}{\exp(\hbar\omega/k_\mathrm{B}T) - 1}\mathrm{d}\omega$$

$$\propto T^{\frac{3}{2}} \int_0^{\hbar\omega_{max}/k_\mathrm{B}T} \frac{x^{\frac{1}{2}}}{\exp x - 1}\mathrm{d}x \quad . \tag{5.103}$$

The second line follows from the subsitution

$$x = \frac{\hbar\omega}{k_\mathrm{B}T} \quad . \tag{5.104}$$

At low temperature, the integral can be replaced by its limiting value for $T \to 0$. A simple result is then obtained for the decrease of the magnetisation: At low temperature, a power law applies

$$\Delta M \propto T^{3/2} \tag{5.105}$$

and this is known as the **Bloch $T^{3/2}$ law**. A good agreement of experimental results with Bloch's law is usually obtained. The temperature dependence of the magnetisation of nickel, which is shown in Fig. 3.8, corresponds at low temperatures to the theoretical curve.

5.1.3 Excitons

Local excitations are never stationary because of the couplings in a crystal. They propagate themselves and are able to transfer the energy gained to a different location. This is not only true for the displacement of an individual atom from its equilibrium position or for the reorientation of an individual spin in a ferromagnet, but also when a crystal experiences a local electronic excitation. **Excitons** are such an example. They represent electronic excitations, which can propagate in the crystal after their generation – this is usually achieved by the absorption of a photon. There are two different types, which are referred to as **Mott–Wannier excitons** and **Frenkel excitons**. The former are encountered in semiconductors, while the latter are found in insulators such as ionic crystals and atomic or molecular van der Waals crystals. Their properties, i.e., the form of the excitation and the associated motional mechanism, will be discussed in the following.

Mott–Wannier Excitons. Figure 5.9 shows the absorption spectrum of GaAs in the frequency region of the energy gap ϵ_g between the valence and the conduction band. A sharp band is observed at a frequency ω_{exc}, which is just below the value of the band gap. The absorption of photons with frequencies $\omega \geq \epsilon_g/\hbar$ leads to the generation of freely mobile electrons and holes, an effect which is evidenced by the appearance of a 'photo current' upon

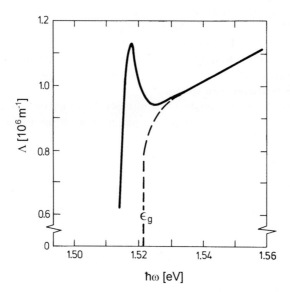

Fig. 5.9. An absorption curve with exciton bands, as measured at 21 K for GaAs in the frequency region of the band gap. In the absence of excitons, the dashed curve would be measured (from Sturge [43]).

applying a voltage. By comparison, no photo current is observed upon the absorption of a photon with frequency ω_{exc}. Instead, an electron-hole pair forms, in which the two particles form a bound state. Such a 'Mott–Wannier exciton' is electrically neutral and does not react to the applied voltage.

The state embodied by an exciton, namely a bonded pair of a positively and a negatively charged particle resembles the ground state of the hydrogen atom, with modifications due to mass differences and differences in the environment. The ground state energy is

$$\epsilon_{\text{exc}} = -\frac{e^2}{4\pi\varepsilon_0\varepsilon r_1}\frac{1}{2n^2} \qquad \text{for} \qquad n = 1 \ , \tag{5.106}$$

where

$$r_1 = \frac{4\pi\varepsilon_0\varepsilon\hbar^2}{e^2 m_{\text{r,eff}}} \tag{5.107}$$

denotes the radius of the ground state orbit and $m_{\text{r,eff}}$ is the reduced mass of the electron-hole system

$$m_{\text{r,eff}}^{-1} = m_{\text{e,eff}}^{-1} + m_{\text{h,eff}}^{-1} \ . \tag{5.108}$$

Effects due to the environment are considered by the incorporation of the dielectric constant $\varepsilon(\simeq 10)$. This takes into account screening effects due to the other electrons and holes in the respective bands, which lead to a reduction in the Coulomb interaction in the exciton. The consequence is a significant increase in the electron-hole separation in an exciton as compared to the Bohr radius of an H atom. For a Ga-As exciton this separation amounts to 4.7 nm. The binding energy is reduced to 3 meV. This energy must be provided if an electron and a hole are to be generated as free charge carriers from an exciton. It is this energy which determines the distance of the exciton band from the onset of the inter-band transitions.

In order to describe the states of a Mott–Wannier exciton, the H atom wavefunctions with quantum numbers n, l, m can be used in a two-particle wavefunction $\psi_2(\boldsymbol{r}_{\text{e}}, \boldsymbol{r}_{\text{h}})$, as

$$\psi_2(\boldsymbol{r}_{\text{e}}, \boldsymbol{r}_{\text{h}}) = \psi_{nlm}(\boldsymbol{r}_{\text{e}} - \boldsymbol{r}_{\text{h}}) \ . \tag{5.109}$$

While, in the example of GaAs, excitons are only generated in the ground state $n = 1$, it is in other semi-conductors also possible for excitons to form in excited states.

Excitons can move freely in an unperturbed crystal in the same way that electrons and holes are freely mobile in their respective bands. The above wavefunction is thus an incomplete description of the state and must be extended as follows

$$\psi_2(\boldsymbol{r}_{\text{e}}, \boldsymbol{r}_{\text{h}}) \propto \exp i\boldsymbol{k}\boldsymbol{r}_{\text{s}} \cdot \psi_{nlm}(\boldsymbol{r}_{\text{e}} - \boldsymbol{r}_{\text{h}}) \ , \tag{5.110}$$

where

$$r_s = \frac{m_{e,eff} \boldsymbol{r}_e + m_{h,eff} \boldsymbol{r}_h}{m_{e,eff} + m_{h,eff}} \tag{5.111}$$

denotes the centre of gravity of the exciton. The exciton possesses as a free particle a total momentum $\hbar\boldsymbol{k}$ with a sharp value and a kinetic energy due to the translational motion

$$\epsilon_{kin} = \frac{\hbar^2 k^2}{2(m_{e,eff} m_{h,eff})^{1/2}} \; . \tag{5.112}$$

In order to represent a localized single exciton, it is necessary to construct a wavepacket from the eigenstates of Eq. (5.110). The dynamics of energy propagation after a local optical excitation is given by the time-dependent development of the wavepacket. Since the momentum carried over from the absorbed photon is negligibly small, this only corresponds to the rate with which the wavepacket disperses. In a perfect crystal, this would be controlled by the dispersion relation Eq. (5.112) and thus determined solely by the exciton mass. Crystals are, because of lattice vibrations and also the ever-present defect sites, never perfect. Therefore, exciton waves, like electron (Bloch) waves, are scattered, and this changes the energy transfer rates. We will discuss that in the treatment of Frenkel excitons.

Excitons only possess a limited lifetime, since they can be destroyed at any time by an electron-hole recombination which is accompanied by the emission of a photon. The dynamics of excitons becomes experimentally visible if the recombination rate is enhanced at defects or foreign atom sites. From the observed time dependence of fluorescence due to the foreign atoms after the generation of excitons by a short pulse of radiation, it is possible to determine how many excitons reach foreign atoms or defects during their lifetime, and hence, by inference, the rate of migration.

Frenkel Excitons. There is, even in insulators, the possibility to generate an electron-hole pair in a bound state by the absorption of a suitable photon, although the two particles are much more strongly coupled together here. Their separation is on the order of the atomic or molecular diameter. Under these conditions, it seems justifiable to speak of a single electronically excited atom or molecule as opposed to an electron-hole two-particle system. This is especially then the case when the absorption spectrum of the crystal does not differ significantly from that of the atoms or molecules in the gas phase. This is indeed found for 'Frenkel excitons' in, for example, KBr, Ar or anthracene crystals.

The fundamental property of excitons, namely that they do not persist in a localised state, but rather propagate themselves, is also a characteristic of an excited atom in a crystal. There is the possibility that the excitation energy is transferred to a different atom by a **resonance transfer**. This occurs by a coupled emission-absorption process or, the exchange of a photon between two equivalent atoms. The transition rate depends on the transition dipole moment and the separation (for larger distances $\propto 1/r^6$). If the matrix element of the dipole transition vanishes in the ground state and the photon-mediated resonance transfer does not exist, there is a further way by which energy transfer can occur. Electrons can change places between neighbours by a tunnelling process; if this happens in a correlated fashion, the excited state can be passed on. **Triplet excitons** use this second mechanism to transport an excited state with a non-vanishing electron spin.

Frenkel excitons are localised at the lattice positions – this is in contrast to Mott–Wannier excitons which are continuously being displaced in space. Using a quantum-mechanical description of the state, this will now be considered for a linear chain with an atom separation of a. The use of creation and annihilation operators $\mathbf{b}_l^+$ and $\mathbf{b}_l$ leads to a concise notation for this many-particle system. The occupation of the lattice position l with an exciton is described by

$$|l\rangle = \mathbf{b}_l^+|\psi_0\rangle \ , \tag{5.113}$$

where ψ_0 denotes the ground state without an exciton. The annihilation operator $\mathbf{b}_{l'}$ has the general effect:

$$\mathbf{b}_{l'}|l''\rangle = \delta_{l'l''}|\psi_0\rangle \tag{5.114}$$

such that the following applies

$$\mathbf{b}_l^+\mathbf{b}_{l'}|l''\rangle = \delta_{l'l''}|l\rangle \ . \tag{5.115}$$

Consider the creation of an exciton and the subsequent distribution of its energy over the linear chain, as described by its probability of being located at a particular position. Such a general state can be described by

$$|\psi\rangle = \sum_l c_l|l\rangle \ . \tag{5.116}$$

In the following, a linear chain comprised of $\mathcal{N}_c$ atoms or molecules is considered, again imposing cyclic boundary conditions. The normalisation condition for the wavefunction is then

$$\langle\psi|\psi\rangle = 1 = \sum_{l=0}^{\mathcal{N}_c-1} |c_l|^2 \ . \tag{5.117}$$

The Hamiltonian of the system has the form:

$$\mathbf{H} = \hbar \sum_{l'=0}^{\mathcal{N}_c-1}\sum_{l''=0}^{\mathcal{N}_c-1} \Omega_{l'l''}\mathbf{b}_{l'}^+\mathbf{b}_{l''} \ . \tag{5.118}$$

The matrix elements $\Omega_{l'l''}$ describe, for $l' \neq l''$, the interaction energy and determine, at the same time, the transition rates between the locations l' and

l'' in the chain – that this is so is apparent from a look at the time-dependent Schrödinger equation:

$$i\hbar \frac{\partial}{\partial t}\psi = \mathbf{H}\psi \ .\tag{5.119}$$

On account of translational symmetry, it follows that

$$\Omega_{l'l''} = \Omega_{l'-l''} \ ,\tag{5.120}$$

i.e., the rates only depend on the distance between the two positions.

It is not difficult to specify the eigenstates of this system. For the existing translational symmetry, they are wave-like and of the form

$$|\psi\rangle = \mathcal{N}_{\mathrm{c}}^{-1/2} \sum_{l=0}^{\mathcal{N}_{\mathrm{c}}-1} \exp(ikal)|l\rangle = |k\rangle \ .\tag{5.121}$$

Since they are characterised by the quantum number k, they are referred to in a shorthand notation as $|k\rangle$; for the imposed cyclic boundary conditions, k assumes, within the first Brillouin zone, the following $\mathcal{N}_{\mathrm{c}}$ values:

$$-\frac{\pi}{a} \leq k = i\frac{2\pi}{\mathcal{N}_{\mathrm{c}}a} \ \ (i \text{ is an integer}) \ < \frac{\pi}{a}\tag{5.122}$$

(cf. Eq. (5.23)). It can be verified by substituting back into the Schrödinger equation that these are indeed eigenfunctions. We obtain

$$\mathbf{H}|k\rangle = \hbar \sum_{l'=0}^{\mathcal{N}_{\mathrm{c}}-1}\sum_{l''=0}^{\mathcal{N}_{\mathrm{c}}-1} \Omega_{l'-l''}\mathbf{b}_{l'}^{+}\mathbf{b}_{l''}\mathcal{N}_{\mathrm{c}}^{-1/2} \sum_{l=0}^{\mathcal{N}_{\mathrm{c}}-1} \exp(ikal)|l\rangle\tag{5.123}$$

$$= \hbar \sum_{l'=0}^{\mathcal{N}_{\mathrm{c}}-1}\sum_{l''=0}^{\mathcal{N}_{\mathrm{c}}-1} \Omega_{l'-l''}\mathcal{N}_{\mathrm{c}}^{-1/2} \exp(ikal'')|l'\rangle$$

$$= \hbar \sum_{l'=0}^{\mathcal{N}_{\mathrm{c}}-1}\sum_{l''=0}^{\mathcal{N}_{\mathrm{c}}-1} \Omega_{l'-l''} \exp[ika(l''-l')]\mathcal{N}_{\mathrm{c}}^{-1/2} \exp(ikal')|l'\rangle$$

$$= \hbar \sum_{l=0}^{\mathcal{N}_{\mathrm{c}}-1} \Omega_{l} \exp(-ikal)|k\rangle \ ,$$

as is required. The energy eigenvalues are, thus,

$$\epsilon(k) = \hbar \sum_{l=0}^{\mathcal{N}_{\mathrm{c}}-1} \Omega_{l} \exp -ikal \ .\tag{5.124}$$

If only transitions between neighbours are possible, the dispersion relation is given by

$$\epsilon(k) = \hbar\Omega_0 + \hbar 2\Omega_1 \cos ka \ .\tag{5.125}$$

Note that a band, whose width is determined by the transition rate between neighbours, Ω_1, has formed.

Using the above expression for the eigenenergy, the following functions are obtained as the stationary solution of the time-dependent Schrödinger equation:

$$|k(t)\rangle = \exp\left(-i\frac{\epsilon(k)t}{\hbar}\right)|k\rangle \tag{5.126}$$

$$= \mathcal{N}_c^{-1/2} \sum_{l=0}^{N_c-1} \exp\left(-i\frac{\epsilon(k)t}{\hbar} + ikal\right)|l\rangle \ .$$

They have wave-like character and describe exciton waves.

The exciton is, as in the case of the Mott–Wannier exciton wave in Eq. (5.110), also delocalised in this state, with a constant probability

$$|\langle l|k(t)\rangle|^2 = \mathcal{N}_c^{-1} \tag{5.127}$$

of being encountered along the chain. In order to determine the energy transfer associated with a local excitation, it is again necessary to begin with a wavepacket. An excitation of the molecule $l = 0$ is described by

$$|l = 0\rangle = \mathcal{N}_c^{-1/2} \sum_{k'} \exp\left(-i\frac{\epsilon(k')}{\hbar}t\right)|k'\rangle \ , \tag{5.128}$$

where the sum is over all wavenumbers in Eq. (5.122). The probability that the exciton is found at the location l after a time t is given by the square of the matrix element:

$$\langle l|\psi(t)\rangle = \mathcal{N}_c^{-1} \sum_{k''} \exp(-ik''al)\langle k''| \sum_{k'} \exp\left(-i\frac{\epsilon(k')}{\hbar}t\right)|k'\rangle$$

$$= \mathcal{N}_c^{-1} \sum_{k'} \exp\left(-i\frac{\epsilon(k')t}{\hbar} - ik'al\right) \ . \tag{5.129}$$

The contributions of all waves are coherently superimposed in the matrix element.

The Transition to Hopping Processes. The wave-like propagation of a local electronic excitation can be described very well by such a process carried by exciton waves. However, it must be stated that this process is rarely encountered under real conditions. Wave-like processes are based on fixed phase relations in time and space. The interaction forces by which exciton waves are carried are very weak in comparison to lattice vibrations or spin waves, such that the required fixed phase relations are all too easily destroyed. The perturbations are due to the lattice vibrations. Transition rates, and in particular the transfer of energy by tunnelling processes associated with triplet excitons, are very sensitive to the separation of the molecules, while the excitation energies are also influenced by changes in the microscopic environment. It is clear that the phase relations in the exciton wavefunction cannot be maintained in the presence of such fluctuations and the propagation mechanism must change. Experiments indicate that there is a change to a process which can be understood as an undirected and temporally random **hopping** of localised Frenkel excitons. There is a model by which the transition from a coherent wave-like mode to an incoherent diffusive hopping mode can be described theoretically. This will be presented here, albeit in a way which foregoes most of the derivation and gives only the main results.

The derivation again starts with the Hamiltonian for a linear chain in Eq. (5.118), with the fluctuations being taken into account by adding a fluctuating part $\omega_{l'l''}(t)$ to the previously constant matrix elements $\Omega_{l'l''}$ to give

$$\Omega_{l'l''} + \omega_{l'l''}(t) \ .$$

In the context of statistics, $\omega_{l'l''}(t)$ is a stochastic function with a vanishing average value

$$\langle \omega_{l'l''}(t) \rangle = 0 \ , \tag{5.130}$$

which is to be understood as the time or ensemble average. It is assumed that the time-dependent auto-correlation function is described by

$$\langle \omega_{l'l''}(t' + t)\omega_{l'l''}(t') \rangle = \langle \omega_{l'l''}^2 \rangle \tau_{\text{vib}}\delta(t) \ , \tag{5.131}$$

i.e., a characteristic time τ_{vib} is assigned to the fluctuation, which is small as compared to the relevant timescale of the motion of the exciton. It is also assumed that there is no correlation between the fluctuations of the matrix elements of different pairs l', l''. In a shorthand form, it is written that

$$\langle \omega_{l'l''}^2 \rangle \tau_{\text{vib}} = 2\gamma_{l'l''} \ . \tag{5.132}$$

The following is true from symmetry reasons

$$\gamma_{l'l''} = \gamma_{l''l'} = \gamma_{|l''-l'|} \tag{5.133}$$

and the assumption can, thus, be made that the coefficients only depend on the atom separations in the chain. The model is thus introduced.

We discuss now the situation of triplet excitons, where only transitions between neighbouring molecules are possible. The model then only contains 3 parameters $\Omega_1 = \Omega_{-1}$, γ_0 and $\gamma_1 = \gamma_{-1}$ if, as is possible, Ω_0 is set equal to zero. In the discussion of exciton dynamics, it can again be asked: What is the probability that an exciton, which is generated at the location $l = 0$, is found at a different location after a time t? The fact that only probabilities can be specified is no longer only a consequence of quantum mechanics, but is now also due to the effect of fluctuations with stochastic character. In such a situation, a transition must be made from a description of the motion by a time-dependent wavefunction

$$\psi(t) = \sum_l c_l(t)|l\rangle$$

to a description in terms of a density matrix $\rho(t)$ with elements

$$\rho_{l'l''}(t) = \langle c_{l'}(t)c_{l''}^*(t)\rangle \quad . \tag{5.134}$$

The probability that an exciton is to be found at position l at a time t is then given by $\rho_{ll}(t)$.

Starting with the general equations, the equation of motion for this model can be given by carrying out the appropriate average-value calculations (see [44]). It is, considering the diagonal and off-diagonal elements separately,

$$\frac{\partial}{\partial t}\rho_{l'l'} = \frac{i}{\hbar}\sum_l(\Omega_{l'l}\rho_{ll'} - \rho_{l'l}\Omega_{ll'})$$
$$-4\gamma_1\rho_{l'l'} + 2\gamma_1\rho_{l'+1,l'-1} + 2\gamma_1\rho_{l'-1,l'-1} \tag{5.135}$$

$$\frac{\partial}{\partial t}\rho_{l'\neq l''} = -\frac{i}{\hbar}\sum_l(\Omega_{l'l}\rho_{ll''} - \rho_{l'l}\Omega_{ll''})$$
$$-2(\gamma_0 + 2\gamma_1)\rho_{l'l''} + (\delta_{l'',l'+1} + \delta_{l'',l'-1})2\gamma_1\rho_{l''l'} \ . \tag{5.136}$$

The elements $\Omega_{l'l''}$ of the time-independent Hamiltonian equation (5.118) are, thus, in this case given by

$$\Omega_{l'l''} = \hbar(\delta_{l'',l'+1} + \delta_{l'',l'-1})\Omega_1 \quad . \tag{5.137}$$

The right-hand side of the equations contain two different parts, the physical meaning of which can be immediately recognised. The part corresponding to the time-independent Hamiltonian would, if only it were present, lead to the energy propagation realised in a rigid lattice by exciton waves, i.e., the previously discussed coherent process is expressed here in the density matrix formalism. The second part of the equation of motion for the diagonal elements describes a hopping process. If only this second part were considered, the equation of motion would be:

$$\frac{\partial}{\partial t}\rho_{l'l'} = -4\gamma_1\rho_{l'l'} + 2\gamma_1\rho_{l'+1,l'+1} + 2\gamma_1\rho_{l'-1,l'-1} \quad . \tag{5.138}$$

This states that the occupation of the position l' is, on the one hand, decreased by jumps to the two neighbours – each with a rate $2\gamma_1$ – and, on the other hand, increased by jumps in the opposite direction, i.e., from the two neighbours. Exactly such a behaviour characterises a hopping process. The incoherent part in the equation of motion for the off-diagonal elements, which only couples $\rho_{l'l''}$ and $\rho_{l''l'}$ in the case where $l'' = l' \pm 1$, cannot be so easily interpreted. It achieves a general time-dependent decay of the values $\rho_{l' \neq l''}$ which is in contrast to the diagonal elements, where the hopping process always only leads to a redistribution of the occupation probabilities. The values $\rho_{l' \neq l''}$ measure the remaining coherence at a time t in the propagation process, and this disappears ever more quickly, the larger the coefficients γ_0 and γ_1 are.

The equation of motion correctly describes the limiting behaviour of a nearly vanishing perturbation as well as predicting a hopping process in the other limiting case of a strong perturbation. According to Eq. (5.138), it would be expected that the jump rate and the associated energy propagation velocity increase with γ_1, and hence with increasing temperature. In fact, this is only observed at higher temperatures. If measurements begin at a low temperature, a decrease in the transition rate is initially observed. Figure 5.10 shows this for the example of the exciton hopping process in crystalline anthracene. In this case, the diffusion coefficient D_s, which is proportional to the jump rate, was determined spectroscopically at different temperatures. The observations can be described by the equations of motion, Eqs. (5.135) and (5.136). If a perturbation theory analysis, starting from the limiting case of only hopping processes described by Eq. (5.138), is carried out in order to determine which changes arise upon 'switching-on' the coherent part for $\Omega_1^2/\gamma_0 \ll 1$, it is found that, while maintaining the hopping character of the propagation process, there is an increase in the jump rate to

$$2\gamma_1 + \frac{\Omega_1^2}{\gamma_0} \quad . \tag{5.139}$$

The additional term increases on cooling because of the decrease in γ_0, thus providing an explanation for the observations.

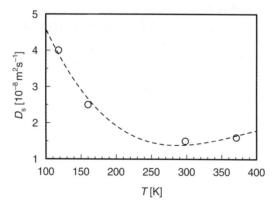

Fig. 5.10. The diffusion coefficient of triplet excitons in anthracene, as measured spectroscopically at different temperatures (from Ern et al. [45]).

5.2 Liquids: Diffusive Motion

Propagating wave-like excitations are also found in liquids in the form of sound waves. In contrast to the case of crystals, it is not to be expected that these waves will continue to exist into the meso- and microscopic region. On these length-scales, liquids lose their macroscopic homogeneous structure. For decreasing wavelengths, this leads, first of all, to an increasing scattering, with an associated reduction in the lifetime of the sound waves, and finally to their complete disappearance. As was mentioned in the introduction to this chapter, only local vibrations of individual molecules in the environment of their neighbours appear in liquids on the microscopic scale, and these are only short-lived because of the constant changes. A qualitatively new type of dynamics is encountered here, namely the diffusion of molecules. This makes the long-range transport of material possible, something which is not the case for crystals. How this can be described will be discussed in the following.

5.2.1 Diffusion Coefficients

The trajectories followed by individual molecules in a liquid or by colloids in solution are subject to statistical laws. As is always the case for processes which do not occur in a deterministic fashion, but rather where there is a large number of different possibilities by which the process may advance, it is necessary to introduce a suitable distribution function in order to describe the time-dependent development. This is, for the case of diffusive motion of individual particles, the **time-dependent auto-correlation function** $g_1(r, t)$. It is defined such that

$$g_1(r, t)\mathrm{d}^3 r$$

describes the probability that a particle moves within a time t from its starting point into a volume element $\mathrm{d}^3 r$ which is r away. g_1 is a probability distribution and, therefore, must be normalised:

$$\int g_1(r, t)\mathrm{d}^3 r = 1 \ . \tag{5.140}$$

Einstein presented a differential equation by which $g_1(r, t))$ can be calculated. The starting point is equation

$$g_1(r, t + \Delta t) = \int g_1(r - r', t)g_1(r', \Delta t)\mathrm{d}^3 r' \ . \tag{5.141}$$

The displacement over a distance r within a time $t + \Delta t$ is broken up into two steps which follow one another. The first step, which is completed during a time t, achieves a displacement of $r - r'$, while a second step, in the remaining time Δt, brings the molecule to the point r. The special feature of this integral notation is that the probability that these two steps follow each other is given by the product of the associated individual probabilities, with an integration over all possible step combinations being then performed. This product notation expresses exactly the basic property of all diffusive motions, namely that they occur in steps, between which there is no correlation at all. The total probability can only be written as the product of the individual probabilities if this is the case.

The integral equation, Eq. (5.141), can be transformed into a differential equation. Carrying out a series expansion with respect to time on the left-hand side and with respect to the position on the right-hand side leads to the following equation

$$g_1(\boldsymbol{r},t) + \Delta t \frac{\partial g_1}{\partial t} \approx \int \left[g_1(\boldsymbol{r},t) - \sum_i r'_i \frac{\partial g_1}{\partial r_i}(\boldsymbol{r},t) \right. $$
$$\left. + \frac{1}{2} \sum_{ij} r'_i r'_j \frac{\partial^2 g_1}{\partial r_i \partial r_j}(\boldsymbol{r},t) \right] g_1(\boldsymbol{r}', \Delta t) \mathrm{d}^3 \boldsymbol{r}' . \quad (5.142)$$

Upon performing the integration on the right-hand side, it is found that several terms disappear, since from symmetry reasons

$$\int g_1(\boldsymbol{r}', \Delta t) r'_i \mathrm{d}^3 \boldsymbol{r}' = 0 \quad (5.143)$$

and also

$$\int g_1(\boldsymbol{r}', \Delta t) r'_i r'_{j \neq i} \mathrm{d}^3 \boldsymbol{r}' = 0 \quad . \quad (5.144)$$

$g_1(\boldsymbol{r},t)$ appears on both sides and can thus be removed. The only term which remains on the right-hand side is

$$\int g_1(\boldsymbol{r}', \Delta t) r'^2_i \mathrm{d}^3 \boldsymbol{r}' = \frac{1}{3} \int g_1(\boldsymbol{r}', \Delta t) |\boldsymbol{r}'|^2 \mathrm{d}^3 \boldsymbol{r}' \quad (5.145)$$

such that we obtain

$$\Delta t \frac{\partial g_1}{\partial t} = \nabla^2 g_1(\boldsymbol{r},t) \cdot \frac{1}{6} \int g_1(\boldsymbol{r}', \Delta t) |\boldsymbol{r}'|^2 \mathrm{d}^3 \boldsymbol{r}' \quad . \quad (5.146)$$

This equation only contains a single parameter, namely the quantity

$$D_\mathrm{s} = \frac{1}{6 \Delta t} \int g_1(\boldsymbol{r}', \Delta t) |\boldsymbol{r}'|^2 \mathrm{d}^3 \boldsymbol{r}' = \frac{\langle |\boldsymbol{r}'|^2 \rangle}{6 \Delta t} \quad , \quad (5.147)$$

which is termed the 'self-diffusion coefficient'. Using D_s, we arrive at a differential equation for the auto-correlation function $g_1(\boldsymbol{r},t)$:

$$\frac{\partial g_1}{\partial t}(\boldsymbol{r},t) = D_\mathrm{s} \nabla^2 g_1(\boldsymbol{r},t) \quad . \quad (5.148)$$

D_s specifies, according to its definition, the mean squared displacement achieved by a diffusing particle per unit time. That this variable indeed exists as a time-independent constant becomes clear from the following consideration. For a diffusive motion, each total displacement $\boldsymbol{r}'$ corresponds to a sum of many uncorrelated individual steps $\boldsymbol{a}_j$. If there are p steps, each in the same time Δt, the overall mean squared displacement of the particle is given by

$$\langle |\boldsymbol{r}'|^2 (p\Delta t)\rangle = \left\langle \left| \sum_{j=1}^{p} \boldsymbol{a}_j \right|^2 \right\rangle = \left\langle \sum_{j,j'=1}^{p} \boldsymbol{a}_j \cdot \boldsymbol{a}_{j'} \right\rangle \ . \tag{5.149}$$

Since individual steps are not correlated, it follows that

$$\langle \boldsymbol{a}_j \cdot \boldsymbol{a}_{j'} \rangle = \langle |\boldsymbol{a}_j|^2 \rangle \delta_{jj'} \tag{5.150}$$

and, therefore,

$$\langle |\boldsymbol{r}'|^2 (p\Delta t)\rangle = p \langle |\boldsymbol{a}_j|^2 \rangle = p \langle |\boldsymbol{r}'|^2 (\Delta t)\rangle \ . \tag{5.151}$$

This means that the mean squared displacement actually increases linearly with time, as the name 'self-diffusion constant' suggests, i.e.,

$$\frac{\langle |\boldsymbol{r}'|^2 (p\Delta t)\rangle}{p\Delta t} = \frac{\langle |\boldsymbol{r}'|^2 (\Delta t)\rangle}{\Delta t} = \text{const} = 6D_s \ . \tag{5.152}$$

Colloids are dispersed particles with diameters in the $10\,\text{nm}$–μm range, which diffuse independently of each other in a liquid – the Brownian motion which they undergo can often be observed directly under a microscope. There is an alternative approach by which their motion can be described. Imagine that a large number of colloids is restricted, at the start, to a very small volume; the restriction is then removed, such that motion sets in. A current of the colloidal particles would then be observed, whose local density is given by Fick's law:

$$\boldsymbol{j} = -D\nabla\rho \ . \tag{5.153}$$

This equation also contains a diffusion coefficient – it is here the constant of proportionality between the particle current density $\boldsymbol{j}$ and the gradient of the particle density $\rho(\boldsymbol{r}, t)$. On combining Fick's law with that for the conservation of mass of the particle,

$$\frac{\partial \rho}{\partial t} + \nabla \boldsymbol{j} = 0 \ , \tag{5.154}$$

a differential equation is obtained for the space and time dependence of the colloid density:

$$\frac{\partial \rho}{\partial t}(\boldsymbol{r}, t) = D\nabla^2 \rho(\boldsymbol{r}, t) \ . \tag{5.155}$$

It is not surprising that this equation for $\rho(\boldsymbol{r}, t)$ has the same form as that in Eq. (5.148) for $g_1(\boldsymbol{r}, t)$: Since each colloid realises a possible trajectory, it follows that the density distribution $\rho(\boldsymbol{r}, t)$ just reproduces the auto-correlation function $g_1(\boldsymbol{r}, t)$.

The self-diffusion coefficient D_s and the diffusion coefficient D in Fick's law are necessarily the same for the case of free, non-interacting colloids. The situation changes when interacting particles are involved. In Fick's law, the gradient of the particle density is then to be replaced by the gradient of the chemical potential, and D becomes a **cooperative diffusion coefficient**, which differs from the self-diffusion coefficient. The definition of the latter remains unchanged; D_s always describes, according to Eq. (5.147), the mean squared displacement per unit time.

In order to determine the auto-correlation function $g_1(\boldsymbol{r}, t)$, it is necessary to solve the differential equation, Eq. (5.148), for the following starting condition:

$$g_1(\boldsymbol{r}, t = 0) = \delta(\boldsymbol{r}) \ . \tag{5.156}$$

The solution of this problem is known as 'Green's function' and has the form of a Gaussian function:

$$g_1(\boldsymbol{r}, t) = \frac{1}{(4\pi D_s t)^{3/2}} \exp{-\frac{|\boldsymbol{r}|^2}{4D_s t}} \ . \tag{5.157}$$

The expression in the denominator of the exponential term is proportional to the mean squared displacement. An evaluation gives, as expected,

$$\langle |\boldsymbol{r}|^2(t) \rangle = \int g_1(\boldsymbol{r}, t) |\boldsymbol{r}|^2 \mathrm{d}^3 \boldsymbol{r} = 6 D_s t \ . \tag{5.158}$$

Exactly the same applies for the time-dependent development of the density distribution $\rho(\boldsymbol{r}, t)$. If $\mathcal{N}$ particles are localised at the origin at the start, this means that

$$\rho(\boldsymbol{r}, 0) = \mathcal{N}\delta(\boldsymbol{r}) \ , \tag{5.159}$$

and the above Gaussian function can again be used to describe the development over time of the density function:

$$\rho(\boldsymbol{r}, t) = \frac{\mathcal{N}}{(4\pi D t)^{3/2}} \exp{-\frac{|\boldsymbol{r}|^2}{4D t}} \ . \tag{5.160}$$

5.2.2 Mobility and the Einstein Relation

It might seem at first that nothing further can be said about the value of a self-diffusion coefficient. In fact, this is not the case. Einstein derived, on the basis of a thought experiment for the case of non-interacting colloidal particles in solution, an expression which can be immediately used. The thought experiment concerns the situation of a colloidal solution in a potential field, which can be simply the gravitational field, or, in the case of charged particles, also an electric field. It is generally observed that a colloid in solution moves, under the influence of an external force, at a constant velocity, which is proportional to the force:

$$\boldsymbol{v} \propto \mathbf{f} \ . \tag{5.161}$$

The proportionality factor

$$\nu = \frac{v}{f} \tag{5.162}$$

is referred to as the **mobility**, while its inverse

$$\zeta = \frac{1}{\nu} = \frac{f}{v} \tag{5.163}$$

has the meaning of a coefficient of friction.

In spite of the motion caused by the field, there is a stationary state, $\rho_{\text{eq}}(\boldsymbol{r})$, for the density distribution of the colloids. It is given by Boltzmann statistics as

$$\rho_{\text{eq}} \propto \exp -\frac{u_{\text{pot}}}{k_{\text{B}}T} \ . \tag{5.164}$$

Einstein explained the existence of this stationarity in terms of the existence of a dynamic equilibrium between two compensating particle currents. There is a particle current density

$$\boldsymbol{j}_f = \rho \boldsymbol{v} = \rho \nu \mathbf{f} \ , \tag{5.165}$$

which is caused to flow by the forces of the potential field. With the force

$$\mathbf{f} = -\nabla u_{\text{pot}} \ , \tag{5.166}$$

it follows that

$$\boldsymbol{j}_f = -\rho \nu \nabla u_{\text{pot}} \ . \tag{5.167}$$

The second particle current in the opposite direction is driven by the gradients in the particle density – Fick's law applies in this case. For non-interacting colloids, it is not necessary to distinguish between D and D_{s}, i.e.,

$$D = D_{\text{s}} \ ,$$

and the diffusive particle current is then given as

$$\boldsymbol{j}_D = -D_{\text{s}} \nabla \rho \ . \tag{5.168}$$

At equilibrium, i.e., for $\rho = \rho_{\text{eq}}$, the two currents compensate each other:

$$\boldsymbol{j}_D + \boldsymbol{j}_f = 0 \ . \tag{5.169}$$

It, thus, follows that

$$D_{\text{s}} \nabla \rho_{\text{eq}} + \rho_{\text{eq}} \nu \nabla u_{\text{pot}} = 0 \ . \tag{5.170}$$

Upon introducing the equilibrium distribution, Eq. (5.164), a generally applicable relation between the self-diffusion coefficient and the mobility is obtained. It is referred to as the **Einstein relation** and is given as

$$D_{\mathrm{s}} = k_{\mathrm{B}} T \nu \ . \tag{5.171}$$

On the right-hand side, the mobility, a reaction parameter, describes the effect exerted by an external force of arbitrary nature on the motion of a colloid in solution. The result of the action is a motion with constant velocity. The left-hand side of the relation contains a parameter which has a purely statistical character – it describes the mean squared displacement of a colloid which would result in the absence of any external force, being solely driven by collisions with the solvent molecules. The Einstein relation states that these very different parameters are linked by the factor $k_{\mathrm{B}} T$.

With the help of the Einstein relation it is possible to make a further step. **Stokes' law**, which is derived using the laws of hydrodynamics, specifies the size of the force which arises when a spherical particle undergoes a motion with a constant velocity $\boldsymbol{v}$ through a liquid with viscosity η. The force depends on the radius of the particle according to

$$\mathbf{f} = 6\pi R \eta \boldsymbol{v} \ . \tag{5.172}$$

Since the viscous force equals the external driving force, Stokes' law provides an explicit expression for the mobility of a colloid. Inserting it into the Einstein relation yields

$$D_{\mathrm{s}} = \frac{k_{\mathrm{B}} T}{6\pi R \eta} \ . \tag{5.173}$$

This is an equation, which can be directly used in order to determine the self-diffusion coefficient of a colloid of radius R in a solution with viscosity η at a temperature T.

The question then arises as to whether this equation also gives sensible values when applied to the case of diffusion of individual molecules in a molecular liquid, using, together with the macroscopic viscosity, an estimate for R, namely the molecular radius, which is a variable in the nm range. In fact, a comparison with experimental results reveals that the so-obtained results are generally correct to at least within an order of magnitude.

5.3 Liquid Crystals: Orientational Fluctuations

All the motional forms discussed above for the liquid state are also encountered in liquid crystals but there is, in addition, a further specific contribution to the dynamics. It is based on two characteristics of nematics, namely, firstly, the restoring elastic forces which arise from a deformation of the director field, and, secondly, the viscous forces which accompany a rotation of the director.

Deformations of the director field result spontaneously through thermal activation. In the following, it will be shown that they can be described as a superposition of wave-like deformation modes. The dynamics is that of a simple relaxator; the viscosity of the medium doesn't permit any vibrational motion.

5.3.1 Splay, Twist and Bend Modes

Consider a uniformly oriented nematic liquid crystal with the director n_0 parallel to the z direction. Small displacements of the director from the equilibrium direction lead to non-vanishing, small transverse components $n_x(r)$, $n_y(r)$. The associated increase in the elastic Helmholtz free energy can be determined using Frank's equation, Eq. (2.8). Integrating over the volume $\mathcal{V}$ gives the expression:

$$\mathcal{F}_{el} = \frac{1}{2} \int_{\mathcal{V}} \left(K_1 \left(\frac{\partial n_x}{\partial x} + \frac{\partial n_y}{\partial y} \right)^2 + K_2 \left(\frac{\partial n_x}{\partial y} - \frac{\partial n_y}{\partial x} \right)^2 + K_3 \left(\frac{\partial n_x}{\partial z} + \frac{\partial n_y}{\partial z} \right)^2 \right) d^3 r.$$

(5.174)

The deformation field can be generally represented as a sum of wave-like contributions:

$$n_i(r) = \mathcal{V}^{-1/2} \sum_k \exp(ikr) n_i(k) \qquad i = x, y \ .$$

(5.175)

The wavevectors k are chosen such that cyclic boundary conditions are fulfilled for a cubic sample with volume $\mathcal{V} = L^3$, i.e.,

$$n_i(x, y, z) = n_i(x + L, y, z) = n_i(x, y + L, z) = n_i(x, y, z + L) \ .$$

(5.176)

This is achieved for the series

$$k_{i_1, i_2, i_3} = \frac{2\pi}{L} (i_1 e_1 + i_2 e_2 + i_3 e_3) \qquad (i_1, i_2, i_3 \text{ are integers})$$

(5.177)

with e_i as the unit vectors of a Cartesian coordinate system. Since the displacements are real, the following applies for the complex amplitudes $n(k)$ of two waves with opposite wavevectors:

$$n_i(-k) = n_i(k)^* \ .$$

(5.178)

Inserting this representation into Eq. (5.174) gives for the first term:

$$\int_{\mathcal{V}} \left(\frac{\partial n_x}{\partial x} + \frac{\partial n_y}{\partial y} \right)^2 d^3 r = \mathcal{V}^{-1} \int_{\mathcal{V}} \sum_{k, k'} [ik_x n_x(k) + ik_y n_y(k)] \cdot [ik'_x n_x(k')$$

$$+ ik'_y n_y(k')] \exp[i(k + k')r] d^3 r \ .$$

(5.179)

Taking into account that

$$\int_{V} \exp[i(\boldsymbol{k} + \boldsymbol{k}')\boldsymbol{r}] \, \mathrm{d}^3\boldsymbol{r} = V\delta_{\boldsymbol{k},-\boldsymbol{k}'} \tag{5.180}$$

leads to the expression

$$\int_{V} \left(\frac{\partial n_x}{\partial x} + \frac{\partial n_y}{\partial y}\right)^2 \mathrm{d}^3\boldsymbol{r} = \sum_{\boldsymbol{k}} |k_x n_x(\boldsymbol{k}) + k_y n_y(\boldsymbol{k})|^2 \ . \tag{5.181}$$

The other terms can be calculated in a similar fashion, and the result is

$$\mathcal{F}_{\mathrm{el}} = \frac{1}{2}\sum_{\boldsymbol{k}} \left(K_1 |k_x n_x(\boldsymbol{k}) + k_y n_y(\boldsymbol{k})|^2 + K_2 |k_y n_x(\boldsymbol{k}) - k_x n_y(\boldsymbol{k})|^2 \right.$$
$$\left. + K_3 k_z^2 (|n_x(\boldsymbol{k})|^2 + |n_y(\boldsymbol{k})|^2)\right) \ . \tag{5.182}$$

A further simplification is achieved if the coordinate system is not considered to be fixed, but rather the y direction is always chosen to be perpendicular to $\boldsymbol{k}$ and $\boldsymbol{n}_0$ for each wavevector $\boldsymbol{k}$. In this individually adjusted coordinate system, it is always true that

$$k_y = 0 \tag{5.183}$$

and thus

$$\mathcal{F}_{\mathrm{el}} = \frac{1}{2}\sum_{\boldsymbol{k}} |n_x(\boldsymbol{k})|^2 (K_3 k_z^2 + K_1 k_x^2) + |n_y(\boldsymbol{k})|^2 (K_3 k_z^2 + K_2 k_x^2) \ . \tag{5.184}$$

This leads to a simple end result for the Helmholtz free energy, namely:

$$\mathcal{F}_{\mathrm{el}} = \frac{1}{2}\sum_{\boldsymbol{k}} \sum_{\alpha=1,2} |n_\alpha(\boldsymbol{k})|^2 K_{\mathrm{eff},\alpha}(\boldsymbol{k}) \ , \tag{5.185}$$

where

$$K_{\mathrm{eff},\alpha}(\boldsymbol{k}) = K_3 k_\parallel^2 + K_\alpha k_\perp^2 \ . \tag{5.186}$$

It can be seen that all couplings between different variables are removed. The Helmholtz free energy consists of independent contributions from the wave-like deformation modes, where there are two independent modes for each wavevector $\boldsymbol{k}$. These have director displacement directions within and perpendicular to the $(\boldsymbol{k}, \boldsymbol{n}_0)$ plane, i.e., their polarisations are perpendicular to each other. $K_{\mathrm{eff},\alpha}(\boldsymbol{k})$, describes, in the sense of a modulus, an elastic stiffness of the mode $(\boldsymbol{k}, \alpha)$ and contains, together with the Frank moduli, the components of the wavevector parallel and perpendicular to the director, $k_\parallel$ and $k_\perp$. Equation (5.186) shows that the two different modes represent a superposition of bend and splay waves, and a superposition of bend and twist deformations respectively. For certain special directions of $\boldsymbol{k}$, there are deformations of only one type. Figure 5.11 shows four such cases, together with the associated lines of the deformed director field.

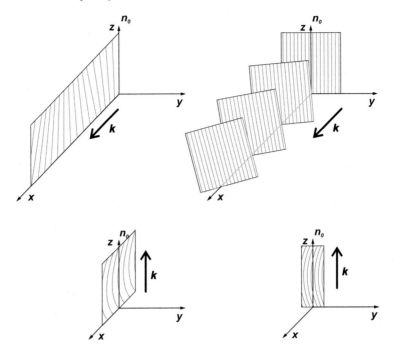

Fig. 5.11. Deformation modes of the director field with the wavevector perpendicular to n_0 (*above*) – splay and twist modes – and with the wavevector k parallel to n_0 (*below*) – two bend modes. A frozen image of the director field in the selected plane is sketched for each of the four modes.

5.3.2 Equilibrium Nematodynamics

Thermal energy excites the modes independently of each other such that the mean squared amplitude $\langle |n_\alpha(k)|^2 \rangle$ can be calculated for all modes using Boltzmann statistics. It is given as

$$\langle |n_\alpha(k)|^2 \rangle = \frac{\int |n_\alpha(k)|^2 \exp -\frac{\delta\mathcal{F}(|n_\alpha(k)|)}{k_\mathrm{B}T} \mathrm{d}|n_\alpha(k)|}{\int \exp -\frac{\delta\mathcal{F}(|n_\alpha(k)|)}{k_\mathrm{B}T} \mathrm{d}|n_\alpha(k)|} \quad , \tag{5.187}$$

where $\delta\mathcal{F}(|n_\alpha(k)|)$ describes the contribution of the mode (k, α) to $\mathcal{F}_\mathrm{el}$. Using Eq. (5.185), it follows that

$$\langle |n_\alpha(k)|^2 \rangle = \frac{k_\mathrm{B}T}{K_{\mathrm{eff},\alpha}(k)} \quad , \qquad \alpha = 1,2 \quad . \tag{5.188}$$

Each deformation mode behaves like an over-damped harmonic oscillator, in which the elastic forces do not act to cause an acceleration, but rather they are dissipated by the viscous forces. Consider as an example the pure twist mode in the top right of Fig. 5.11, which corresponds to the following director deformation field:

$$n_y(x) = n_\perp \cos(k_x x) \ . \tag{5.189}$$

Re-expressing it in the form of the Fourier representation of Eq. (5.175), i.e., as

$$n_y(x) = \mathcal{V}^{-1/2} \left[\frac{\mathcal{V}^{1/2} n_\perp}{2} \exp(ik_x x) + \frac{\mathcal{V}^{1/2} n_\perp}{2} \exp(-ik_x x) \right] \tag{5.190}$$

yields the value for the associated Fourier amplitude. Equation (5.185) can be used to determine the elastic energy associated with the twist wave – it is given by

$$\mathcal{F}_{el} = \frac{\mathcal{V} K_{eff,2} n_\perp^2}{4} \ . \tag{5.191}$$

An increase in the amplitude $dn_\perp$ leads to an increase

$$d\mathcal{F}_{el} = \frac{\mathcal{V} K_{eff,2} n_\perp}{2} dn_\perp \tag{5.192}$$

in the Helmholtz free energy and, at the same time, to an increase in the restoring local torque $dT_x/d^3\mathbf{r}$. The work per unit volume dw, which is performed by the torque in turning the director, amounts to

$$dw = \frac{dT_x}{d^3\mathbf{r}} \frac{dn_y}{dt} dt \ . \tag{5.193}$$

Using Eq. (2.42) which relates this torque to the rotational viscosity γ_1, one obtains

$$dw = \gamma_1 \left(\frac{dn_y}{dt} \right)^2 dt \ . \tag{5.194}$$

The integration over the sample volume, taking into account Eq. (5.189), yields for the total work $\mathcal{W}$

$$d\mathcal{W} = \frac{\gamma_1 \mathcal{V}}{2} \left(\frac{dn_\perp}{dt} \right)^2 dt \ . \tag{5.195}$$

Since this work is performed using the stored elastic energy, it follows that

$$-\frac{\mathcal{V} K_{eff,2} n_\perp}{2} \frac{dn_\perp}{dt} = \frac{\mathcal{V} \gamma_1}{2} \left(\frac{dn_\perp}{dt} \right)^2 \ . \tag{5.196}$$

The equation of motion for $n_\perp$ is thus given to be

$$\frac{dn_\perp}{dt} = -\frac{K_{eff,2} n_\perp}{\gamma_1} \ . \tag{5.197}$$

The solution is

$$n_\perp(t) \propto \exp -\frac{t}{\tau} \quad , \tag{5.198}$$

where

$$\tau = \frac{\gamma_1}{K_{\mathrm{eff},2}} \quad . \tag{5.199}$$

It can be seen that the considered twist mode decays exponentially after a thermal excitation, within a time, which depends on the rotational viscosity and the associated effective modulus

$$K_{\mathrm{eff},2} = K_2 k^2 \quad . \tag{5.200}$$

A similar dynamics, corresponding to a simple relaxation, is also found for all the other modes. The relaxation time
is always given by an expression of the form

$$\tau_\alpha(\boldsymbol{k}) = \frac{\eta_\alpha(\boldsymbol{k})}{K_{\mathrm{eff},\alpha}(\boldsymbol{k})} \quad , \tag{5.201}$$

where $\eta_\alpha(\boldsymbol{k})$ denotes the viscosity which is effective for the mode $(\boldsymbol{k}, \alpha)$.

Experimental results indicate, generally, for the case where there is not a pure twist mode, that it is not completely correct to describe the dynamics as only a reorientation of the director. The bend and splay modes shown in Fig. 5.11 actually set flow processes free, such that modes arise, which are to be considered as coupled reorientation-flow processes. A corresponding extension of the theory leads to the following expressions for the effective viscosities associated with splay and bend modes:

$$\eta_{\mathrm{splay}} = \gamma_1 - \frac{\alpha_3^2}{\eta_{\mathrm{b}}} \tag{5.202}$$

and

$$\eta_{\mathrm{bend}} = \gamma_1 - \frac{\alpha_2^2}{\eta_{\mathrm{c}}} \quad . \tag{5.203}$$

The terms, which appear together with the rotational viscosity γ_1, contain, in addition to the Miesowicz viscosities η_{b}, η_{c} (see Fig. 2.5), two further parameters associated with the flow processes, α_2 and α_3, which are referred to as the **Leslie viscosity coefficients**. As is apparent from the negative sign of the additional terms, they cause a reduction in the effective viscosity. What this means is indicated in Fig. 5.12, which shows the velocity fields $\boldsymbol{v}(\boldsymbol{r})$ for a bend and a splay mode. The effect is evident from the zero crossing points: The flow vortex supports the return of the director orientation to equilibrium, i.e., makes it easier.

The described **nematodynamics** is associated with a variation in the orientation of the optical index ellipsoid and the dielectric tensor, and thus leads to the scattering of light. An explanation is thus provided for the reduction,

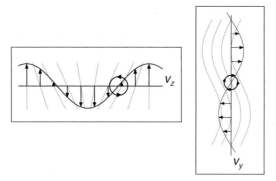

Fig. 5.12. The velocity field resulting from the director field deformation (*dashed lines*) for the case of a splaying mode (*left*) and a bend mode (*right*). The arrows denote the local flow velocities. The flow vortexes (two are shown) act in support of the reorientation back into equilibrium.

as compared to liquids, in the transparency of liquid-crystalline samples, even when they are uniformly oriented. Light scattering experiments can be used to exactly analyse the nematodynamics. If a scattering experiment is performed with a scattering vector $\boldsymbol{q}$, both the modes with $\boldsymbol{k} = \boldsymbol{q}$ are selected. The scattering function $S(\boldsymbol{q})$ is then associated with their mean squared amplitudes, i.e., it is found that

$$S(\boldsymbol{q}) \propto (\Delta\varepsilon)^2 \sum_{\alpha=1,2} \frac{F_{\mathrm{p},\alpha}}{K_3 q_\parallel^2 + K_\alpha q_\perp^2} \quad . \tag{5.204}$$

$\Delta\varepsilon$ is the anisotropy of the dielectric constant (see Eq. (1.65)) and $F_{\mathrm{p},\alpha}$ is a polarisation factor, which depends on the chosen polarisation direction of the incident and scattered light as well as the director orientation. Dynamic light scattering experiments provide, as will be described in the final section of this chapter, insight into the dynamics by the determination of the time-dependent scattering function $S(\boldsymbol{q}, t)$. For liquid crystals, this directly provides the relaxation time associated with the selected mode:

$$S(\boldsymbol{q}, t) = S(\boldsymbol{q}) \exp -\frac{t}{\tau_\alpha(\boldsymbol{q})} \quad . \tag{5.205}$$

Figure 5.13 shows, as an example, typical experimental results, which were obtained for a standard liquid crystal. It is found, in agreement with Eqs. (5.186), (5.201) and (5.204), that both I^{-1} and τ^{-1} are proportional to the square of the wavevector.

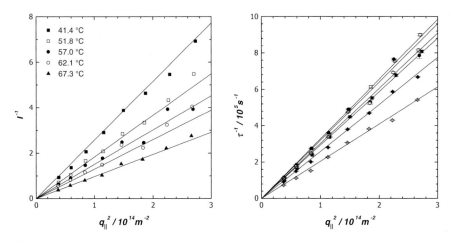

Fig. 5.13. The results of a light scattering experiment on a standard liquid crystal (sample ZLI1132 from Merck). The q dependence of the inverse intensity (*left*) and the inverse relaxation time (*right*) of the light scattering caused by the bending modes.

5.4 Polymer Melts: Chain Dynamics

Polymers in the melt are constantly undergoing changes in their conformational state. Transitions occur between the manifold different forms which a coil can adopt, such that all possible conformations are passed through over the course of time. The question then arises as to how this type of dynamics can be described, and this is the subject of this section.

It is clear that inertia forces can be neglected in the description of the conformational dynamics of a polymer chain. It is only necessary to consider the following two factors:

- Each monomer experiences a frictional force during its motion through the melt because of the interaction with other monomers. This is proportional to the velocity of the monomer, as in the case of colloids in a liquid.
- In contrast to the case of diffusion of free colloids discussed above, it is not possible for the monomers in a chain to move independently from each other. The connectivity generates additional forces within the chain. It is these forces which lead to restoring forces in rubber and other cross-linked polymers.

5.4.1 Rubber-Elastic Forces and the Rouse Model

At first, the restoring forces will be described. The situation for an individual macromolecule is shown in Fig. 5.14. The illustration indicates that it is necessary to apply a force f if the two chain ends are to be held at

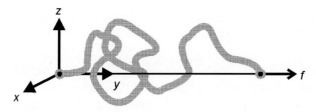

Fig. 5.14. A force f is required if the two ends of a macromolecule are to be held at a fixed separation.

a fixed separation, which is chosen here to be along the y axis. Expressed in a different way: If the ends were released, the chain, which is initially somewhat stretched out, would tie itself up again. This trend is felt for a rubber as an elastic force. The origin of this force is easy to appreciate. It is entropic in nature, exactly like the pressure of an ideal gas. A stretching of the chain, under conditions of constant temperature, lowers the entropy, like in the case of the isothermal compression of an ideal gas, since the chain can adopt less conformations in this stretched state as compared to the situation where the chain ends lie near to each other. A chain always has the tendency to adopt a state with a smaller end-to-end distance, in the same way that a gas spontaneously fills the available volume upon expansion. If it is desired to hinder this by fixing the chain ends, a force appears.

The strength of this force can be calculated starting from the generally applicable expression for an isothermal stretching process:

$$f = \frac{\partial \mathcal{F}}{\partial y} = \frac{\partial \mathcal{U}}{\partial y} - T \frac{\partial \mathcal{S}}{\partial y} \qquad (5.206)$$

(cf. Eq. (A.6) which gives an analogous expression for the stress σ). As was shown in Sect. 1.4.2, it is possible to describe polymers in the liquid state as ideal coils with Gaussian distributions for all monomer-monomer distance vectors. It is necessary to calculate how the number of conformations changes as a function of the end-to-end distance y. This will yield an expression for the dependence on y of the entropy of the macromolecule shown in Fig. 5.14. Energetic contributions play no role in this model representation. For an ideal coil, the total energy $\mathcal{U}$ depends, like in an ideal gas, solely on the kinetic energy, and this does not change upon isothermal stretching. Since the restoring force is thus solely of entropic origin, it follows that the following shortened expression can be used for the external force:

$$f = \frac{\partial \mathcal{F}}{\partial y} = -T \frac{\partial \mathcal{S}}{\partial y} \quad . \qquad (5.207)$$

The calculation of the entropy $\mathcal{S}$ of the macromolecule in the stretched state begins with the general equation

$$S = k_B \ln \mathcal{Z}(y) \quad , \tag{5.208}$$

which links $\mathcal{S}$ with the partition function $\mathcal{Z}$. The latter can be directly derived from the statistical properties of an ideal coil. It was shown in Sect. 1.4.2 that the probability that the two ends of a polymer chain are separated from each other by x, y, z is given by the Gaussian function (Eq. (1.73))

$$w(x, y, z) = \left(\frac{3}{2\pi R_0^2} \right)^{3/2} \exp - \frac{3(x^2 + y^2 + z^2)}{2R_0^2} \quad ,$$

where R_0^2 denotes the mean squared end-to-end distance

$$R_0^2 = \langle x^2 \rangle + \langle y^2 \rangle + \langle z^2 \rangle \quad . \tag{5.209}$$

The partition function, which counts the number of conformations which are associated with a particular fixed end-to-end distance, is proportional to this probability. For the chain in Fig. 5.14, it thus follows that

$$\mathcal{Z}(y) \propto w(0, y, 0) \propto \exp - \frac{3y^2}{2R_0^2} \quad . \tag{5.210}$$

The y dependence of the entropy is thus given by

$$S(y) = S(0) - \frac{3k_B y^2}{2R_0^2} \quad . \tag{5.211}$$

Differentiating then leads to a simple result for the force:

$$f = \frac{3k_B T}{R_0^2} y = by \quad , \tag{5.212}$$

i.e., there is a linear force law, exactly as in the case of a spring with a spring constant b. One can recognise the parameters which affect the **rubber-elastic restoring force**, namely, as ever for forces of purely entropic nature, the absolute temperature and the mean squared end-to-end distance R_0^2.

It could now be asked as to why a coil, in which these entropically elastic forces are active, does not simply collapse, i.e., the chain ends are brought together with a zero separation, which is the state for which no forces act. That this is not the case is a consequence of the thermal energy. The mean squared displacement at thermal equilibrium, for an entropically elastic spring, is calculated using the following expression, which is valid for all elastic springs:

$$\frac{b}{2} \langle y^2 \rangle = \frac{k_B T}{2} \quad . \tag{5.213}$$

Upon inserting the spring constant, it is found that

$$\langle y^2 \rangle = \frac{R_0^2}{3} \quad , \tag{5.214}$$

i.e., a non-vanishing value corresponding exactly to that which is to be expected for an isotropic coil.

Rouse developed a model to describe the chain dynamics, which brings the frictional forces and the entropically elastic forces together in a suitable form. Figure 5.15 illustrates the representation of the polymer chain according to the **Rouse model**: It consists of a series of spheres, which interact with each other by spring-like forces. The spheres represent the targets for the frictional forces, while the springs carry the entropically elastic restoring forces. If a particular polymer chain is to be represented by the Rouse model, it is necessary to break it up into **Rouse sequences**. The length of the Rouse sequences is in the mesoscopic region, but it is not defined any more precisely. The sole condition is that the Rouse sequences are long enough that the entropically elastic forces, which arise upon extending the sequences, can be correctly described by the simple expression in Eq. (5.212). If a definite subdivision is chosen, the mean-squared end-to-end distance $\langle \Delta r^2 \rangle$ of each Rouse sequence is fixed; this is expressed by

$$\langle \Delta r^2 \rangle = a_\mathrm{R}^2 \quad . \tag{5.215}$$

The following force law applies for each spring in the Rouse model:

$$\mathbf{f} = b_\mathrm{R} \Delta \mathbf{r} \quad , \tag{5.216}$$

where the spring constant b_R is given by Eq. (5.212) as

$$b_\mathrm{R} = \frac{3k_\mathrm{B}T}{a_\mathrm{R}^2} \quad . \tag{5.217}$$

In order to describe the frictional forces, which act on the mobile spheres, an expression corresponding to Eq. (5.163) is chosen:

$$\mathbf{f} = \zeta_\mathrm{R} \mathbf{v} \quad , \tag{5.218}$$

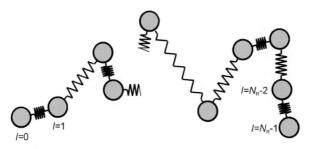

Fig. 5.15. A representation of a polymer chain according to the Rouse model: N_R inertia-free spheres are coupled by means of elastic springs.

where ζ_R is the coefficient of friction. Both parameters, ζ_R and b_R, depend on the length of the selected Rouse sequence.

In this way, all variables, which are required in order to formulate the equations of motion of the **Rouse chain**, have been introduced. Since, as mentioned above, it is not necessary to consider inertia effects, the equation of motion is simply given by

$$\zeta_R \frac{d\mathbf{r}_l}{dt} = b_R(\mathbf{r}_{l+1} - \mathbf{r}_l) + b_R(\mathbf{r}_{l-1} - \mathbf{r}_l) \ . \tag{5.219}$$

The coordinates $\mathbf{r}_l$ refer to the positions of the N_R spheres with indices $l = 0, \ldots, N_R - 1$. The equation states that the viscous force, which pulls a sphere through the melt, comes from the entropically elastic forces which act between neighbours in the chain.

It can be seen that the three spatial directions are separated from each other in the equation of motion. The z direction can thus be chosen as a generally representative example, and the associated dynamics analysed on the basis of the equation

$$\zeta_R \frac{dz_l}{dt} = b_R(z_{l+1} - z_l) + b_R(z_{l-1} - z_l) \ . \tag{5.220}$$

It is easy to find a suitable trial solution, namely:

$$z_l \propto \exp{-\frac{t}{\tau}} \cdot \exp(il\delta) \ . \tag{5.221}$$

It describes a simple relaxation process with a wave-like displacement pattern for the spheres, with δ denoting the phase difference between neighbours. Inserting Eq. (5.221) into Eq. (5.220) proves that the trial solution solves the equation and does it indeed exactly when the following relationship between the relaxation rate τ^{-1} and the phase difference δ applies:

$$\tau^{-1} = \frac{b_R}{\zeta_R}(2 - 2\cos\delta) = \frac{4b_R}{\zeta_R}\sin^2\frac{\delta}{2} \ . \tag{5.222}$$

Macromolecules always have a definite molecular weight and hence a finite total length. The ends are force-free, which leads to the following boundary condition:

$$z_1 - z_0 = z_{N_R-1} - z_{N_R-2} = 0 \ . \tag{5.223}$$

On account of the large number of monomer units in the macromolecule, it is possible to replace the discrete index l by a continuous variable. The boundary condition can then be written in the following form:

$$\frac{dz}{dl}(l = 0) = \frac{dz}{dl}(l = N_R - 1) = 0 \ . \tag{5.224}$$

Both the real and imaginary parts of the complex trial solution in Eq. (5.221) represent a solution:

$$z_l \propto \cos(l\delta) \cdot \exp-\frac{t}{\tau} \tag{5.225}$$

$$z_l \propto \sin(l\delta) \cdot \exp-\frac{t}{\tau} \ . \tag{5.226}$$

Only the former fulfils the boundary condition at the position $l = 0$. At the other end, the boundary condition is satisfied for this solution when

$$\frac{\mathrm{d}z_l}{\mathrm{d}l}(l = N_R - 1) \propto \sin((N_R - 1)\delta) = 0 \ . \tag{5.227}$$

This yields a series of N_R discrete values for the phase difference δ, namely:

$$\delta_p = \frac{\pi}{N_R - 1}p \ , \quad p = 0, 1, 2, \cdots, N_R - 1 \ , \tag{5.228}$$

which is in agreement with the number of spheres and thus the number of degrees of freedom (a motion of the whole chain without inner deformation is included for $p = 0$).

The longest relaxation time, τ_1, is found for the **Rouse fundamental mode** with the smallest phase difference, δ_1. It amounts to

$$\tau_1^{-1} \approx \frac{b_R}{\zeta_R} \frac{\pi^2}{(N_R - 1)^2} \ . \tag{5.229}$$

The associated displacement pattern is shown on the right-hand edge of Fig. 5.16. Starting from the collapsed rest state, in which all the spheres are at $z = 0$, the end points of the chain move to be far away from each other. The largest stretching is found in the centre at the unmovable centre of gravity. At the two free ends, the spheres stay together such that no forces

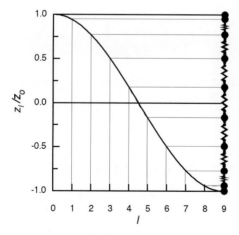

Fig. 5.16. The displacement pattern of the Rouse fundamental mode, starting from a collapsed rest state with all spheres at $z = 0$.

arise. It is to be noted that higher-order Rouse modes can exhibit force-free positions also in the centre.

The Rouse model is based on a linear equation of motion. Therefore, an arbitrary motional state of the chain can be described as a superposition of Rouse modes, which are the eigenmodes of the system. There are, corresponding to the number of degrees of freedom, N_R modes in each spatial direction, and thus $3N_R$ Rouse modes overall. Each mode executes a relaxator-like motion with a definite relaxation time, i.e., it is a straightforwardly describable process.

All Rouse-modes are thermally excited in a melt, whereby the amplitudes of the modes decrease with increasing order p. A short consideration shows how this decrease occurs. At first, it is necessary to consider how the Helmholtz free energy of the Rouse chain changes with the excitation amplitude for the Rouse mode with order p. The amplitude Z_p determines the displacements of all beads according to

$$z_l = Z_p \cos(l\delta_p) \ . \tag{5.230}$$

The increase in the Helmholtz free energy $\Delta\mathcal{F}$ associated with a displacement is given by

$$\begin{aligned}
\Delta\mathcal{F} &= \frac{b_R}{2} \sum_{l=0}^{N_R-2} (z_{l+1} - z_l)^2 \\
&= \frac{b_R}{2} Z_p^2 \sum_{l=0}^{N_R-2} (\cos[(l+1)\delta_p] - \cos[l\delta_p])^2 \\
&= \frac{b_R}{2} Z_p^2 \delta_p^2 \sum_{l=0}^{N_R-2} \sin^2(\delta_p l) = \frac{b_R}{2} Z_p^2 \delta_p^2 \frac{N_R-1}{2} \ .
\end{aligned} \tag{5.231}$$

A quadratic dependence is, thus, found as in the case of a simple spring:

$$\Delta\mathcal{F} \propto Z_p^2 \ . \tag{5.232}$$

Since the following must apply for the take-up of energy by a harmonic oscillator at thermal equilibrium:

$$\langle \Delta\mathcal{F} \rangle = \frac{k_B T}{2} \tag{5.233}$$

the mean quadratic amplitude $\langle Z_p^2 \rangle$ can be calculated. It is found that

$$\frac{b_R}{2} \cdot \frac{N_R-1}{2} \delta_p^2 \langle Z_p^2 \rangle = \frac{3k_B T}{2a_R^2} \cdot \frac{N_R-1}{2} \delta_p^2 \langle Z_p^2 \rangle = \frac{k_B T}{2} \tag{5.234}$$

such that

$$\langle Z_p^2 \rangle = \frac{2a_R^2}{3(N_R-1)\delta_p^2} \propto \frac{1}{p^2} \ , \tag{5.235}$$

independent of temperature. The rapid decrease in the amplitudes of the Rouse modes with increasing p means that it is the 3 Rouse fundamental modes which are the main contributions to the relaxator-like motion of the polymer chains in a melt. They are always thermally excited and relax with a time, which has, according to Eq. (5.229) i.e., it is a quadratic dependence on the number of Rouse sequences and, thus, a quadratic dependence on the molecular weight.

5.4.2 Entanglement Effects and the Tube Model

Although the Rouse model provides a straightforward way to describe the dynamics of macromolecules in the melt, its applicability is limited. If the molecular weight M of a polymer exceeds a critical value, entanglement effects set in, which hinder the free mobility assumed by the Rouse model. The encountered situation is shown schematically in Fig. 5.17. The interpenetration of the macromolecules in the melt means that entanglements form, which are more long-lived, the larger the molecular weight is. They create limitations to the mobility, which qualitatively change the dynamics of the molecule. The **tube model** which was introduced by Edwards allows the important features of the situation to be correctly described. As indicated in the figure, the tube is constructed from those loops which hinder the sideways motion of the selected chain. The loops limit the local lateral motion to distances of a few nanometres, and this value then corresponds to the diameter of the tube. While Rouse modes continue to exist, they are restricted to within the pipe, which means that they have a higher order p and correspondingly make smaller displacements and have shorter relaxation times. The long-wave Rouse modes are suppressed by the tube and are replaced by other motional forms.

What is the nature of this new motional form? It is similar to that of a reptile, which crawls through the tube, and is, therefore, following a sug-

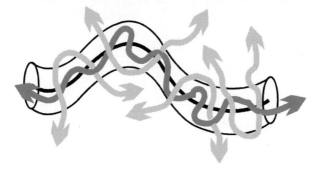

Fig. 5.17. The tube model of the motion of a polymer chain in an entangled melt. Fluctuations of the chain (*dark grey*) about the primitive path in the centre of the tube (*black*).

gestion from de Gennes, referred to as **reptation**. It is possible to analyse the kinetics of the reptation motion in the framework of the tube model, and thus, in a simple way, gain some fundamental insights. As well as the structural parameters of the individual chain inherent to the Rouse model, namely the number of sequences N_R and the squared end-to-end distance of each segment a_R^2, there are two further parameters for the entangled melt, which characterise the structure of the tube. One of them, namely the diameter of the tube, has already been introduced. In introducing the second parameter, it is, first of all, necessary to note that, while the tube strongly influences the dynamics of the chain, it leaves the spatial properties unchanged. The expression which is applicable for Gaussian chains continues to apply for the end-to-end separation:

$$R_0^2 = N_R a_R^2 = l_c a_R \quad . \tag{5.236}$$

In the second step, the number N_R of Rouse sequences is replaced by the contour length of the Rouse chain, which is given by

$$l_c = N_R a_R \quad . \tag{5.237}$$

Within the tube, the chain fluctuates about an average path which, as indicated in the diagram, runs through the centre of the tube. Edward introduced the term **primitive path** for it. The primitive path also links the end points of the chain like in a curve, which is Gaussian in its statistical character, but does it on a shorter way. Together with Eq. (5.236), the following expression thus also applies for the end-to-end distance of the chain:

$$R_0^2 = l_{pr} a_{pr} \quad . \tag{5.238}$$

The equation contains a new parameter, a_{pr} and it is evident that it characterises the stiffness of the tube with respect to bending; the larger a_{pr} is, the smaller the degree to which the tube is bent.

How can the motion in the tube be described? The reptation of a polymer chain is still diffusive, however, it is no longer in three dimensions, but one-dimensional and thereby 'curvilinear'. It can be described with the help of a curvilinear diffusion coefficient $\hat{D}$. The value of $\hat{D}$ is given by the Einstein relation, Eq. (5.171). Using a coefficient of friction instead of the mobility, it is written as

$$\hat{D} = \frac{k_B T}{\zeta_p} \quad . \tag{5.239}$$

The coefficient of friction ζ_p is that of all N_R Rouse spheres in the polymer chain together. Since all spheres move at the same time, it follows that ζ_p and the friction coefficient of one Rouse sphere, ζ_R, are related by

$$\zeta_p = N_R \zeta_R \quad . \tag{5.240}$$

It is now possible to estimate the time required for a chain to completely remove itself from the tube, which it occupied at time $t = 0$, and by which it was also stabilised. Figure 5.18 indicates how this process occurs. A piece of the tube is lost with each sliding out step, until, at the end, the chain finds itself in a completely new environment, i.e., in a different tube. Only the fundamental property of diffusive motions is used here, namely that the mean squared displacement, now of the centre of the chain, increases proportionally with time. Together with Eqs. (5.239) and (5.240) as well as the proportionality $l_{\mathrm{pr}} \propto N_{\mathrm{R}}$, this leads to

$$\tau_{\mathrm{d}} \simeq \zeta_{\mathrm{R}} N_{\mathrm{R}}^3 \ . \tag{5.241}$$

This is an important result, since it states that the disentangling time increases very rapidly with an increasing molecular weight, namely with M cubed. The necessary time τ_{d} is on the order of

$$\tau_{\mathrm{d}} \simeq \frac{l_{\mathrm{pr}}^2}{\hat{D}} \ . \tag{5.242}$$

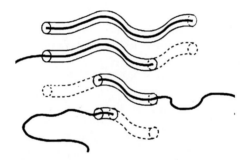

Fig. 5.18. Reptation model: Variation in time of the primitive path associated with a chain and decay of the initial tube.

A large number of experiments confirm that there is indeed such an increase of the disentangling time with the molecular weight. As a first example, Fig. 5.19 presents the results of investigations of the dielectric relaxation of polyisoprene samples with different molecular weights. As explained in Sect. 2.2.2, dipole reorientation times can be derived from frequency-dependent dielectric measurements. In this case, the time which is investigated is that required by a chain to completely change its overall orientation – this corresponds exactly to the disentangling time τ_{d}. It is evident that above a molecular weight $M = 10^4 \, \mathrm{g \, mol^{-1}}$, the time increases rapidly, with a power which is even somewhat greater than three. In the lower molecular weight region, where chain entanglements do not occur, a quadratic dependence is found. The reorientation occurs there by means of the 3 Rouse fundamental modes, whose molecular weight dependence follows a quadratic law – this can be seen from combining Eqs. (5.229), (5.217), (5.236), and (5.240):

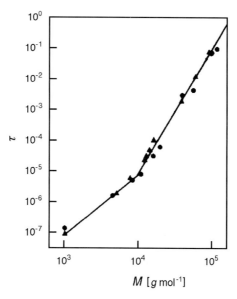

Fig. 5.19. The molecular weight dependence of the disentangling time of different polyisoprene samples, as derived from measurements of the dielectric relaxation time (from Boese and Krämer [46]).

$$\tau_1 \propto \frac{\zeta_R (N_R - 1)^2}{b_R} \propto a_R^2 \zeta_R (N_R - 1)^2$$

$$\approx R_0^2 \zeta_p \propto M^2 \quad . \tag{5.243}$$

A particularly sensitive test is provided by viscosity measurements. A macromolecule cannot flow away as long as it is held within a tube. The disentangling time τ_d, thus, also determines the flow velocity; the larger τ_d is, the larger the viscosity is. Figure 5.20 shows the dependence of the viscosity η on the molecular weight for a series of different polymers. An increase with a power of about three is also found here above the critical value of the molar mass, at which entanglement effects set in.

The reptation motion of macromolecules can even be directly observed in a microscope, provided that, first, the chain is very stiff and long, and, second, the tube diameter is large. Such conditions are met for concentrated solutions of biopolymers. An example is shown in Fig. 5.21, namely a DNA chain, which has been labelled with a fluorescent material, in a concentrated solution of the same type of unlabelled chains. The spherical end of the chain was pulled rapidly to the bottom at the beginning of the experiment using a pair of optical tweezers. It is evident that the so established entropically elastic force causes a diffusive motion of the chain along its own contour – exactly this represents the characteristics of reptation.

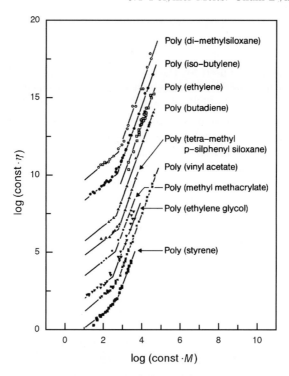

Fig. 5.20. The molecular weight dependence of the viscosity for different polymers. To allow a better comparison, the curves were shifted in both the horizontal and vertical directions (data from Berry and Fox [47]).

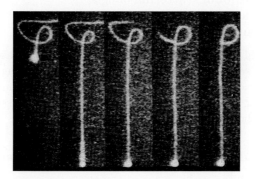

Fig. 5.21. Images of a DNA chain, which has been labelled with a fluorescent material, in a concentrated solution of the same type of unlabelled chains: The original form of the chain (*left*), the stretching out using optical tweezers, and the subsequent pulling through the tube (from Perkins et al. [48]).

5.5 Time-Resolved Scattering Experiments

In the last section of the first chapter, it was explained how and why scattering experiments can be used for the analysis of condensed matter structures. The structural information is contained in the interference of the scattered waves arising from the individual particles in the sample. The amplitude of the total scattered radiation, which results from the sum of all interfering scattered waves, is determined by the positions of the particles in the sample. If the particles move, and this is, with a few exceptions, generally the case, there is a corresponding variation in the scattering amplitude. In the previously described 'static scattering experiments', whose only aim was a structural analysis, there is, for the determination of the scattered intensity, an average taken over these fluctuations. It was shown that these experiments provide, correspondingly, information about the average structural properties, which can, for example, be expressed in the pair distribution function $g_2(r)$. If the experiment is now set up such that the time-dependent variations are registered, insight into the dynamics of the system is gained. This will be discussed in this section, and it will be shown that a space–time analysis of the structural-dynamic properties in the mesoscopic and microscopic region of the sample can be carried out in this way. At first, some general concepts which will be used in the evaluation of the experiments will be introduced. Subsequently, three examples of motions, which can be investigated by means of dynamic scattering experiments, will be considered:

- the diffusion of individual colloids in solution, as investigated by **dynamic light scattering**
- thermally excited sound and heat waves in liquids, to be analysed by **Rayleigh–Brillouin scattering** of light
- lattice vibrations, the dispersion curves of which can be determined by **inelastic neutron scattering experiments**

The evaluation in the two light-scattering experiments is based on a classical time-dependent scattering theory. Inelastic scattering experiments are best performed with neutrons, in which case quantum effects must be considered in the evaluation, with the corresponding scattering processes being treated as quantum-mechanical collision processes.

5.5.1 Time- and Frequency-Dependent Structure Factors

In the most common case of scattering at a one component system, which may be liquid or solid, it is possible to express the amplitude of the scattered radiation as a sum of phase factors, completely independent of whether electromagnetic waves or neutrons are used. Equation (1.98) represented this and formulates the relationship between the locations at which the particles are found. Let the motion of the particles now be explicitly included in the equation by writing

$$C(\boldsymbol{q}, t) = \sum_{j=1}^{\mathcal{N}} \exp[-\mathrm{i}\boldsymbol{q}\boldsymbol{r}_j(t)] \quad . \tag{5.244}$$

From the motion of the particles, there results a variation with time of the phase factor and consequently also of the scattering amplitude.

The sum can be replaced by an integral since a huge number of particles are involved in a scattering experiment. The integral notation links the scattering amplitude with the spatio-temporal fluctuations of the local particle density $\rho(\boldsymbol{r}, t)$, according to

$$C(\boldsymbol{q}, t) = \int_{V} \exp(-\mathrm{i}\boldsymbol{q}\boldsymbol{r})\rho(\boldsymbol{r}, t)\mathrm{d}^3\boldsymbol{r} \quad . \tag{5.245}$$

A subtraction of the average density, ρ, from the local density is achieved by the introduction of a second integral:

$$C(\boldsymbol{q}, t) = \int_{V} \exp(-\mathrm{i}\boldsymbol{q}\boldsymbol{r})(\rho(\boldsymbol{r}, t) - \rho)\mathrm{d}^3\boldsymbol{r} + \int_{V} \exp(-\mathrm{i}\boldsymbol{q}\boldsymbol{r})\rho\mathrm{d}^3\boldsymbol{r} \quad . \tag{5.246}$$

As was noted earlier, the second integral corresponds to a contribution, which is very near to the incident direction of the primary radiation; this is not detected and the term can thus be omitted. The remaining first integral shows that scattered radiation in the experimentally accessible region is only found if there are fluctuations in the particle density within the sample. For a homogeneous sample no scattering effects result.

The scattering amplitude $C(\boldsymbol{q}, t)$ is a time-dependent complex variable which consists of a modulus and a phase. While it is not possible to directly determine it experimentally, certain statistical properties can be measured. Of central importance in dynamic scattering experiments is the temporal auto-correlation function because it can be determined and evaluated. This function is denoted as $S(\boldsymbol{q}, t)$ and is defined as the average

$$S(\boldsymbol{q}, t) = \frac{1}{\mathcal{N}}\langle C(\boldsymbol{q}, t' + t)C^*(\boldsymbol{q}, t')\rangle \quad . \tag{5.247}$$

The averaging, as expressed by the triangular brackets, has the experimental meaning of an average value over all possible starting times t', for which the product of two scattering amplitudes, measured at two time points separated by t, is to be calculated. $S(\boldsymbol{q}, t)$ becomes, upon dividing by the total number of particles in the sample, $\mathcal{N}$, a volume-independent and precisely defined variable. It is referred to as the **time-dependent structure factor**, or, for reasons which will become clear later, the **intermediate scattering law**.

Two experiments allow the determination of $S(\boldsymbol{q}, t)$, first, dynamic light scattering performed with the help of digitally operating correlators, and second, neutron scattering experiments with a 'spin-echo spectrometer'. More precisely expressed, in the case of dynamic light scattering, the temporal auto-correlation function of the scattering intensity, which fluctuates with the amplitude, is calculated and $S(\boldsymbol{q}, t)$ is derived. This is in contrast to the neutron spin-echo spectrometer, which directly gives the intermediate scattering law.

If $C(\boldsymbol{q}, t)$ is again expressed by the sum in Eq. (5.244), a second general expression for $S(\boldsymbol{q}, t)$ is obtained:

$$S(\boldsymbol{q}, t) = \frac{1}{N} \sum_{j,k=1}^{N} \langle \exp[-i\boldsymbol{q}(\boldsymbol{r}_j(t+t') - \boldsymbol{r}_k(t'))] \rangle \ . \tag{5.248}$$

The choice of whether to use Eq. (5.248) or Eq. (5.247) together with Eq. (5.245) depends on the particular question under consideration.

In static scattering experiments, the pair distribution function $g_2(\boldsymbol{r})$ is derived from the structure factor by a Fourier analysis. Time-resolved scattering experiments investigate the **van Hove function** $g(\boldsymbol{r}, t)$, which is defined as follows:

$$g(\boldsymbol{r}, t)\mathrm{d}^3\boldsymbol{r}$$

denotes how many particles are found on average at a time t in a volume element $\mathrm{d}^3\boldsymbol{r}$ which is a distance $\boldsymbol{r}$ away from the origin at which a particle was found at $t = 0$. The van Hove function is composed of two basic parts:

$$g(\boldsymbol{r}, t) = g_1(\boldsymbol{r}, t) + g_2(\boldsymbol{r}, t) \ . \tag{5.249}$$

The first part, $g_1(\boldsymbol{r}, t)$, is associated with the probability that a particular particle moves by $\boldsymbol{r}$ in the time t, while the second part, $g_2(\boldsymbol{r}, t)$, is the temporal extension of the pair distribution function $g_2(\boldsymbol{r})$ used above and describes the space–time correlations between different particles. A short consideration reveals that Eq. (5.248) can, with the help of the van Hove function, be expressed as an integral with the following form:

$$S(\boldsymbol{q}, t) = \int_V \exp(-i\boldsymbol{q}\boldsymbol{r})(g(\boldsymbol{r}, t) - \rho)\mathrm{d}^3\boldsymbol{r} \ . \tag{5.250}$$

As in Eq. (5.246), the asymptotic value ρ attained by the van Hove function at both large distances $\boldsymbol{r}$ and large time intervals t is incorporated into the formulation of the integral.

Even though the intermediate scattering law can be determined by the two named techniques, i.e., the use of a digital correlator in light scattering and measurements with a neutron spin-echo spectrometer, dynamic scattering experiments are usually carried out differently. Instead of the determination of the temporal correlation function a frequency analysis of the scattered radiation is usually performed with the help of a suitable monochromator. The result of such a measurement is the **double differential scattering cross section**

$$\frac{\mathrm{d}^2\sigma}{\mathrm{d}\omega'\mathrm{d}\Omega}(\omega', \Omega) \ .$$

Ω denotes, as ever, the spatial direction of the detector, $\mathrm{d}\Omega$ is the spatial angle covered by the detector and $\mathrm{d}\omega'$ is the interval selected by the monochromator during the recording of the scattered radiation, with its centre at the frequency ω'.

What is the relationship between the double differential scattering cross section and the previously introduced temporal auto-correlation function of the scattering amplitude, i.e., between the results of frequency- and time-resolved experiments? In order to answer this question, it is necessary to incorporate the frequency of the incident primary radiation, ω_0, which has not been considered up until now. In the case of electromagnetic radiation, a time-dependent fluctuating field $E'(t)$ reaches the monochromator, where it is subject to a Fourier analysis:

$$E'(\omega') \propto \int_{t=-\infty}^{\infty} \exp(\mathrm{i}\omega't)E'(t)\mathrm{d}t \ . \tag{5.251}$$

The intensity determination at the detector behind the monochromator yields the **power spectrum** $\langle|E'(\omega')|^2\rangle$, such that the following applies for the double differential scattering cross section:

$$\frac{\mathrm{d}^2\sigma}{\mathrm{d}\omega'\mathrm{d}\Omega}(\omega', \Omega) \propto \langle|E'(\omega')|^2\rangle \ . \tag{5.252}$$

$\langle|E'(\omega')|^2\rangle$ is in a Fourier relationship with the time correlation function $\langle E'(t'+t)E'^*(t')\rangle$, since

$$\langle|E'(\omega')|^2\rangle \propto \int_{t'=-\infty}^{\infty}\int_{t''=-\infty}^{\infty} \exp(\mathrm{i}\omega't' - \mathrm{i}\omega't'')\langle E'(t')E'^*(t'')\rangle\mathrm{d}t'\mathrm{d}t''$$

$$\propto \int_{t=-\infty}^{\infty} \exp(\mathrm{i}\omega't)\langle E'(t'+t)E'^*(t')\rangle\mathrm{d}t \ . \tag{5.253}$$

In the above, it was solely assumed that $E'(t)$ is a statistically fluctuating, but stationary function, i.e., all mean values are time independent, and thus, in particular, $\langle E'(t'+t)E'^*(t')\rangle$ is independent of t'. Equation (5.253) is often referred to as the **Wiener–Khinchin theorem**. If $\langle E'(t'+t)E'^*(t')\rangle$ is linked to $S(\boldsymbol{q}, t)$ according to

$$\langle E'(t'+t)E'^*(t')\rangle \propto \langle C(\boldsymbol{q}, t'+t)\exp[-\mathrm{i}\omega_0(t'+t)]C^*(\boldsymbol{q}, t')\exp(\mathrm{i}\omega_0 t')\rangle$$

$$\propto \exp(-\mathrm{i}\omega_0 t)S(\boldsymbol{q}, t) \ , \tag{5.254}$$

the following is obtained from Eqs. (5.252) and (5.253) employing a basic property of convolutions

$$\frac{d^2\sigma}{d\omega' d\Omega} \propto \int_{t=-\infty}^{\infty} \exp(i\omega' t) \, \exp(-i\omega_0 t) \, dt \otimes \int_{t=-\infty}^{\infty} \exp(i\omega' t) \, S(\boldsymbol{q}, t) dt$$

$$\propto \delta(\omega' - \omega_0) \otimes S(\boldsymbol{q}, \omega') \quad . \tag{5.255}$$

In the above, the frequency-dependent **dynamic structure factor** $S(\boldsymbol{q}, \omega)$ has been introduced according to

$$S(\boldsymbol{q}, \omega) = \int_{t=-\infty}^{\infty} \exp(i\omega t) S(\boldsymbol{q}, t) dt \quad . \tag{5.256}$$

With this result it has been found that:

- Frequency-resolved scattering experiments yield the dynamic structure factor $S(\boldsymbol{q}, \omega)$, where the frequency ω denotes the experimental deviation from the frequency ω_0 of the primary irradiation.
- There is a Fourier relationship between the results of a time- and a frequency-dependent measurement, namely the scattering functions $S(\boldsymbol{q}, t)$ and $S(\boldsymbol{q}, \omega)$ are related by Eq. (5.256) and the reciprocal relation:

$$S(\boldsymbol{q}, t) = \frac{1}{2\pi} \int_{\omega=-\infty}^{\infty} \exp(-i\omega t) S(\boldsymbol{q}, \omega) d\omega \quad . \tag{5.257}$$

Combining Eqs. (5.256) and (5.250) yields a general relationship between the dynamic structure factor and the van Hove function:

$$S(\boldsymbol{q}, \omega) = \int_V \int_{t=-\infty}^{\infty} \exp(-i\boldsymbol{q}\boldsymbol{r} + i\omega t)(g(\boldsymbol{r}, t) - \rho) d^3\boldsymbol{r} dt \quad . \tag{5.258}$$

It states that a space–time Fourier analysis of the van Hove function is achieved by performing a frequency-dependent scattering experiment. If only the spatial Fourier transformation and not that with respect to time is performed, 'the intermediate scattering law' corresponding to Eq. (5.250) is obtained.

It is useful to make clear the relationship between the results of a time-resolved or a frequency-dependent scattering experiment and that of a static experiment on the same sample. It is obvious that

$$S(\boldsymbol{q}) = S(\boldsymbol{q}, t = 0) \tag{5.259}$$

and

$$S(\boldsymbol{q}) = \frac{1}{2\pi} \int\limits_{\omega=-\infty}^{\infty} S(\boldsymbol{q},\omega)\mathrm{d}\omega \quad . \tag{5.260}$$

The latter equation states that a static experiment proceeds without an analysis of the changed frequency distribution of the radiation which arises generally through the scattering process.

5.5.2 Dynamic Light Scattering in Liquids

Colloidal Diffusion. Figure 5.22 shows the result of a dynamic light scattering experiment performed on a dilute solution of polystyrene in toluene. The presented data were obtained after a subtraction of the scattering due to pure toluene from that observed for the solution, such that they represent only the dynamic scattering effects due to the polymer molecules. It was stated

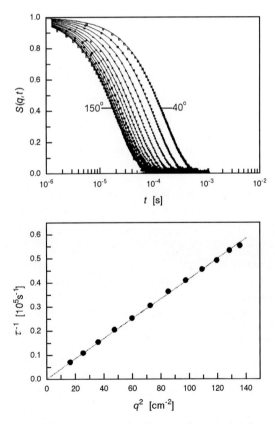

Fig. 5.22. Dynamic light scattering of a dilute solution of polystyrene in toluene. Time-dependent scattering function for different scattering vectors (*above*). The dependence of the decay constant on the scattering vector (*below*).

above that the dynamic light scattering experiment yields the intermediate scattering law $S(q, t)$. The upper part of the figure shows the time-dependence of $S(q, t)$ as measured for different scattering angles. Each of the curves can be described by a simple exponential function, with it being apparent that the characteristic time τ becomes ever smaller as the scattering angle increases. In the lower part of the figure, it is seen that a square law $\tau^{-1} \propto q^2$ is obeyed.

The cause and significance of this result are readily understood. A polymer molecule in dilute solution exhibits a diffusive motion which is independent from that of the other polymer molecules. The time-dependent auto-correlation function $g_1(r, t)$ which was introduced in Sect. 5.2 can be used to describe this motion. According to Eq. (5.249), $g_1(r, t)$ is a part of the van Hove function, indeed, in the absence of pair-correlation functions, it is the only contribution. $S(q, t)$ can be calculated using Eq. (5.250), making use of the calculated expression (Eq. (5.157)) for the auto-correlation function

$$g_1(r, t) = \frac{1}{(4\pi D_s t)^{3/2}} \exp -\frac{|r|^2}{4 D_s t} \ .$$

Upon realising that the pair correlation term g_2 is given by

$$g_2 = \rho \tag{5.261}$$

in the case of a dilute solution in the absence of interactions between the polymers, it follows that

$$S(q, t) = \int_V \exp(-iqr)g_1(r, t)\mathrm{d}^3 r = \exp(-D_s q^2 t) \ . \tag{5.262}$$

This corresponds exactly to the observed experimental behaviour, $\tau^{-1} \propto q^2$. Moreover, the theoretical analysis has revealed that the self-diffusion coefficient D_s can be determined from the slope of the line in the figure.

Density Fluctuations. Figure 5.23 shows, for the example of CCl_4, the results of a frequency-resolved 'Rayleigh–Brillouin scattering experiment' on a liquid. The liquid, which was contained in a round cuvette, was irradiated by a laser with a frequency ω_0 in the visible region. The scattered light was recorded for different scattering angles 2θ and subjected to a frequency analysis using a Fabry–Pérot monochromator. Three spectra are shown; in each case, a central **Rayleigh** line, with its maximum at the excitation frequency, is observed together with a **Brillouin** doublet, for which the frequency shift increases with increasing scattering angle.

How are these spectra to be understood? In a scattering experiment density fluctuations are detected, as was explained above and expressed by Eq. (5.246). Scattering experiments break up the density fluctuations into wave-like components and select a particular wavevector by means of the scattering angle. Correspondingly, the spectra in the figure represent the dynamics

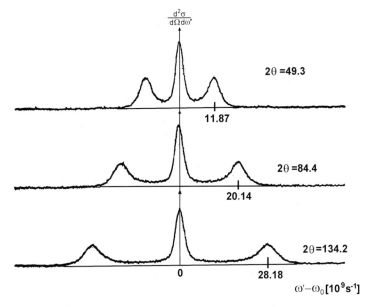

Fig. 5.23. Rayleigh–Brillouin spectra of CCl₄ at room temperature, as measured for different scattering angles 2θ.

of particular density waves. For liquids and for the considered frequency region, it is evident that there are then two different types. It is not difficult to recognise their particular characters. The Brillouin doublet is associated with sound waves – in every liquid, they are thermally excited with all possible wavelengths, with the associated frequencies being determined by the speed of sound. The density fluctuations which give rise to the central Rayleigh line are of a different nature. They do not cause a frequency shift, and, thus, in contrast to the sound waves, do not propagate themselves. In this case, fluctuations in the entropy density are involved and their kinetics relates to the heat diffusion. It is possible to observe these fluctuations because of the thermal expansion which leads to corresponding density fluctuations.

The combined contribution of **sound waves** and **heat waves** lead to a mean squared density fluctuation given by

$$\langle \Delta\rho^2 \rangle = \left(\frac{\partial\rho}{\partial p}\right)_s^2 \langle \Delta p^2 \rangle_s + \left(\frac{\partial\rho}{\partial s}\right)_p^2 \langle \Delta s^2 \rangle_p \ . \tag{5.263}$$

The pressure fluctuations due to the sound waves occur adiabatically with an amplitude $\langle \Delta p^2 \rangle_s$, while the isobaric entropy fluctuations have an amplitude $\langle \Delta s^2 \rangle_p$. Corresponding to the two components, the dynamic structural factor is also composed of two contributions:

$$S(\boldsymbol{q},\omega) = S_p(\boldsymbol{q},\omega) + S_s(\boldsymbol{q},\omega) \ . \tag{5.264}$$

The two parts will be discussed one after the other.

The pressure fluctuations move as sound waves

$$\Delta p(\boldsymbol{r}, t) = \Delta p_0(t) \exp(\pm i\omega_p t + i\boldsymbol{k}_p \boldsymbol{r}) \tag{5.265}$$

through the liquid, whereby the following dispersion relation applies:

$$\omega_p(k_p) = c_s k_p \quad . \tag{5.266}$$

c_s is the speed of sound. The experiment detects, by means of the choice of the scattering vector, a particular sound wave, namely that with the wavevector

$$\boldsymbol{k}_p = \boldsymbol{q} \quad . \tag{5.267}$$

A measurement of the time-dependent structure factor, for a scattering vector $\boldsymbol{q}$, would give the result:

$$S_p(\boldsymbol{q}, t) \propto (\exp[i\omega_p(q)t] + \exp[-i\omega_p(q)t] \langle \Delta p_0(t + t')\Delta p_0^*(t') \rangle \quad . \tag{5.268}$$

The amplitude of the sound wave, Δp_0, is thermally excited again and again, each time being followed by a decay because of scattering processes. It is therefore to be expected that the following applies for the correlation function of the amplitude of the sound wave:

$$\langle \Delta p_0(t + t')\Delta p_0^*(t') \rangle \propto \exp -\frac{|t|}{\tau_p(q)} \quad . \tag{5.269}$$

The time constant τ_p describes the lifetime of the sound wave, i.e., the mean time for which its propagation is unperturbed. It can be assumed that the lifetime decreases upon decreasing the wavelength, i.e., with increasing q. A Fourier transformation achieves the transition from the time- to the frequency-dependent structure factor:

$$S_p(\boldsymbol{q}, \omega) \propto \mathrm{Ftr}(\exp[i\omega_p(q)t] + \exp[-i\omega_p(q)t]) \otimes \mathrm{Ftr} \exp -\frac{|t|}{\tau_p(q)} \quad . \tag{5.270}$$

The second term in the convolution is given by

$$\mathrm{Ftr} \exp -\frac{|t|}{\tau_p(q)} = \int_0^\infty \exp(i\omega t) \exp -\frac{t}{\tau_p(q)} dt + \int_0^\infty \exp(-i\omega t) \exp -\frac{t}{\tau_p(q)} dt$$

$$= \frac{1}{-i\omega + \tau_p^{-1}(q)} + \frac{1}{i\omega + \tau_p^{-1}(q)} = \frac{2\tau_p^{-1}(q)}{\omega^2 + \tau_p^{-2}(q)} \quad . \tag{5.271}$$

Overall, it thus follows that

$$S_p(\boldsymbol{q}, \omega) \propto [\delta(\omega + \omega_p(q)) + \delta(\omega - \omega_p(q))] \otimes \frac{2\tau_p^{-1}(q)}{\omega^2 + \tau_p^{-2}(q)}$$

$$= \frac{2\tau_p^{-1}(q)}{(\omega - \omega_p(q))^2 + \tau_p^{-2}(q)} + \frac{2\tau_p^{-1}(q)}{(\omega + \omega_p(q))^2 + \tau_p^{-2}(q)} \quad . \tag{5.272}$$

Exactly this is the description of a Brillouin doublet, with the frequency shift corresponding to the frequency of the selected sound wave, while the linewidth is associated with the lifetime of the wave.

For the analysis of the Rayleigh line, it is to be noted, first of all, that the entropy fluctuation is proportional to the temperature fluctuation:

$$\Delta s = \frac{c_p}{T} \Delta T \ , \tag{5.273}$$

where c_p denotes the heat capacity at constant pressure

$$c_p = \frac{1}{\mathcal{V}} \frac{\partial \mathcal{H}}{\partial T} \ . \tag{5.274}$$

It is, thus, possible to discuss temperature fluctuations $\Delta T(\boldsymbol{r}, t)$ instead of entropy fluctuations. Their dynamics are described by the differential equation

$$\frac{\partial \Delta T}{\partial t} = D_T \nabla^2 \Delta T \ . \tag{5.275}$$

D_T is the **thermal diffusion coefficient** which depends on c_p and the heat conductivity λ_Q:

$$D_T = \frac{\lambda_Q}{c_p} \ . \tag{5.276}$$

In order to analyse the scattering experiment, a description of the dynamics of the temperature waves is required. Choosing a corresponding Fourier representation, we write

$$\frac{\partial}{\partial t} \frac{1}{(2\pi)^3} \int \Delta T(\boldsymbol{k}, t) \exp(\mathrm{i}\boldsymbol{k}\boldsymbol{r}) \, \mathrm{d}^3\boldsymbol{k} = D_T \nabla^2 \frac{1}{(2\pi)^3} \int \Delta T(\boldsymbol{k}, t) \exp(\mathrm{i}\boldsymbol{k}\boldsymbol{r}) \, \mathrm{d}^3\boldsymbol{k}$$

$$= -D_T \frac{1}{(2\pi)^3} \int k^2 \Delta T(\boldsymbol{k}, t) \exp(\mathrm{i}\boldsymbol{k}\boldsymbol{r}) \, \mathrm{d}^3\boldsymbol{k}$$

which leads to a differential equation for the amplitude of the heat wave with a wavevector $\boldsymbol{k}$:

$$\frac{\partial}{\partial t} \Delta T(\boldsymbol{k}, t) = -D_T k^2 \Delta T(\boldsymbol{k}, t) \ . \tag{5.277}$$

The solution is

$$\Delta T(\boldsymbol{k}, t) \propto \exp -\frac{t}{\tau_s(k)} \ , \tag{5.278}$$

with

$$\tau_s = \frac{1}{D_T k^2} \ . \tag{5.279}$$

Heat waves, therefore, possess a relaxator-like dynamics, with a relaxation time τ_s, which is determined by the thermal diffusion coefficient. The temporal correlation function of the amplitude of the temperature waves, which are being spontaneously excited in thermal equilibrium, is thus given by

$$\langle \Delta T(\boldsymbol{k}, t' + t) \Delta T(\boldsymbol{k}, t') \rangle \propto \exp - \frac{|t|}{\tau_s(k)} \quad . \tag{5.280}$$

The light scattering experiment selects the wavevector of the thermal wave according to

$$\boldsymbol{k} = \boldsymbol{q} \quad ,$$

such that the temperature wave part of the dynamic structure factor S_s is given by Fourier transformation to be

$$S_s(\boldsymbol{q}, \omega) \propto \frac{2\tau_s^{-1}(q)}{\omega^2 + \tau_s^{-2}(q)} \quad . \tag{5.281}$$

The result, which accounts for the Rayleigh line, shows that the central line is broadened as compared to the laser light, with the broadening being determined by τ_s. The information content is in the linewidth, from which the thermal diffusion coefficient can be extracted.

Rayleigh–Brillouin spectra can be obtained for crystals as well as for liquids. Figure 5.24 shows, as an example, a spectrum recorded for quartz. In contrast to the case of the liquid, three pairs of frequency-shifted lines are now observed. Their assignment is evident: In crystals, there are three different acoustic oscillations – in quartz, one is longitudinally polarised and two are transversely polarised. The question then arises as to why the transverse acoustic vibrations are visible, since these are due to shear deformations and, therefore, do not generate density fluctuations. In fact, light reacts, upon scattering, to all the fluctuations of the optical index ellipsoid, and, in the general

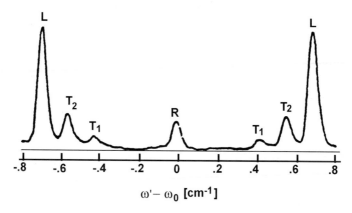

Fig. 5.24. The Rayleigh–Brillouin spectrum of a quartz crystal with 3 bands, which are assigned to longitudinal and transverse acoustic vibrations (from Shapiro et al. [49]).

case, shear deformations change its direction. This effect, which only appears in scattering experiments with light, is not included in the equations given in this section.

5.5.3 Inelastic Neutron Scattering in Crystals

Up to this point, scattering experiments have only been described classically. This was the case for the introduction of the fundamental scattering functions $S(q, t)$ and $S(q, \omega)$, as well as for the examples of the diffusion of colloids and density fluctuations in liquids. The prerequisites for a classical treatment were met in both cases. The changes in the frequency of light caused by the scattering processes are negligibly small in comparison to the initial frequency, and sound waves are populated by a very large number of phonons at room temperature; quantum effects are not observed under these conditions. A completely different situation, however, is encountered in the investigation of acoustic lattice vibrations with wavevectors, which are far away from the origin or belong to the optical branches. Quantum effects are then definitely not negligible and must be taken account of.

Scattering experiments with neutrons are very well suited to the analysis of the vibrational behaviour of crystals. The frequency of the neutrons can be chosen such that it is on the order of that of the lattice vibrations, in which case there is a high and easily measurable energy transfer.

The discussion of scattering experiments, which are inherently quantum-mechanical, requires an approach, which is different to the classical picture of a generation of scattered waves. It is here appropriate to change to a **particle picture** and to describe the scattering process as a quantum-mechanical collision with an associated creation and annihilation of phonons. This is represented schematically in Fig. 5.25. A certain initial state is transformed into an end state by the interaction process. The initial state depends on the wavevector k_n of the neutron and the phonon occupation numbers n_{ij} of all lattice vibrations of the body at which the neutrons are scattered. Both are changed by the interaction process. In general, after a scattering event, there is a different neutron wavevector, k'_n, and different occupation numbers n'_{ij} for the lattice vibrations. To be in the particle picture means that, firstly, for the neutron wavevector k_n, the associated momentum

$$p_n = \hbar k_n \tag{5.282}$$

is considered, and, secondly, for the frequency of the neutron wave, the kinetic energy of the neutron

$$\epsilon_n = \hbar \omega_n \tag{5.283}$$

is introduced in the treatment. There is the following relationship between ω_n and the modulus of the wavevector k_n, and correspondingly between ϵ_n and p_n:

$$\hbar \omega_n = \epsilon_n = \frac{p_n^2}{2m_n} = \frac{\hbar^2 k_n^2}{2m_n} \ , \tag{5.284}$$

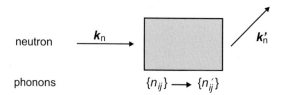

Fig. 5.25. The description of the scattering of neutrons at a crystal as a quantum-mechanical collision process.

where m_n is the mass of the neutron.

The main effect of the interaction of the neutron with the body at which scattering occurs is the creation or the annihilation of phonons, predominantly in the form of a one-phonon process. The total momentum and the total energy must be conserved. If a phonon with a wavevector $\boldsymbol{k}_i$ and a frequency $\omega_j(\boldsymbol{k}_i)$ is annihilated, it follows that

$$\hbar\boldsymbol{k}_n + \hbar\boldsymbol{k}_i = \hbar\boldsymbol{k}'_n \tag{5.285}$$

$$\hbar\omega_n + \hbar\omega_j = \hbar\omega'_n \ . \tag{5.286}$$

In an inelastic process, the scattering vector $\boldsymbol{q}$ can still be used as the central variable in the geometrical description of the scattering event, with the same definition as for elastic scattering:

$$\boldsymbol{k}'_n - \boldsymbol{k}_n = \boldsymbol{q} \ . \tag{5.287}$$

In the general case, it has the meaning of a momentum transfer – Eq. (5.285) can be expressed as

$$\boldsymbol{q} = \boldsymbol{k}_i \ . \tag{5.288}$$

If a collision leads to the creation of a phonon, the requirement for the conservation of momentum and energy are expressed by

$$\hbar\boldsymbol{k}_n = \hbar\boldsymbol{k}'_n + \hbar\boldsymbol{k}_i \tag{5.289}$$

$$\hbar\omega_n = \hbar\omega'_n + \hbar\omega_j \ . \tag{5.290}$$

In contrast to the case of a phonon annihilation, it is now found that

$$\boldsymbol{q} = -\boldsymbol{k}_i \ , \tag{5.291}$$

i.e., the scattering vector corresponds to the negative wavevector of the created phonon. Scattering processes which lead to the creation of phonons are referred to as **Stokes processes**, while those in which a phonon disappears are called **anti-Stokes processes**.

The question arises as to whether it is possible to have a change in the momentum of a neutron without the creation or annihilation of a phonon, i.e., without a simultaneous change in the energy. This would correspond to an **elastic scattering process**, i.e., a process which only changes the direction of the wavevector and leaves the frequency unchanged. In fact, this does occur, with the requirement being that

$$k'_n - k_n = G_{hkl} \quad , \tag{5.292}$$

i.e., the scattering vector must correspond to a vector of the reciprocal lattice. This is exactly the condition, formulated by the Laue equation, Eq. (1.132), for which a Bragg reflex is observed in a crystal. That it is actually so that the Bragg reflections occur in a purely elastic fashion becomes clear when, recalling the representation of the lattice vibrations in the first Brillouin zone in the reduced scheme, it is noted that all wavevectors which agree with vectors of the reciprocal lattice while initially lying outside coincide with $k = 0$. At this point, however, a vanishing frequency is found for the three acoustic branches:

$$\omega_j = 0 \quad . \tag{5.293}$$

One might say that here the neutron collides with the whole crystal, i.e., a macroscopic object.

Inelastic scattering experiments with neutrons can be performed in different ways. In the simplest case, a fixed scattering direction Ω is chosen and the frequency distribution of the scattered neutrons is analysed using a monochromator. Figure 5.26 shows a typical result from such an experiment, carried out here on a sample of germanium. A spectrum with a series of bands is

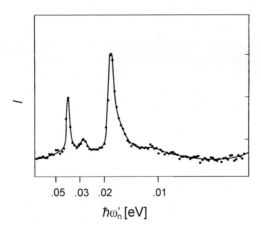

$$\hbar\omega'_n \, [\text{eV}]$$

Fig. 5.26. Inelastic neutron scattering by germanium. The results of a measurement of the scattering intensity in a fixed direction obtained by varying the setting of the monochromator (from Pelah [50]).

observed. A band appears exactly when both energy and momentum are conserved. How this is achieved and how such a spectrum can be analysed is explained in more detail in Fig. 5.27. Along the arrow in the Ω direction, there is a change, which is given by Eq. (5.284), in the momentum k'_n and the energy ω'_n of the scattered neutron. The energy is reduced and amplified in the Stokes and the anti-Stokes region, respectively. Each point fixes the momentum of a potentially created or annihilated phonon. A band in the spectrum appears exactly when the associated frequency change of a neutron corresponds to a phonon frequency in any branch. Therefore, each band can be assigned, in an unambiguous fashion, to the wavevector and frequency of a phonon. As was mentioned above, elastic scattering is only found for the exceptional case where the Bragg condition $q = G_{hkl}$ is fulfilled, as is also shown in the figure.

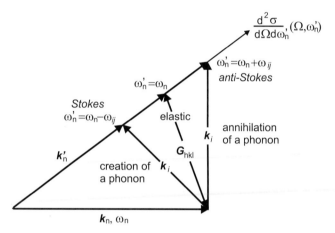

Fig. 5.27. The kinematics of the scattering process for a constant scattering direction. The neutrons scattered in the Ω direction are subject to a frequency analysis. The presence of a band at a frequency ω'_n fixes the wavevector of the scattered neutrons as well as the wavevector and the frequency of the created (for $\omega'_n < \omega_n$) or annihilated (for $\omega'_n > \omega_n$) phonon. Examples for this are given. In addition, the case of elastic scattering at the Bragg reflex G_{hkl} is shown.

In fact, experiments are not normally carried out in this way. In order to read off a dispersion relation equidistantly, it is necessary to fix the scattering vector and then change it in equal steps. Figure 5.28 illustrates how such a measurement 'with constant q' can be carried out for a given energy of the incident neutrons. All the shown pairs of wavevectors of incident (k_n) and scattered (k'_n) neutrons are linked by the same scattering vector. For each pair, there is a particular arrangement of the crystal relative to the incident neutron beam, a particular direction in which the scattered radiation is measured, and a particular aperture frequency of the monochromator. A suitable

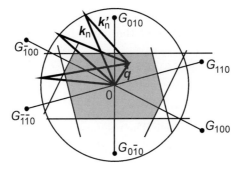

Fig. 5.28. The measurement for a fixed scattering vector q and a fixed frequency ω_n of the incident neutrons, upon varying the energy transfer. This can be achieved by a rotation of the crystal about an axis perpendicular to q together with an associated change in the arrangement of the detector and the monochromator.

control mechanism allows such a sequence to be exactly realised. The dispersion relations of aluminium, which are shown in Fig. 5.29, were obtained in this way. They correspond to the lattice vibrations for wavevectors along two selected directions in k space, as is indicated in the centre of the figure, together with the boundaries of the first Brillouin zone.

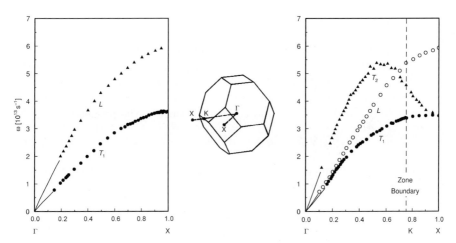

Fig. 5.29. Phonon dispersion relations of aluminium, obtained by inelastic neutron scattering along the [100] (*left*) and [110] (*right*) direction. k space with the first Brillouin zone of the body-centred cubic lattice (*centre*); the two directions as well as the special points Γ (the origin), K and X are indicated. The wavenumbers along the two selected directions are given as relative values. The two transverse acoustic branches are degenerate along ΓX for symmetry reasons, while they are separated along $\Gamma K X$ (from Yarnell in [51]).

Each individual measurement, carried out in a particular direction for a particular aperture frequency of the monochromator, selects a scattering vector

$$q = k'_n - k_n$$

and a frequency

$$\omega = \omega'_n - \omega_n \ , \tag{5.294}$$

with which the system is probed. The response is described, as in the classical case, by the dynamic structure factor $S(q, \omega)$. The above discussion showed how this can be calculated for classical systems, namely by the spatio-temporal Fourier transformation of the van-Hove function. The theoretical approach and the meaning of $S(q, \omega)$ changes for the case of a quantum-mechanical collision process. The dynamic structure factor is now associated with quantum-mechanical transition frequencies, in particular with the probability of those processes with a change in the momentum q and an energy change ω. Transition probabilities of quantum-mechanical collision processes are described, in the general case, by **Fermi's golden rule**. In order to keep the notation simple, a particular lattice vibration ij (wavevector k_i, jth branch) in a well-defined quantum state, i.e., populated with n_{ij} phonons, is selected. We then ask the following question: Does the annihilation of a phonon result in a contribution to the dynamic structure factor? The application of Fermi's golden rule here gives

$$S(q, \omega) \propto | \langle k_n, n_{ij} | V | k_n + q, n_{ij} - 1 \rangle |^2 \delta(\omega - \omega_j) \ . \tag{5.295}$$

The equation contains the square of the magnitude of the matrix element describing a change, caused by the interaction potential V, in the state of the body undergoing scattering with the neutron. The δ function is an expression of the conservation of energy. The conservation of momentum is contained in the transition matrix element and is a consequence of the property

$$\langle k_n, n_{ij} | V | k_n + q, n_{ij} - 1 \rangle \propto \delta(q - k_i + G_{hkl}) \ , \tag{5.296}$$

which is due to symmetry (its derivation cannot be presented here). In the case of the creation of a photon we have correspondingly

$$S(q, \omega) \propto | \langle k_n, n_{ij} | V | k_n + q, n_{ij} + 1 \rangle |^2 \delta(\omega + \omega_j) \tag{5.297}$$

and

$$\langle k_n, n_{ij} | V | k_n + q, n_{ij} + 1 \rangle \propto \delta(q + k_i + G_{hkl}) \ . \tag{5.298}$$

As stated above, an individual lattice vibration with a particular phonon occupation number was, in the first instance, selected here. The full dynamic structure factor is given by a summation, taking into account the appropriate weighting factors, over all occupation numbers and all lattice vibrations. This will not be done here, instead only one important property will be highlighted: While the observation of equal intensity for the two components of the Brillouin doublet in Figs. 5.23 and 5.24 is typical for the classical case, the situation changes for an inelastic neutron scattering experiment, with the band on the anti-Stokes side usually being clearly weaker than that on the Stokes side. The intensity ratio is determined, in the general case, by the Boltzmann expression, as

$$\frac{I_{AS}}{I_S} = \exp -\frac{\hbar\omega_j}{k_B T} \ .$$
(5.299)

For optical phonons with high frequencies, it describes the probability that a phonon which can be annihilated is present in the sample at the time of the collision.

5.6 Exercises

1. Consider the vibrational modes of a linear chain of atoms of mass m, which are connected to each other at a distance a by springs with a spring constant b. Let the interaction to the next-but-one neighbours be described by an additional spring with a spring constant b'.

   (a) Derive the dispersion relation.
   (b) Plot the dispersion curve for $b' = b/2$.

2. Thermal expansion can be understood in a straightforward way by considering a classical oscillator with anharmonic terms in the potential $u(x)$. Calculate the average expansion making use of the Boltzmann distribution:

$$\langle x \rangle = \frac{\int\limits_{-\infty}^{\infty} x \ \exp -\frac{u(x)}{k_B T} \ \mathrm{d}x}{\int\limits_{-\infty}^{\infty} \exp -\frac{u(x)}{k_B T} \ \mathrm{d}x} \qquad \text{with} \qquad u(x) = c_2 x^2 - c_3 x^3 - c_4 x^4 \ .$$

   Express the exponential function as a series expansion with respect to the anharmonic part and calculate an approximate value of $\langle x \rangle$ for the case where the anharmonic term is small as compared to $k_B T$.
   Hint: The above expression for $u(x)$ correctly describes the local potential only. Neglect the anharmonic part of the potential in the denominator of the above expression, and use

$$\int\limits_{-\infty}^{\infty} \exp(-\alpha x^2) \ \mathrm{d}x = \left(\frac{\pi}{\alpha}\right)^{1/2} \ .$$

3. Calculate the heat capacity $c(T) = \partial \mathcal{U}/\partial T$ of a system of $\mathcal{N}$ harmonic oscillators with the same eigenfrequency ω. In addition, determine the limiting behaviour of $c(T)$ for low temperatures $T \ll \hbar\omega/k_B$ and compare this to the corresponding behaviour predicted by the Debye theory of specific heats.

4. At which temperature T_0 does the heat capacity of the free electrons in a metal become larger than that of the lattice vibrations? Express this temperature as a function of the Debye temperature T_D and the electron concentration. Calculate T_0 for copper.

5. An electric field of strength $E = 750\,\mathrm{Vm^{-1}}$ is applied on colloids in a liquid ($T = 20\,^\circ\mathrm{C}$) which have a self-diffusion coefficient $D_s = 8 \times 10^{-11}\,\mathrm{m^2 s^{-1}}$ and an elementary electric charge $+e$.

   (a) With what velocity do the colloids move in the direction of the cathode?

   (b) The viscosity of the liquid is $\eta = 10^{-3}\,\mathrm{Pa\,s}$. How large is the colloid radius?

6. A colloidal solution of particles with mass density ρ_m in a liquid with mass density $\rho_{m,f} < \rho_m$ is subjected to a high centrifugal acceleration in a centrifuge with a rotation at an angular velocity w. The particles have the form of spheres with radius a. The solution is in a volume with a maximum distance r_{max} from the rotation axis. What is the resulting radial distribution of the particle density $\rho(r)$ in the centrifuge?

7. The upper end of a polymer molecule in solution with $R_0 = 10\,\mathrm{nm}$ is fixed at $z = 0$. A gold particle ($\rho_{m,Au} \approx 19\,\mathrm{g\,cm^{-3}}$) with a radius of $1\,\mathrm{\mu m}$ hangs on the lower end. What is the average value of z ($T = 300\,\mathrm{K}$, neglect the buoyancy force)? What happens when the temperature is increased?

8. Show that the conservation of energy and momentum means that neutrons with very low incident energies ($\hbar w_n \to 0, \hbar k_n \to 0$) can absorb photons from all branches of the lattice vibration spectrum (anti-Stokes lines).

Appendix

A

Thermodynamic Potentials

Thermodynamics provides equations for different thermodynamic potentials which are of use for the description of the properties of condensed matter. They are often applied in this book, and are summarised here.

Since many processes of interest involve the effect of the action of an external field, be it mechanical, electric, or magnetic, it is necessary to have the thermodynamic potentials in a form which contains the field variables. It is well known that certain variables are always associated with a particular thermodynamic potential. Indeed, the following variables must be considered:

$$s = \frac{S}{\mathcal{V}} , \; \frac{\Delta \mathcal{V}}{\mathcal{V}_0} , \; e(e_{zz} \text{ or } e_{zx}) , \; P , \; D , \; M , \; B$$

and

$$T , \; p , \; \sigma(\sigma_{zz} \text{ or } \sigma_{zx}) , \; E , \; H .$$

The variables in the second group are, without exception, intensive variables, while the first group contains, with the exception of D and B, variables which correspond to densities and, thus, by multiplication with the volume, become extensive variables. The density of the internal energy

$$u = \frac{\mathcal{U}}{\mathcal{V}} ,$$

the primary thermodynamic potential, has the following dependence:

$$u(s, \Delta \mathcal{V}/\mathcal{V}_0, e, B, D) .$$

u contains, in general, a material part together with the energy density of the macroscopic electromagnetic field.

A 'reduced internal energy' $\hat{u}$, which is defined as follows, is sometimes used instead of u:

$$\hat{u} = u - \frac{\mu_0}{2} H^2 - \frac{\varepsilon_0}{2} E^2 . \tag{A.1}$$

It is given by the subtraction from u of terms due to the applied external electromagnetic field, such as that which exists before bringing the sample into a capacitor, where the term depends on the fixed potential difference, or in a coil, where the term depends on the fixed current in the coil.

The following applies for the change of the density of the internal energy for infinitesimal variations in the variables:

$$du = Tds - p\frac{d\mathcal{V}}{\mathcal{V}_0} + \sigma de + HdB + EdD \quad , \tag{A.2}$$

while the work per unit volume is given by the following differential equation:

$$dw = -p\frac{d\mathcal{V}}{\mathcal{V}_0} + \sigma de + HdB + EdD \quad . \tag{A.3}$$

Using Eqs. (A.1) and (A.2), it follows that

$$d\hat{u} = Tds - p\frac{d\mathcal{V}}{\mathcal{V}_0} + \sigma de + \mu_0 HdM + EdP \quad . \tag{A.4}$$

For the case of isothermal systems, s and T must be exchanged and the transition from the internal energy to the density of the *Helmholtz free energy* f must be achieved by means of the Legendre transformation:

$$f(T, \Delta\mathcal{V}/\mathcal{V}_0, e, B, D) = u - \frac{\partial u}{\partial s}s = u - Ts \quad . \tag{A.5}$$

The change df is then given by

$$df = -sdT - p\frac{d\mathcal{V}}{\mathcal{V}_0} + \sigma de + HdB + EdD \quad . \tag{A.6}$$

A reduced form of the Helmholtz free energy is analogously given by

$$\hat{f}(T, \Delta\mathcal{V}/\mathcal{V}_0, e, M, P) = \hat{u} - \frac{\partial \hat{u}}{\partial s}s = \hat{u} - Ts \quad , \tag{A.7}$$

for which the following differential equation finds many applications:

$$d\hat{f} = -sdT - p\frac{d\mathcal{V}}{\mathcal{V}_0} + \sigma de + \mu_0 HdM + EdP \quad . \tag{A.8}$$

In order to treat processes which occur in the presence of an applied mechanical, electric, or magnetic field with a strength σ, E, or H, a potential is required which contains the field strength as a variable. This is the density of the *Gibbs free energy* g, with the exchange of the variables being achieved by a multiple Legendre transformation of the Helmholtz free energy to give

$$g(T, p, \sigma, H, E) = f + p\frac{\mathcal{V}}{\mathcal{V}_0} - \sigma e - BH - ED \quad . \tag{A.9}$$

The change $\mathrm{d}g$ is now given by

$$\mathrm{d}g = -s\mathrm{d}T + \frac{\mathcal{V}}{\mathcal{V}_0}\mathrm{d}p - e\mathrm{d}\sigma - B\mathrm{d}H - D\mathrm{d}E \ . \qquad (\text{A.10})$$

A reduced form of the Gibbs free energy $\hat{g}$ can also be introduced as

$$\hat{g}(T, p, \sigma, H, E) = \hat{f} + p\mathcal{V}/\mathcal{V}_0 - \sigma e - \mu_0 HM - EP \ , \qquad (\text{A.11})$$

with the differential dependence

$$\mathrm{d}\hat{g} = -s\mathrm{d}T + \frac{\mathcal{V}}{\mathcal{V}_0}\mathrm{d}p - e\mathrm{d}\sigma - \mu_0 M\mathrm{d}H - P\mathrm{d}E \ . \qquad (\text{A.12})$$

It is to be noted that while the thermodynamic potentials introduced above are often used, an unambiguous naming convention is frequently lacking. In particular, the literature often makes no distinction between $f, g, \hat{f}$ and $\hat{g}$, simply referring to any of them as a 'free energy density'.

In the formulation of the above expressions, a fixed number of particles $\mathcal{N}$ has been assumed. The Gibbs free energy can, of course, also be used for the treatment of open systems with variable particle numbers, such as in the discussion of phase equilibria. For such a case, the chemical potential can be introduced as

$$\mu = \frac{\partial \mathcal{G}}{\partial \mathcal{N}} \ . \qquad (\text{A.13})$$

The corresponding molar variable is

$$\tilde{\mu} = N_\mathrm{A}\mu \ . \qquad (\text{A.14})$$

B

Solutions to the Exercises

Chapter 1

1. Unit cell with lattice parameter a, sphere radius b, n spheres per cell:
   $\phi = n\frac{4\pi}{3}b^3/a^3$.

   (a) $a = 2b$, $\quad \phi = \dfrac{\pi}{6} = 0.52$, $\quad z = 6$.

   (b) $a = 4b/\sqrt{3}$, $\quad \phi = \dfrac{\pi\sqrt{3}}{8} = 0.68$, $\quad z = 8$.

   (c) $a = 2\sqrt{2}b$, $\quad \phi = \dfrac{\pi}{3\sqrt{2}} = 0.74$, $\quad z = 12$.

2. (a) $\mathrm{C_{2v}}$:   1 two-fold rotation axis $\mathrm{C_2}$,
   2 vertical mirror planes σ_v.

   (b) $\mathrm{D_{\infty h}}$:   1 infinite rotation axis $\mathrm{C_\infty}$,
   ∞ vertical mirror planes σ_v,
   1 horizontal mirror plane σ_h,
   1 centre of inversion i.

   (c) $\mathrm{C_{3v}}$:   1 three-fold rotation axis $\mathrm{C_3}$,
   3 vertical mirror planes σ_v.

   (d) $\mathrm{T_d}$:   3 four-fold rotation-reflection axes $\mathrm{S_4}$,
   4 three-fold rotation axis $\mathrm{C_3}$,
   6 mirror planes σ_d.

   (e) $\mathrm{D_{6h}}$:   1 six-fold rotation axis $\mathrm{C_6}$,
   6 two-fold rotation axes $\mathrm{C_2}$,
   1 horizontal mirror plane σ_h,
   6 vertical mirror planes σ_v,
   1 centre of inversion i.

(f) D_{3d} :   1 three-fold rotation axis C_3,
        3 two-fold rotation axes C_2,
        3 mirror planes σ_d,
        1 centre of inversion i.

      C_{2v} :   1 two-fold rotation axis C_2,
        2 vertical mirror planes σ_v.

(g) C_{2h} :   1 two-fold rotation axis C_2,
        1 horizontal mirror plane σ_h,
        1 centre of inversion i.

(h) D_{4h} 1 four-fold rotation axis C_4,
        4 two-fold rotation axes C_2,
        4 vertical mirror planes σ_v,
        1 horizontal mirror plane σ_h,
        1 centre of inversion i.

3. See Fig. B.1

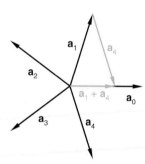

Fig. B.1. Let a_0 be a vector of minimal length and $a_1 \ldots a_4$ be vectors which arise from rotations of whole-number multiples of $2\pi/5$. In this way, $a_1 + a_4$ is parallel to a_0, but shorter.

4. (a) Molecular axes parallel to the reference axis: $\vartheta = 0$,

$$S_2 = \left\langle \frac{3\cos^2 \vartheta - 1}{2} \right\rangle = \frac{3-1}{2} = 1 \ .$$

(b) Molecular axes perpendicular to the reference axis: $\vartheta = \pi/2$,

$$S_2 = \left\langle \frac{3\cos^2 \vartheta - 1}{2} \right\rangle = -\frac{1}{2} \ .$$

(c) Isotropic distribution: $P(\vartheta) = \text{const.}$

$$S_2 = \frac{\int_0^\pi \dfrac{3\cos^2\vartheta - 1}{2} P(\vartheta)\sin\vartheta\, d\vartheta}{\int_0^\pi P(\vartheta)\sin\vartheta\, d\vartheta}$$

$$= \frac{\frac{3}{2}\int_0^\pi \cos^2\vartheta \sin\vartheta\, d\vartheta - \dfrac{1}{2}\int_0^\pi \sin\vartheta\, d\vartheta}{\int_0^\pi \sin\vartheta\, d\vartheta}$$

$$= \frac{\frac{3}{2}\left[\frac{1}{3}\cos^3\vartheta\right]_0^\pi - \dfrac{1}{2}[-\cos\vartheta]_0^\pi}{[-\cos\vartheta]_0^\pi} = \frac{-1+1}{2} = 0 \ .$$

5. The expectation values result from the Poisson distribution

$$w(N) = \exp(-\bar{N})\frac{\bar{N}^N}{N!} \quad :$$

$$\sigma^2 \ = \langle (N - \langle N\rangle)^2\rangle = \langle N^2\rangle - \langle N\rangle^2 \ ,$$

$$\langle N\rangle \ = \sum_{N=1}^{\infty} w(N)N = \sum_{N=1}^{\infty} N \exp(-\bar{N})\frac{\bar{N}^N}{N!}$$

$$= \exp(-\bar{N})\bar{N}\sum_{N=1}^{\infty}\frac{\bar{N}^{(N-1)}}{(N-1)!} = \bar{N}\exp(-\bar{N})\exp(\bar{N}) = \bar{N} \ ,$$

$$\langle N^2\rangle \ = \bar{N}^2 + \bar{N} \ ,$$

$$\sigma^2 \ = \bar{N} \ .$$

The mean molecular weights M_n and M_w can be determined from the molecular weight M_m of the monomer:

$$M_\mathrm{n} = M_\mathrm{m}\langle N\rangle \ ,$$

$$M_\mathrm{w} = M_\mathrm{m}^2\langle N^2\rangle / M_\mathrm{m}\langle N\rangle \ ,$$

$$U \ = \frac{M_\mathrm{w}}{M_\mathrm{n}} - 1 = \frac{\langle N^2\rangle}{\langle N\rangle\langle N\rangle} - 1 = \frac{\sigma^2}{\langle N\rangle^2} = \frac{1}{\bar{N}} \ .$$

For $\bar{N} = 100$ it follows that

$$U = \frac{1}{100} = 0.01 \ .$$

6. The series expansion in the quadratic approximation is given by:

$$I(q) \propto \sum_{j,k=1}^{N} f^2 \langle \exp(iq \cdot (r_j - r_k)) \rangle$$

$$= \sum_{j,k=1}^{N} f^2 \left\langle 1 + iq \cdot (r_j - r_k) - \frac{1}{2}(q \cdot (r_j - r_k))^2 + \ldots \right\rangle$$

$$\approx f^2 N^2 \left(1 - \frac{q^2}{3} \frac{1}{2N^2} \sum_{j,k=1}^{N^2} \langle (r_j - r_k)^2 \rangle \right) = f^2 N^2 \left(1 - \frac{R_g^2 q^2}{3} \right) .$$

$$R_g^2 = \frac{1}{2N^2} \sum_{j,k}^{N} \langle (r_j - r_k)^2 \rangle = \frac{1}{2N^2} \sum_{j,k}^{N} \langle [(r_j - r_c) - (r_k - r_c)]^2 \rangle$$

$$= \frac{1}{N} \sum_{j}^{N} \langle (r_j - r_c)^2 \rangle .$$

7. The basis vectors of the Bravais lattice are

$$a_1 = (1/2 \, \text{nm}, 0, 0) ,$$

$$a_2 = (\cos 120°, \sin 120°, 0) \, \text{nm} = (-1/2 \, \text{nm}, \sqrt{3}/2 \, \text{nm}, 0) ,$$

$$a_3 = (0, 0, 3/2 \, \text{nm}) .$$

The basis vectors of the reciprocal lattice are given by

$$\hat{a}_1 = 2\pi \frac{a_2 \times a_3}{(a_1 \times a_2) \cdot a_3} \qquad \text{etc. as :}$$

$$\hat{a}_1 = \frac{4\pi}{\sqrt{3}} \begin{pmatrix} \sqrt{3} \\ 1 \\ 0 \end{pmatrix} \text{nm}^{-1} , \qquad \hat{a}_2 = \frac{4\pi}{\sqrt{3}} \begin{pmatrix} 0 \\ 1 \\ 0 \end{pmatrix} \text{nm}^{-1}$$

$$\hat{a}_3 = \frac{4\pi}{\sqrt{3}} \begin{pmatrix} 0 \\ 0 \\ \frac{1}{\sqrt{3}} \end{pmatrix} \text{nm}^{-1} .$$

Using the Miller indices, the scattering vector, the plane-to-plane distance, and the scattering angle are given by:

$$\boldsymbol{q}_{321} = 3\hat{\boldsymbol{a}}_1 + 2\hat{\boldsymbol{a}}_2 + \hat{\boldsymbol{a}}_3 = \frac{4\pi}{\sqrt{3}} \begin{pmatrix} 3\sqrt{3} \\ 5 \\ 1 \\ \sqrt{3} \end{pmatrix} \, \text{nm}^{-1} \quad ,$$

$$|\boldsymbol{q}_{321}| = \frac{4\pi}{3} \sqrt{157} \, \text{nm}^{-1} = 52.48 \, \text{nm}^{-1} \quad ,$$

$$d_{321} = \frac{2\pi}{q_{321}} = 0.1197 \, \text{nm} \quad ,$$

$$2\theta = 2 \arcsin\left(\frac{q\lambda}{4\pi}\right) = 2 \arcsin\left(\frac{\sqrt{157}}{3} \cdot 0.154\right) = 80.06° \quad .$$

8. The cell structure factor is given by the positions of the particles in the cell according to:

$$f_c(hkl) = f \sum_{j=1}^{4} \exp\left[\mathrm{i}2\pi \left(hx_1^j + kx_2^j + lx_3^j\right)\right] \quad ,$$

$$\{x_i^1\} = (0,0,0) \quad , \quad \{x_i^2\} = \left(\frac{1}{2}, \frac{1}{2}, 0\right) \quad , \quad \{x_i^3\} = \left(\frac{1}{2}, 0, \frac{1}{2}\right) \quad ,$$

$$\{x_i^4\} = \left(0, \frac{1}{2}, \frac{1}{2}\right),$$

$$f_c(hkl) = f\left\{1 + \exp\left[\mathrm{i}\pi(h+k)\right] + \exp\left[\mathrm{i}\pi(h+l)\right] + \exp\left[\mathrm{i}\pi(k+l)\right]\right\}$$

$$= \begin{cases} 4 \text{ for: } h,k,l \text{ all even} \\ 4 \text{ for: } h,k,l \text{ all odd} \\ 0 \text{ for: only one index even} \\ 0 \text{ for: only one index odd} \end{cases} \quad .$$

Chapter 2

1. The elastic element: The viscous element:

$$\sigma_{zx} = G \cdot \tan\gamma \approx G \cdot \gamma \qquad \sigma_{zx} = \eta \frac{\mathrm{d}}{\mathrm{d}t} \tan\gamma \approx \eta\dot{\gamma}$$

$$\gamma = \frac{\sigma_{zx}}{G} \qquad\qquad\qquad \dot{\gamma} = \frac{\sigma_{zx}}{\eta}$$

Maxwell element: $\gamma = \gamma_{\mathrm{el}} + \gamma_{\mathrm{vis}}$,

Kelvin element: $\sigma = \sigma_{\mathrm{el}} + \sigma_{\mathrm{vis}}$.

(a) Creep experiment

Maxwell element: Kelvin element:

$$\gamma = \frac{\sigma_0}{G} + \int_0^t \frac{\sigma_0}{\eta} dt \qquad\qquad \sigma = G\gamma + \eta\dot{\gamma} = \sigma_0 \ ,$$

$$= \frac{\sigma_0}{G} + \frac{\sigma_0}{\eta} t \ . \qquad\qquad G\dot{\gamma} + \eta\ddot{\gamma} = 0$$

$$\Rightarrow \gamma(t) = \frac{\sigma_0}{G}\left(1 - \exp(-\frac{G}{\eta}t)\right) \ .$$

(b) Stress relaxation experiment for the Maxwell element:

$$\gamma_0 = \frac{\sigma}{G} + \int_0^t \frac{\sigma}{\eta} dt' \ , \quad \dot{\gamma} = 0 = \frac{\dot{\sigma}}{G} + \frac{\sigma}{\eta} \ ,$$

$$\sigma = \sigma(0)\exp(-\frac{G}{\eta}t) \quad \text{with } \sigma(0) = \gamma_0 G \ .$$

(c) Dynamic experiment

trial solution: $\sigma(t) = \sigma_0 \exp(-\mathrm{i}\omega t), \quad \gamma(t) = \gamma_0 \exp(-\mathrm{i}\omega t)$

Maxwell element: $\dot{\gamma} = \dfrac{\dot{\sigma}}{G} + \dfrac{\sigma}{\eta} \ ,$

insert : $\Rightarrow$ $\qquad \gamma_0 = \dfrac{\sigma_0}{G} + \mathrm{i}\dfrac{\sigma_0}{\eta\omega}$

$$|\gamma_0|^2 = \sigma_0^2\left(\frac{1}{G^2} + \frac{1}{\eta^2\omega^2}\right) \ , \quad \tan\delta = \frac{G}{\eta\omega} \ .$$

Kelvin element: $\sigma = G\gamma + \eta\dot{\gamma} \ ,$

insert : $\Rightarrow$ $\qquad \gamma_0 = \dfrac{\sigma_0}{G - \mathrm{i}\eta\omega}$

$$|\gamma_0|^2 = \frac{\sigma_0^2}{G^2 + \eta^2\omega^2} \ , \quad \tan\delta = \frac{\eta\omega}{G} \ .$$

2. (a) The velocity gradient and the shear stress σ_s are constant for the given geometry such that

$$\sigma_\mathrm{s} = \frac{\mathrm{d}f}{\mathrm{d}A} = \eta\frac{\mathrm{d}v}{\mathrm{d}z} \ ,$$

$$\frac{\mathrm{d}v}{\mathrm{d}z} = \frac{\omega \cdot r}{r\tan\alpha} = \frac{\omega}{\tan\alpha} = \frac{\dot{\varphi}}{\tan\alpha} \ .$$

The torque is given by integrating over the total surface:

$$T = \int r \frac{df}{dA} \, dA = \frac{2\pi\eta\omega R^3}{3\tan\alpha} \quad .$$

(b) For oscillatory rotation, T is given by differentiating the rotation angle with respect to time:

$$T(t) = \frac{2\pi\eta R^3}{3\tan\alpha} \frac{d}{dt} [\varphi_0 \cos(\omega t)]$$

$$= -\frac{2\pi\eta R^3}{3\tan\alpha} \varphi_0 \omega \sin(\omega t) \quad .$$

3. The following is obtained for the real and imaginary regions:

$$\chi(\omega) = \chi_0 \frac{1 + i\omega\tau}{1 + \omega^2\tau^2}; \quad 0 \le \Re\left(\frac{\chi}{\chi_0}\right) \le 1; \quad 0 \le \Im\left(\frac{\chi}{\chi_0}\right) \le \tfrac{1}{2}$$

A semi-circle with its centre at $(\tfrac{1}{2}, \ 0)$, since :

$$\left(\frac{1}{1+\omega^2\tau^2} - \frac{1}{2}\right)^2 + \left(\frac{\omega\tau}{1+\omega^2\tau^2}\right)^2 = \frac{(\tfrac{1}{2} - \tfrac{1}{2}\omega^2\tau^2)^2 + \omega^2\tau^2}{(1+\omega^2\tau^2)^2}$$

$$= \frac{\left(\tfrac{1}{2}\right)^2 (1+\omega^2\tau^2)^2}{(1+\omega^2\tau^2)^2} = \left(\frac{1}{2}\right)^2 \quad .$$

4. $\langle r^2 \rangle = \dfrac{1}{\pi a_0^3} \displaystyle\int r^2 e^{-2r/a_0} d\Omega r^2 dr = 3a_0^2$,

$\kappa_{\text{dia}} = -\mu_0 \dfrac{N_A e^2}{6m_e} 3a_0^2$,

$\kappa_{\text{dia}} = 2.98 \times 10^{-5} \, \text{cm}^3 \, \text{mol}^{-1}$.

5. (a) The occupation numbers are proportional to the Boltzmann factors:

$$\mathcal{N}_\uparrow \propto \exp\left(\frac{\mu_B B}{k_B T}\right) \quad ,$$

$$\mathcal{N}_\downarrow \propto \exp\left(-\frac{\mu_B B}{k_B T}\right) \quad ,$$

$$\frac{\mathcal{N}_\uparrow - \mathcal{N}_\downarrow}{\mathcal{N}} = \frac{\mathcal{N}_\uparrow - \mathcal{N}_\downarrow}{\mathcal{N}_\uparrow + \mathcal{N}_\downarrow} = \frac{\exp x - \exp(-x)}{\exp x + \exp(-x)} = \tanh(x) \quad , \quad x = \frac{\mu_B B}{k_B T}$$

$$\mu_B = \frac{e\hbar}{2m_e} = \frac{e h}{4\pi m_e} = 9.273 \times 10^{-24}\,\mathrm{Am^2} \quad,$$

$$x = \frac{\mu_B B}{k_B T} = 2.238 \times 10^{-3} \quad,$$

$$\frac{\mathcal{N}_\uparrow - \mathcal{N}_\downarrow}{\mathcal{N}} = \tanh(x) \approx x = 2.238 \times 10^{-3} \quad.$$

(b) $M = (\mathcal{N}_\uparrow - \mathcal{N}_\downarrow)\mu_B = \mathcal{N}\mu_B \tanh(x) \approx \mathcal{N}\mu_B x$

$$= 207.6\,\mathrm{A/m} \quad.$$

6. (a) The additional field due to the neighbouring magnetic moment $\boldsymbol{m}$ at a distance $\boldsymbol{r}$ (with $\boldsymbol{B}_0 \parallel \boldsymbol{z}$):

$$\Delta\boldsymbol{B} = \frac{\mu_0}{4\pi} \frac{3(\boldsymbol{m}\cdot\boldsymbol{r})\boldsymbol{r} - \boldsymbol{m}r^2}{r^5} \quad,$$

$$\Delta B_z = \frac{\mu_0 m_z}{4\pi r^3}(3\cos^2\vartheta - 1) \quad,$$

$$\vartheta_{\max} = 0 \quad.$$

(b) $\Delta B_{z,\max} = \dfrac{\mu_0 m_z}{2\pi r^3} \quad,$

$$m_z = g_n \mu_n m \quad, \qquad m = \pm 1/2 \quad,$$

$$\Delta B_{z,\max} = \pm \frac{\mu_0\, g_n \mu_n}{4\pi r^3} \quad,$$

$$g_n = 5.588 \quad, \quad \mu_n = \frac{e\hbar}{2m_p} = 5.049 \times 10^{-27}\,\mathrm{Am^2} \quad,$$

$$\Delta E = g_n \mu_n \Delta m (B_0 + \Delta B_{z,\max}) = h\nu = g\mu_n(B_0 + \Delta B_{z,\max}) \quad.$$

A symmetric splitting of the frequency about ν_0:

$$h(\nu_0 \pm \Delta\nu/2) = g_n \mu_n \left(B_0 \pm \frac{\mu_0 g_n \mu_n}{4\pi r^3}\right) \quad,$$

$$\Delta\nu = \frac{\mu_0 g_n^2 \mu_n^2}{2\pi r^3 h} = 3.003 \times 10^4\,\mathrm{s^{-1}} \quad.$$

(c) The condition for a vanishing splitting is:

$$3\cos^2\vartheta - 1 = 0 \quad,$$

$$\vartheta = \arccos\frac{1}{\sqrt{3}} = 54.74° \quad.$$

Chapter 3

1. (a) The dipole and the inter-atomic axis parallel: $E(a) = \frac{2p_1}{4\pi\varepsilon_0 a^3}$.

   Self-stabilisation criterion: $p_2 = \beta E(a) \geq p_1 \Rightarrow \beta \geq 2\pi\varepsilon_0 a^3$.

   (b) An anti-ferroelectric arrangement would be preferred upon orienting the dipole perpendicular to the inter-atomic axis. A calculation analogous to that in (a) shows that the ferroelectric self-stabilisation sets in earlier when the atoms approach each other.

   (c) The dipole and the inter-atomic axis parallel: $E(a) = \frac{2p_1}{4\pi\varepsilon_0 a^3} \sum \frac{2}{n^3}$.

   Self-stabilisation criterion: $p_2 = \beta E(a) > p_1 \Rightarrow \beta > \pi\varepsilon_0 a^3 / \sum \frac{1}{n^3}$.

2. electric dipole: $\quad \Delta u = \dfrac{4p_0^2}{4\pi\varepsilon_0 r^3} = 3.21 \times 10^{-21}\,\text{J}$,

   magnetic dipole: $\quad \Delta u =$

   $$\frac{4\mu_0 m_0^2}{4\pi r^3} = 2.75 \times 10^{-25}\,\text{J} \ ,$$

   cf. thermal energy at $300\,\text{K}$: $\quad k_B T = 4.1 \times 10^{-21}\,\text{J}$.

3. Equilibrium condition:

   $$\frac{\partial}{\partial M} g = 2b(T - T_c)M + 4c_4 M^3 = 0 \ .$$

   1st solution $M = 0$ (equilibirium for $T > T_c$) ,

   2nd solution $M = \pm\sqrt{\dfrac{b}{2c_4}(T_c - T)}$ (equilibrium for $T < T_c$) .

   The Gibbs free energy density g:

   $$T > T_c: \quad g = g_0 \ ,$$

   $$T < T_C: \quad g = g_0 - \frac{b^2}{4c_4}(T - T_c)^2 \ .$$

   The entropy density s:

   $$T > T_c: \quad s = s_0 = -\frac{\partial}{\partial T} g_0 \ ,$$

   $$T < T_c: \quad s = s_0 - \frac{\partial}{\partial T}\left(-\frac{b^2}{4c_4}(T - T_c)^2\right) = s_0 - \frac{b^2}{2c_4}(T_c - T) \ .$$

   The heat capacity c_v:

   $$T > T_c: \quad c_v = c_{v0} = T\frac{\partial s_0}{\partial T} \ ,$$

$$T < T_{\mathrm{c}} :\quad c_v = c_{v0} + T\frac{b^2}{2c_4}\quad.$$

Chapter 4

1. The Fermi-Dirac particle density

   at $T = 0$: for $\epsilon < 0 \Rightarrow w(\epsilon) = 1$,

   $\qquad\qquad$ for $\epsilon > 0 \Rightarrow w(\epsilon) = 0$;

   at $T \neq 0, k_{\mathrm{B}}T \ll \mu_{\mathrm{e}}$:

   Expanding about $\epsilon = \mu_{\mathrm{e}}$,

   $$w|_{\epsilon=\mu_{\mathrm{e}}} = \frac{1}{2}\quad,$$

   $$\frac{\mathrm{d}}{\mathrm{d}\epsilon}w\,\big|_{\epsilon=\mu_{\mathrm{e}}} = \frac{-\exp\dfrac{\epsilon-\mu_{\mathrm{e}}}{k_{\mathrm{B}}T}}{\left(\exp\dfrac{\epsilon-\mu_{\mathrm{e}}}{k_{\mathrm{B}}T}+1\right)^2}\cdot\frac{1}{k_{\mathrm{B}}T}\,\big|_{\epsilon=\mu_{\mathrm{e}}} = -\frac{1}{4k_{\mathrm{B}}T}\quad.$$

   The width of the deviation region, using a linear approximation:

   $$\frac{1}{2} - \frac{\Delta\epsilon}{4k_{\mathrm{B}}T} = 0 \Rightarrow \Delta\epsilon = 2k_{\mathrm{B}}T\quad.$$

2. The Helmholtz free energy of the Fermi electron gas:

   $$\mathcal{F} = \frac{3}{5}\mathcal{N}_{\mathrm{e}}\epsilon_{\mathrm{F}} = \frac{3}{5}\mathcal{N}_{\mathrm{e}}\frac{\hbar^2}{2m_{\mathrm{e}}}\left(\frac{3\pi^2\mathcal{N}_{\mathrm{e}}}{\mathcal{V}}\right)^{2/3} = \frac{3\hbar^2\mathcal{N}_{\mathrm{e}}^{5/3}(3\pi^2)^{2/3}}{10m_{\mathrm{e}}}\mathcal{V}^{-2/3}\quad,$$

   $$\frac{\partial\mathcal{F}}{\partial\mathcal{V}} = -\frac{2}{3}\frac{3\hbar^2\mathcal{N}_{\mathrm{e}}^{5/3}(3\pi^2)^{2/3}}{10m_{\mathrm{e}}}\mathcal{V}^{-5/3}\quad,$$

   $$\frac{\partial^2\mathcal{F}}{\partial\mathcal{V}^2} = +\frac{10}{9}\frac{3\hbar^2\mathcal{N}_{\mathrm{e}}^{5/3}(3\pi^2)^{2/3}}{10m_{\mathrm{e}}}\mathcal{V}^{-8/3}\quad,$$

   $$K = \mathcal{V}\frac{\partial^2\mathcal{F}}{\partial\mathcal{V}^2} = \frac{\hbar^2(3\pi^2)^{2/3}}{3m_{\mathrm{e}}}\left(\frac{\mathcal{N}_{\mathrm{e}}}{\mathcal{V}}\right)^{5/3} = \frac{\pi^{4/3}\hbar^2}{3^{1/3}m_{\mathrm{e}}}\left(\frac{N_{\mathrm{A}}\rho m}{M}\right)^{5/3}$$
   $$= 8.55\times10^9\ \mathrm{Pa}\quad.$$

3. (a) Trial solution for the Schrödinger equation $-\frac{\hbar^2}{2m}\Delta\psi = \epsilon\psi$
   for the boundary condition $\psi(x,y,z) = 0$ for $|x| \geq a$:

$$\psi(x, y, z) = C \exp[i(k_y y + k_z z)] \cos(k_x x) \quad \text{for} \ |x| \leq a \ ,$$

$$k_x a = n_x \pi + \pi/2 \ ,$$

$$k_y L = n_y 2\pi \ ,$$

$$k_z L = n_z 2\pi \ ,$$

$$\Delta \psi = - \left(k_x^2 + k_y^2 + k_z^2 \right) \psi = k^2 \psi \ ,$$

$$\epsilon = \frac{\hbar^2 k^2}{2m_e} \ .$$

For $n_x = 0$:

$$k_x = \frac{\pi}{2a} \ ,$$

ϵ is now only dependent on k_y and k_z.

Integrating in k space in the y, z plane:

$$\mathcal{D}(\epsilon)\mathrm{d}\epsilon = \mathcal{D}(\epsilon)\frac{\mathrm{d}\epsilon}{\mathrm{d}k}\mathrm{d}k = \mathcal{D}(k)\mathrm{d}k = \frac{A}{(2\pi)^2} 2\pi k \mathrm{d}k \ ,$$

$$\mathcal{D}(\epsilon) = \frac{A}{2\pi} k \frac{\mathrm{d}k}{\mathrm{d}\epsilon} \ ,$$

$$k = \frac{\sqrt{2m_e \epsilon}}{\hbar} \ ,$$

$$\frac{\mathrm{d}k}{\mathrm{d}\epsilon} = \frac{1}{2\hbar}\sqrt{\frac{2m_e}{\epsilon}} = \frac{1}{\hbar}\sqrt{\frac{m_e}{2\epsilon}} \ ,$$

$$k\frac{\mathrm{d}k}{\mathrm{d}\epsilon} = \frac{m_e}{\hbar^2} \ ,$$

$$\mathcal{D}(\epsilon) = \frac{A m_e}{2\pi \hbar^2} \ .$$

(b) Trial solution for the boundary condition

$$\psi(x, y, z) = 0 \ \text{for} \ |x| \geq a \ \text{and} \ |y| \geq b \ :$$

$$\psi(x, y, z) = C \exp(ik_z z) \cos(k_x x) \cos(k_y y) \ \text{for} \ |x| \leq a \ \text{and} \ |y| \leq b \ ,$$

$$k_x a = n_x \pi + \pi/2 \ ,$$

$$k_y b = n_y \pi + \pi/2 \ ,$$

$$k_z L = n_z 2\pi \ .$$

For $n_x = n_y = 0$:

$$k_x = \frac{\pi}{2a} \quad, \quad k_y = \frac{\pi}{2b} \quad,$$

ϵ is now only dependent on k_z.

Integrating in k space in the z direction:

$$\mathcal{D}(\epsilon)d\epsilon = \mathcal{D}(\epsilon)\frac{d\epsilon}{dk}dk = \mathcal{D}(k)dk = \frac{L}{2\pi}dk \quad,$$

$$\mathcal{D}(\epsilon) = \frac{L}{2\pi}\frac{dk}{d\epsilon} = \frac{L}{2\pi\hbar}\sqrt{\frac{m_e}{2\epsilon}} \quad.$$

4. Electrons, which move with angles $\vartheta_1 \leq \vartheta \leq \vartheta_2$ with respect to the layer normal, do not reach the interface before completing the distance λ_f:

$$\left.\begin{array}{l} \cos\vartheta_1 = \dfrac{d-x}{\lambda_f} \\[2ex] \cos\vartheta_2 = -\dfrac{x}{\lambda_f} \end{array}\right\} \quad \text{for} \quad d \leq \lambda_f \ .$$

The path length λ is a function of x and ϑ:

$$\lambda(x,\vartheta) = \begin{cases} \dfrac{d-x}{\cos\vartheta} & \text{for } 0 \leq \vartheta < \vartheta_1 \\[2ex] \lambda_f & \text{for } \vartheta_1 \leq \vartheta \leq \vartheta_2 \\[2ex] -\dfrac{x}{\cos\vartheta} & \text{for } \vartheta_2 < \vartheta \leq \pi \end{cases} \ .$$

The mean path length as the integral over the whole layer and all spatial angles:

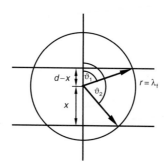

Fig. B.2. Angle-dependent limitation of the free path length.

$$\lambda_{\mathrm{f}}(d) = \frac{1}{2d} \int_0^d \int_0^\pi \lambda(x, \vartheta) \sin \vartheta \mathrm{d}\vartheta \mathrm{d}x$$

$$= \frac{1}{2d} \int_0^d \left[(d - x) \int_0^{\vartheta_1} \tan \vartheta \mathrm{d}\vartheta + \lambda_{\mathrm{f}} \int_{\vartheta_1}^{\vartheta_2} \sin \vartheta \mathrm{d}\vartheta - x \int_{\vartheta_2}^\pi \tan \vartheta \mathrm{d}\vartheta \right] \mathrm{d}x$$

$$= \frac{1}{2d} \int_0^d \left[(d - x)(\ln |\cos 0| - \ln |\cos \vartheta_1|) + \lambda_{\mathrm{f}}(\cos \vartheta_1 - \cos \vartheta_2) \right.$$

$$\left. - x(\ln |\cos \vartheta_2| - \ln |\cos \pi|) \right] \mathrm{d}x$$

$$= \frac{1}{2d} \int_0^d \left[-(d - x) \ln \frac{d - x}{\lambda_{\mathrm{f}}} + d - x \ln \frac{x}{\lambda_{\mathrm{f}}} \right] \mathrm{d}x \quad,$$

$$u = \frac{d - x}{\lambda_{\mathrm{f}}} \quad, \quad \mathrm{d}u = -\frac{\mathrm{d}x}{\lambda_{\mathrm{f}}} \quad, \quad \mathrm{d}x = -\lambda_{\mathrm{f}} \mathrm{d}u \quad,$$

$$I_1 = \int_0^d -(d - x) \ln \frac{d - x}{\lambda_{\mathrm{f}}} \mathrm{d}x = \lambda_{\mathrm{f}}^2 \int_{d/\lambda_{\mathrm{f}}}^0 u \ln u \mathrm{d}u = -\lambda_{\mathrm{f}}^2 \int_0^{d/\lambda_{\mathrm{f}}} u \ln u \mathrm{d}u \quad,$$

$$v = \frac{x}{\lambda_{\mathrm{f}}} \quad, \quad \mathrm{d}x = \lambda_{\mathrm{f}} \mathrm{d}v \quad,$$

$$\int_0^d -x \ln \frac{x}{\lambda_{\mathrm{f}}} \mathrm{d}x = -\lambda_{\mathrm{f}}^2 \int_0^{d/\lambda_{\mathrm{f}}} v \ln v \mathrm{d}v = I_1 \quad,$$

$$\int u \ln u \mathrm{d}u = u^2 \left(\frac{\ln u}{2} - \frac{1}{4} \right) \quad,$$

$$I_1 = -d^2 \left(\frac{\ln \frac{d}{l_{\mathrm{f}}}}{2} - \frac{1}{4} \right) \quad,$$

$$\lambda_{\mathrm{f}}(d) = \frac{1}{2d}(2I_1 + d^2) = d \left(\frac{1}{4} - \frac{1}{2} \ln \frac{d}{\lambda_{\mathrm{f}}} + \frac{1}{2} \right) = d \left(\frac{3}{4} + \frac{1}{2} \ln \frac{\lambda_{\mathrm{f}}}{d} \right) \quad,$$

$$\frac{\sigma_{\mathrm{el}}(d)}{\sigma_0} = \frac{\lambda_{\mathrm{f}}(d)}{\lambda_{\mathrm{f}}} = \frac{d}{\lambda_{\mathrm{f}}} \left(\frac{3}{4} + \frac{1}{2} \ln \frac{\lambda_{\mathrm{f}}}{d} \right) \quad.$$

5. (a) Plasma frequency: $\omega_{pl}^2 = \dfrac{\rho_e e^2}{\varepsilon_0 m_e} = \dfrac{N_A \rho_m e^2}{M \varepsilon_0 m_e} = (8.99 \times 10^{15} \mathrm{s}^{-1})^2$,

$$k = \frac{\omega}{c_1} \quad , \qquad \lambda = \frac{2\pi}{k} = 2\pi \frac{c_1}{\omega} = 2.095 \times 10^{-7} \, \mathrm{m} = 209.5 \, \mathrm{nm} \quad .$$

(b) $\varepsilon(\omega) = 1 + \chi(\omega) = 1 - \dfrac{\rho_e e^2}{\varepsilon_0 m_e \omega^2}$,

$$\chi(\omega) = -\frac{N_A \rho_m e^2 \hbar^2}{M \varepsilon_0 m_e \varepsilon^2} = -3.5 \times 10^{-7} \quad ,$$

$$\varepsilon = 1 + \chi = 0.99999965 \quad , \qquad n = \sqrt{\varepsilon} \approx 1 + \frac{1}{2}\chi = 0.999999825 \quad .$$

$n < 1$: Total reflection for incidence from a vacuum is possible.

6. Electric field for

$z > 0:$ $E_z^+ = kV_0 \cos(kx) \exp(-kz)$, $E_x^+ = kV_0 \sin(kx) \exp(-kz)$,

$z < 0:$ $E_z^- = -kV_0 \cos(kx) \exp(kz)$, $E_x^- = kV_0 \sin(kx) \exp(kz)$.

The tangential components of E are continuous.

Continuity of the normal components of D:

$$\varepsilon_0 \varepsilon(\omega) E_{z=0}^+ = \varepsilon_0 E_{z=0}^- \quad \Rightarrow \quad \varepsilon(\omega) = -1 \quad ,$$

$$\varepsilon(\omega) \approx 1 - \frac{\rho_e e^2}{\varepsilon_0 m_e \omega^2} = 1 - \frac{\omega_{pl}^2}{\omega^2} \quad \Rightarrow \quad \omega^2 = \frac{1}{2}\omega_{pl}^2 \quad .$$

7. Required formulae:

$$\int \cos^2 x \, dx = \frac{1}{2} \sin x \cos x + \frac{x}{2} + C \quad ,$$

$$\int \sin^2 x \, dx = -\frac{1}{2} \sin x \cos x + \frac{x}{2} + C \quad ,$$

$$\cos^2 x = \frac{1}{2}(1 + \cos 2x) \quad ,$$

$$\sin^2 x = \frac{1}{2}(1 - \cos 2x) \quad .$$

Normalisation:

$$\int_0^a A^2 \cos^2\left(\frac{\pi}{a}x\right) dx = A^2 \frac{a}{\pi} \int_0^\pi \cos^2 u\, du = A^2 \frac{a}{2} = 1 \Rightarrow A = \sqrt{2/a} \;;$$

The normalisation for ψ_- is analogous.

For $\psi_+, k_1 = \dfrac{\pi}{a}$:

$$\epsilon = \frac{2}{a} \int_0^a \cos\left(\frac{\pi}{a}x\right) \left[\frac{-\hbar^2}{2m} \frac{d^2}{dx^2} - u_0 \cos\left(\frac{2\pi}{a}x\right)\right] \cos\left(\frac{\pi}{a}x\right) dx$$

$$= \frac{2}{a} \frac{\hbar^2 k_1^2}{2m} \int_0^a \cos^2\left(\frac{\pi}{a}x\right) dx - \frac{2}{a} u_0 \int_0^a \cos\left(\frac{2\pi}{a}x\right) \cos^2\left(\frac{\pi}{a}x\right) dx$$

$$= \frac{\hbar^2 k_1^2}{2m} - \frac{2}{a} u_0 \int_0^a \cos\left(\frac{2\pi}{a}x\right) \frac{1}{2}\left[1 + \cos\left(\frac{2\pi}{a}x\right)\right] dx$$

$$= \frac{\hbar^2 k_1^2}{2m} - \frac{u_0}{a} \int_0^a \cos^2\left(\frac{2\pi}{a}x\right) dx = \frac{\hbar^2 k_1^2}{2m} - \frac{u_0}{a}\frac{a}{2\pi} \int_0^{2\pi} \cos^2 u\, du$$

$$= \frac{\hbar^2 k_1^2}{2m} - \frac{u_0}{2} \; .$$

For $\psi_-, k_1 = \dfrac{\pi}{a}$:

$$\epsilon = \frac{2}{a} \int \sin\left(\frac{\pi}{a}x\right) \left[\frac{\hbar^2}{2m}\frac{d^2}{dx^2} - u_0 \cos\left(\frac{2\pi}{a}x\right)\right] \sin\left(\frac{\pi}{a}x\right) dx$$

$$= \frac{\hbar^2 k_1^2}{2m} - \frac{2}{a} u_0 \int_0^a \cos\left(\frac{2\pi}{a}x\right) \sin^2\left(\frac{\pi}{a}x\right) dx$$

$$= \frac{\hbar^2 k_1^2}{2m} - \frac{2}{a} u_0 \int_0^a \cos\left(\frac{2\pi}{a}x\right) \frac{1}{2}\left[1 - \cos\left(\frac{2\pi}{a}x\right)\right] dx$$

$$= \frac{\hbar^2 k_1^2}{2m} + \frac{u_0}{2} \; .$$

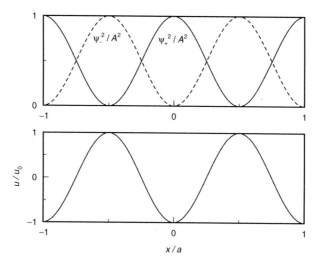

Fig. B.3. The solution ψ_- has maximum probabilities of occupation at the maxima of the potential, and thus has the highest energy.

8. Energy of the electron: $\epsilon = \dfrac{\hbar^2 k^2}{2m_e}$,

$$\Rightarrow \frac{\mathrm{d}}{\mathrm{d}\boldsymbol{k}}\epsilon = \frac{\hbar^2}{m_e}\boldsymbol{k} ,$$

$$\tau = \frac{\hbar^2}{e} \oint \frac{|\mathrm{d}\boldsymbol{k}|}{\left|\dfrac{\mathrm{d}}{\mathrm{d}\boldsymbol{k}}\epsilon \times B\right|} = \frac{\hbar^2}{e} \oint \frac{|\mathrm{d}\boldsymbol{k}|}{\left|\dfrac{\hbar^2}{m_e}\boldsymbol{k} \times B\right|} = \frac{m_e}{eB}\frac{2\pi k_\perp}{k_\perp} = 2\pi\frac{m_e}{eB} ,$$

$$\Rightarrow \frac{2\pi}{\tau} = \frac{eB}{m_e} = \omega_c .$$

Chapter 5

1. (a) The equation of motion for atom l with displacement z_l from the rest position:

$$m\ddot{z}_l = b(z_{l+1} - z_l + z_{l-1} - z_l) + b'(z_{l+2} - z_l + z_{l-2} - z_l)$$

$$= b(z_{l+1} + z_{l-1} - 2z_l) + b'(z_{l+2} + z_{l-2} - 2z_l) .$$

Trial solution: $z_l = Z\exp[\mathrm{i}(\omega t - kal)]$,

$$m\ddot{z}_l = -m\omega^2 z_l = -m\omega^2 Z \exp[i(\omega t - kal)]$$

$$= Z\{b\exp[i(\omega t - ka(l+1))] + b\exp[i(\omega t - ka(l-1))]$$

$$+ b'\exp[i(\omega t - ka(l+2))] + b'\exp[i(\omega t - ka(l-2))]$$

$$- 2(b+b')\exp[i(\omega t - kal)]\} \quad,$$

$$\omega^2 = \frac{1}{m}\{2(b+b') - b[\exp(ika) + \exp(-ika)] - b'[\exp(2ika)$$

$$+ \exp(-2ika)]\}$$

$$= \frac{b}{m}\left\{2 - 2\cos(ka) + \frac{b'}{b}[2 - 2\cos(2ka)]\right\}$$

$$= \frac{2b}{m}\left\{1 - \cos(ka) + \frac{b'}{b}[1 - \cos(2ka)]\right\} \quad,$$

$$\omega = \sqrt{\frac{2b}{m}} \cdot \sqrt{1 - \cos(ka) + \frac{b'}{b}[1 - \cos(2ka)]\}} \quad.$$

(b) Plot for $b' = b/2$

Fig. B.4 plots the expression

$$\omega\sqrt{\frac{m}{b}} = \sqrt{2}\sqrt{1 - \cos(ka) + \frac{1}{2}[1 - \cos(2ka)] \}} \quad.$$

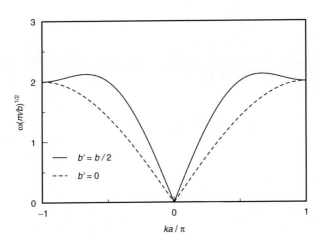

Fig. B.4. The dispersion relation for a linear chain with interactions between next-but-one neighbours.

2. Integration of the expansion for small displacements:

$$\int x\, e^{-\beta u}\mathrm{d}x \approx \int e^{-\beta c_2 x^2}(x + \beta c_3 x^4 + \beta c_4 x^5)\mathrm{d}x =$$

$$\int e^{-\beta c_2 x^2}\beta c_3 x^4 \mathrm{d}x = \beta c_3 \frac{\mathrm{d}^2}{\mathrm{d}(\beta c_2)^2}\int e^{-\beta c_2 x^2}\mathrm{d}x = \frac{3\sqrt{\pi}\,c_3}{4 c_2^{5/2}\beta^{3/2}} \quad,$$

$$\int e^{-\beta u}\mathrm{d}x \approx \int e^{-\beta c_2 x^2}\mathrm{d}x = \left(\frac{\pi}{\beta c_2}\right)^{1/2}$$

$$\Rightarrow \langle x\rangle = \frac{3 c_3}{4 c_4^2} k_{\mathrm{B}}T \quad.$$

3. The internal energy is given by the expectation value $\langle n\rangle$ of the phonon occupation number of the oscillator:

$$U = \mathcal{N}\hbar\omega\left(\langle n\rangle + \frac{1}{2}\right) \quad , \quad \langle n\rangle = \frac{1}{\exp\left(\frac{\hbar\omega}{k_{\mathrm{B}}T}\right) - 1}$$

$$\Rightarrow U = \mathcal{N}\hbar\omega\left[\frac{1}{\exp\left(\frac{\hbar\omega}{k_{\mathrm{B}}T}\right) - 1} + \frac{1}{2}\right]$$

$$c_v = \frac{\partial U}{\partial T} = \mathcal{N}\hbar\omega\left[\frac{-\exp\left(\frac{\hbar\omega}{k_{\mathrm{B}}T}\right)\left(-\frac{\hbar\omega}{k_{\mathrm{B}}T^2}\right)}{[\exp\left(\frac{\hbar\omega}{k_{\mathrm{B}}T}\right) - 1]^2}\right]$$

$$= \mathcal{N}\frac{(\hbar\omega)^2}{k_{\mathrm{B}}T^2}\frac{\exp\left(\frac{\hbar\omega}{k_{\mathrm{B}}T}\right)}{\left[\exp\left(\frac{\hbar\omega}{k_{\mathrm{B}}T}\right) - 1\right]^2}$$

$$= \mathcal{N}k_{\mathrm{B}}\left(\frac{\hbar\omega}{k_{\mathrm{B}}T}\right)^2\frac{\exp\left(\frac{\hbar\omega}{k_{\mathrm{B}}T}\right)}{\left[\exp\left(\frac{\hbar\omega}{k_{\mathrm{B}}T}\right) - 1\right]^2} \quad.$$

For $T \ll \hbar\omega/k_{\mathrm{B}}$:

$$c_v \approx \mathcal{N}k_{\mathrm{B}}\left(\frac{\hbar\omega}{k_{\mathrm{B}}T}\right)^2 \lim_{x\to\infty}\frac{e^x}{(e^x - 1)^2} = \mathcal{N}k_{\mathrm{B}}\left(\frac{\hbar\omega}{k_{\mathrm{B}}T}\right)^2 \exp\left(-\frac{\hbar\omega}{k_{\mathrm{B}}T}\right)$$

: exponential decay with $\exp\left(-\dfrac{\hbar\omega}{k_{\mathrm{B}}T}\right)$.

cf. the Debye law: $c_v \propto T^3$.

4. The lattice vibration part of the internal energy:

$$\mathcal{U}_{\text{vib}} = 9\mathcal{N}\frac{k_{\text{B}}^4 T^4}{\hbar^3 \omega_{\text{D}}^3}\frac{\pi^4}{15} = \frac{3\pi^4}{5}\mathcal{N}k_{\text{B}}\frac{T^4}{T_{\text{D}}^3} \quad ,$$

$$c_{v,\text{vib}} = \frac{\text{d}\mathcal{U}_{\text{vib}}}{\text{d}T} = \frac{12\pi^4}{5}\mathcal{N}k_{\text{B}}\left(\frac{T}{T_{\text{D}}}\right)^3 .$$

The electronic part of the internal energy:

$$c_{v,\text{e}} = \frac{\pi^2}{2}\mathcal{N}k_{\text{B}}\frac{T}{T_{\text{F}}} \quad .$$

Equivalence at:

$$\frac{12\pi^4}{5}\mathcal{N}k_{\text{B}}\left(\frac{T_0}{T_{\text{D}}}\right)^3 = \frac{\pi^2}{2}\mathcal{N}k_{\text{B}}\frac{T_0}{T_{\text{F}}} \quad ,$$

$$T_0^2 = \frac{5T_{\text{D}}^3}{24\pi^2 T_{\text{F}}} \quad ,$$

$$T_{\text{F}} = \frac{\hbar^2}{2m_{\text{e}}k_{\text{B}}}\left(\frac{3\pi^2\mathcal{N}}{\mathcal{V}}\right)^{2/3} \quad ,$$

$$T_0 = \frac{T_{\text{D}}^{3/2}}{\pi\hbar}\left(\frac{5m_{\text{e}}k_{\text{B}}}{12}\right)^{1/2}\left(3\pi^2\frac{\mathcal{N}}{\mathcal{V}}\right)^{-1/3} \quad ,$$

in the case of copper:

$$\rho_{\text{m}} = 8.96\,\text{g}\,\text{cm}^{-3} \quad , \quad M = 63.54\,\text{g}\,\text{mol}^{-1}$$

$$T_{\text{D}} = 343\,\text{K} \quad , \quad \frac{\mathcal{N}}{\mathcal{V}} = \frac{\rho_{\text{m}}N_{\text{A}}}{M} \quad ,$$

$$T_0 = 3.23\,\text{K} \quad .$$

5. (a) Mobility, Einstein relation:

$$v = \nu eE = \frac{eE}{k_{\text{B}}T}D_s \quad ,$$

$$v = 2.38 \cdot 10^{-6}\,\frac{\text{m}}{\text{s}} \quad .$$

(b) $a = \dfrac{k_{\text{B}}T}{6\pi\eta D} = 2.68\,\text{nm}.$

6. Particle current balance:

$$j = v\frac{4\pi}{3}a^3(\rho_{\text{m}} - \rho_{\text{m,f}})\omega^2 r\rho(r) - D\frac{\text{d}\rho}{\text{d}r} = 0 \quad ,$$

$$\frac{d\rho}{\rho} = \frac{\nu 4\pi a^3 (\rho_m - \rho_{m,f})\omega^2 r}{3D} dr$$

$$= \frac{4\pi a^3 (\rho_m - \rho_{m,f})\omega^2}{3k_B T} r dr \quad,$$

$$\ln \frac{\rho(r_{max})}{\rho(r)} = \frac{4\pi a^3 (\rho_m - \rho_{m,f})\omega^2}{3k_B T} \int_r^{r_{max}} r' dr'$$

$$= \frac{4\pi a^3 (\rho_m - \rho_{m,f})\omega^2 (r_{max}^2 - r^2)}{6k_B T} \quad,$$

$$\rho(r) = \rho(r_{max}) \exp\left[\frac{2\pi a^3 (\rho_m - \rho_{m,f})\omega^2 (r^2 - r_{max}^2)}{3k_B T}\right] \quad.$$

7. The Helmholtz free energy at equilibrium:

$$\mathcal{F} = mgz + kT\frac{3z^2}{2R_0^2}$$

$$\frac{d\mathcal{F}}{dz} = mg + kT\frac{3z}{R_0^2} = 0 \quad \Rightarrow z = -\frac{1}{3}\frac{mgR_0^2}{kT} \quad,$$

$$a = 1\,\mu m : \quad m = 8.0 \times 10^{-14}\,kg \quad, \quad z = -6.3\,nm \quad.$$

8. Kinetic energy of the neutrons:

$$\epsilon_n = \frac{(\hbar k_n)^2}{2m_n} \quad.$$

The conservation of energy and momentum:

The neutron gains the energy $\hbar\omega$ and the momentum $\hbar k$ of the phonon. As is shown in Fig. B.5, this is possible for all ks, for which the condition $\epsilon_n(k) = \hbar\omega(k)$ is fulfilled.

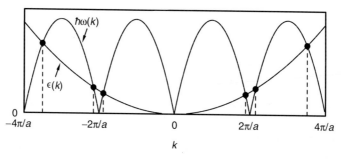

Fig. B.5. The dispersion relations of the neutron and the phonon.

C

A Small Selection of Further Reading

Physics of Condensed Matter

P.M. Chaikin and T.C. Lubensky: *Principles of Condensed Matter Physics*. Cambridge, New York: Cambridge University Press 1995

Crystalline Solids

N.W. Ashcroft and N.D. Mermin: *Solid State Physics*. Orlando: Saunders College 1976

J.R. Christman: *Fundamentals of Solid State Physics*. New York: John Wiley & Sons 1988

H. Ibach and H. Lüth: *Solid State Physics*, 2nd ed. Berlin, Heidelberg: Springer 2003

J.M. Ziman: *Principles of the Theory of Solids*. Cambridge, New York: Cambridge University Press 1979

Superconductors

W. Buckel: *Superconductivity: Fundamentals and Applications*. New York: John Wiley & Sons 1991

Simple Liquids

P.A. Egelstaff: *An Introduction to the Liquid State*. Oxford, New York: Oxford University Press 1992

J.-P. Hansen and I.R. McDonald: *Theory of Simple Liquids*. New York, London: Academic Press 1986

Electrolytes

J.O.M. Bockris and A.K.N. Reddy: *Modern Electrochemistry*. New York: Plenum Press 2000

Liquid Crystals

S. Chandrasekhar: *Liquid Crystals*. Cambridge, New York: Cambridge University Press 1992

P.G. de Gennes and J. Prost: *The Physics of Liquid Crystals*. Oxford, New York: Oxford University Press 1993

G. Vertogen and W. de Jeu: *Thermotropic Liquid Crystals*. Berlin, Heidelberg: Springer 1988

Polymers

P.G. de Gennes: *Scaling Concepts in Polymer Physics*. Ithaca, London: Cornell University Press 1979

A.Y. Grosberg and A.R. Khokhlov: *Statistical Physics of Macromolecules*. New York: AIP Press 1994

G. Strobl: *The Physics of Polymers*. Berlin, Heidelberg: Springer 1997

Light, X-Ray, and Neutron Scattering

B. Berne and R. Pecora: *Dynamic Light Scattering*. New York: Dover 2000

A. Guinier: *X-ray Diffraction*. New York: Dover 1994

S.W. Lovesey: *Theory of Neutron Scattering from Condensed Matter*. Oxford, New York: Oxford University Press 1984

D

Nomenclature

a_0	effective length per monomer (Eq. (1.80))
a_I	ionic radius
a_R	size of a Rouse-sequence (Eq. (5.215))
a_T	temperature shift factor
$a(t)$	general time-dependent modulus
$\boldsymbol{a}_i$	basis vectors of Bravais-lattice (Eq. (1.1))
$\widehat{\boldsymbol{a}}_i$	basis vectors of reciprocal lattice (Eq. (1.129))
A	atomic weight
$\boldsymbol{A}(\boldsymbol{r})$	vector potential
b_R	Rouse spring constant (Eq. (5.217))
$\mathbf{b}_l^+, \mathbf{b}_l$	exciton-creation and -annihilation operator
$\boldsymbol{B}, B$	magnetic field strength
c_l	speed of light
c_s	sound velocity
c_a	heat capacity per mole of atoms
c_v	heat capacity per unit volume
$c_{v,e}$	electronic part of c_v
$c_{v,\text{vib}}$	contribution of lattice vibrations to c_v

$c_{\boldsymbol{k}}$	momentum representation of a quantum state
$\tilde{c}$	molar concentration
$C(\boldsymbol{q})$	scattering amplitude (Eqs. (1.98) and (5.244))
d_{hkl}	separation of the set of lattice planes hkl
D	diffusion coefficient (Eq. (5.153))
D_{s}	self-diffusion coefficient (Eq. (5.147))
D_T	thermal diffusion coefficient (Eq. (5.276))
$D_{\mathrm{t}}, D_{\mathrm{t}}(t), D_{\mathrm{t}}(\omega)$	tensile compliance (Eqs. (2.4), (2.45), (2.49))
$\mathcal{D}(\omega)$	density of levels (Eq. (4.36))
$\boldsymbol{D}$	electric displacement
e	electron charge
e_{zz}	strain (Eq. (2.2))
$E_{\mathrm{t}}, E_{\mathrm{t}}(t), E_{\mathrm{t}}(\omega)$	Young's modulus (Eqs. (2.3), (2.46), (2.51))
$\boldsymbol{E}, E$	electric field strength
f	density of the Helmholtz free energy (Eq. (A.5))
$\hat{f}$	density of the reduced Helmholtz free energy (Eq. (A.7))
f_{c}	cell structure factor (Eq. (1.120))
f_j	atomic or molecular form factor (Eq. (1.89))
$\mathbf{f}, f$	force
F_{p}	polarisation factor (Eqs. (1.89) and (5.204))
$\mathcal{F}$	Helmholtz free energy of a sample
g	density of the Gibbs free energy (Eq. (A.9))
$\hat{g}$	density of the reduced Gibbs free energy (Eq. (A.11))
g	g-factor of an electron shell (Eq. (2.149))
g_{n}	g-factor of a nucleus (Eq. (2.168))
$g_2(\boldsymbol{r})$	pair distribution function
$g(\boldsymbol{r}, t)$	van Hove function
$g_1(\boldsymbol{r}, t)$	auto-correlation function
$G, G(t), G(\omega)$	shear modulus (Eq. (2.5))

$\mathcal{G}$	Gibbs free energy of a sample
$\Delta\mathcal{G}_{\mathrm{mix}}$	Gibbs free energy of mixing
$\Delta\mathcal{G}_{\mathrm{r}}$	Gibbs reaction free energy
$\boldsymbol{G}_{hkl}, G_h$	vector of the reciprocal lattice (Eq. (1.130))
$\tilde{h}$	enthalpy per mole
H	Hamiltonian function
H_{c}	critical magnetic field for a superconductor (Fig. 4.25)
$\mathcal{H}$	enthalpy of a sample
$\mathbf{H}$	Hamiltonian operator
$\boldsymbol{H}, H$	strength of the applied magnetic field
I	electric current
I	radiation intensity
I	quantum number of nuclear spin
$\boldsymbol{I}$	nuclear spin
$\boldsymbol{j}, j, j_i$	particle current density (for species i)
$\boldsymbol{j}_\eta, j_\eta$	charge current density
j_Q	heat current density
J	quantum number of the total angular momentum of an electron shell
$J, J(t), J(\omega)$	shear compliance
$\boldsymbol{J}$	total angular momentum of an electron shell
k_{B}	Boltzmann constant
k_{F}	radius of the Fermi sphere (Fig. 4.5)
$\boldsymbol{k}$	wavevector
K_j	Frank moduli of a nematic (Eq. (2.8))
l_{c}	contour length of a polymer chain
L	quantum number of the total orbital angular momentum of an electron shell
m_{e}	electron mass
$m_{\mathrm{e,eff}}$	effective mass of electrons (Eq. (4.147))

$m_{h,eff}$	effective mass of holes (Eq. (4.152))
m_p	proton mass
m_I	quantum number of the z component of a nuclear spin
m_J	quantum number of the z component of the total angular momentum of the electron shell
m_S	quantum number of the z component of the total spin of the electron shell
$\boldsymbol{m}, m, m_z$	magnetic dipole moment
M	molecular weight
$\boldsymbol{M}, M$	magnetisation
n	refractive index
n_B	effective magneton number (Eq. (2.162))
n_L	quantum number of Landau levels (Eq. (4.221))
n_{ij}	number of phonons occupying the lattice vibration ij
$\tilde{n}$	molar number
$\boldsymbol{n}$	nematic director
N	degree of polymerisation
N_A	Avogadro number
N_s	number of segments in a polymer chain
$\mathcal{N}$	number of atoms, molecules, cells in a sample
p	pressure
$\boldsymbol{p}$	momentum operator
$\boldsymbol{p}, p$	electric dipole moment
$\boldsymbol{p}, p_i$	momentum
$\boldsymbol{P}, P$	polarisation
$\boldsymbol{q}$	scattering vector (Eq. (1.93))
Q	charge
$\mathcal{Q}$	heat
r_e	classical electron radius (Eq. (1.88))
R_0	mean size of a polymer coil (Eq. (1.80))
R_H	Hall coefficient (Eq. (4.208))

R_Ω	Ohm resistance
$\tilde{R}$	universal gas constant
$\boldsymbol{R}_{uvw}$	vectors of a Bravais lattice (Eq. (1.1))
s	entropy density
$\tilde{s}$	entropy per mole
S_2	nematic order parameter
S	quantum number of the total spin of the electron shell
S_D	Debye scattering function (Eq. (1.117))
S_L	interference function of an ideal lattice
$S(\boldsymbol{q})$	scattering function, structure factor
$S(\boldsymbol{q}, t)$	time-dependent structure factor
$S(\boldsymbol{q}, \omega)$	dynamic structure factor
$\mathcal{S}$	entropy of a sample
$\boldsymbol{S}$	spin
T	temperature
T_1, T_2	longitudinal and transverse relaxation time in magnetic resonance experiments (Eq. (2.179))
T_A	activation temperature (Eq. (1.82))
T_c	critical temperature
T_D	Debye temperature (Eq. (5.61))
T_F	Fermi temperature (Eq. (4.55))
T_g	glass transition temperature
T_m	melting point temperature
T_V	Vogel temperature (Eq. (1.82))
$\boldsymbol{T}$	torque
u	density of the internal energy
$\hat{u}$	density of the reduced internal energy (Eq. (A.1))
u_{kin}	kinetic energy of a particle
u_{pot}	potential energy of a particle
$u(r)$	pair interaction energy

$u(\varphi)$	rotation potential
U_b	lattice energy (Eq. (1.5))
U_{e0}	ground state energy of the electrons in a sample
$\mathcal{U}$	internal energy of a sample
$\mathcal{U}_{vib}$	total vibrational energy of a crystal
$\mathcal{U}_e$	total electronic energy of a crystal
$\tilde{v}$	molar volume
$\boldsymbol{v}, v$	velocity
$\boldsymbol{v}_g, v_g$	group velocity
$V, \Delta V$	electrostatic potential, potential difference
V_c	volume of the unit cell
$\widehat{V}_c$	volume of the unit cell of the reciprocal lattice
$\mathcal{V}$	volume of a sample
$\mathbf{V}$	interaction potential operator
w	probability density
w_i	occupation probability for state i
w	work per unit volume
$\mathcal{W}$	work
X	general deformation variable
z	complex frequency (Eq. (2.208))
$\mathcal{Z}$	partition function
$\alpha(\omega)$	general susceptibility
$\alpha_p(t)$	general pulse response function
$\alpha_c(t)$	general compliance
β	polarisability
β_M	Madelung constant (Eq. (1.13))
β_{ex}	spin–spin interaction coefficient (Eq. (3.30))
γ	gyromagnetic ratio
γ	shearing angle
γ_1	rotational viscosity of a nematic (Eq. (2.42))

γ_1	exciton hopping rate (Eq. (5.138))
γ_i	activity coefficient (Eq. (4.297))
ϵ_i	energy of eigenstate i
ϵ_F	Fermi energy
ϵ_g	energy gap in a semiconductor (Fig. 4.14)
$\Delta\epsilon_s$	energy gap of a superconductor
ε	dielectric constant
$\Delta\varepsilon$	dielectric anisotropy (Eq. (1.65))
ε_0	electric constant
ζ	friction coefficient of a particle (Eq. (5.163))
ζ_R	friction coefficient of a Rouse sequence (Eq. (5.218))
η	charge density
η	viscosity
$\eta_{a,b,c}$	Miesowicz viscosity coefficients (Eq. (2.41))
θ	Bragg angle (Abb. 1.35)
κ	magnetic susceptibility
κ_n	magnetic susceptibility of nuclear spins (Eq. (2.173))
λ	wavelength
λ_f	mean free path
λ_L	London penetration depth (Eq. (4.257))
λ_Q	heat conductivity (Eq. (4.15))
μ	magnetic permeability(Eq. (2.135))
μ	chemical potential
$\tilde{\mu}$	molar chemical potential (Eq. (A.14))
μ_e	(electro-)chemical potential of electrons
μ_B	Bohr magneton
μ_0	magnetic constant
μ_n	nuclear magneton (Eq. (2.169))
ν	frequency [s^{-1}]

ν	mobility (Eq. (5.162))
ν_{el}	electric mobility (Eq. (4.158))
$\nu_{el,e}, \nu_{el,h}$	electric mobility of electrons and holes(Eq. (4.158))
$\nu_{el,+}, \nu_{el,-}$	electric mobility of anions and cations (Eq. (4.301))
ξ	general force
ξ_D	Debye screening length (Eq. (4.329))
ξ_G	Ginzburg–Landau coherence length (Fig. 4.36)
ξ_H	magnetic coherence length in nematics (Eq. (2.32))
ξ_ϕ	correlation length for concentration fluctuations (Eq. (3.64))
ρ	mean particle density
ρ_e	mean electron density
ρ_h	mean hole density
ρ_c	number of unit cells per unit volume
$\rho(\boldsymbol{r})$	local particle density
$\rho_e(\boldsymbol{r})$	local electron density
$\boldsymbol{\rho}, \rho_{l'l''}$	density matrix
σ	scattering cross section
σ_i	surface charge density
σ_{el}	electrical conductivity (Eqs. (4.5)), (4.291))
$\tilde{\sigma}_{el}$	molar electrical conductivity (Eq. (4.292))
σ_{ij}	component of stress tensor
τ	relaxation time
τ_f	collosion time
ϕ_i	volume fraction of species i
Φ	magnetic flux
Φ_0	quantum of magnetic flux (Eq. (4.259))
χ_F	Flory–Huggins parameter (Eq. (3.70))
χ	dielectric susceptibility
$\psi(\boldsymbol{r})$	wavefunction

$\psi_2(\boldsymbol{r}_1, \boldsymbol{r}_2)$	2-particle wavefunction (Cooper pairs, excitons)
$\Psi(\boldsymbol{r})$	wavefunction of superconducting particles
ω	frequency $[\mathrm{rad\,s}^{-1}]$
ω_c	cyclotron frequency (Eq. (4.200))
ω_D	Debye cut-off frequency (Fig. 5.5)
ω_L	Larmor frequency (Eq. (2.139))
ω_pl	plasma frequency (Eq. (4.94))
Ω	direction in space
$\Omega_{l',l''}$	matrix element of a Hamiltonian (Eq. (5.118))
$\boldsymbol{\Omega}$	rotation matrix

References

1. F. Kohler: *The Liquid State*, p. 5. Weinheim: Verlag Chemie 1972
2. G. Vertogen and W. de Jeu: *Thermotropic Liquid Crystals, Fundamentals*, p. 24. Berlin, Heidelberg: Springer 1988
3. P. Cladis, M. Kleman and P. Pieranski: *C. R. Acad. Sci. Paris B* **273**, 275, (1971)
4. P. Chatelain and M. Germain: *C. R. Acad. Sci. Paris* **259**, 127 (1964)
5. G. Vertogen and W. de Jeu: *Thermotropic Liquid Crystals, Fundamentals*, p. 175. Berlin, Heidelberg: Springer 1988
6. R. Eppe, E.W. Fischer and H.A. Stuart: *J. Polym. Sci.* **34**, 721 (1959)
7. B. Kanig: *Progr. Colloid Polym. Sci.* **57**, 176 (1975)
8. A.J. Kovacs: *Fortschr. Hochpolym. Forsch.* **3**, 394 (1966)
9. R. Kirste, W.A. Kruse and K. Ibel: *Polymer* **16**, 120 (1975)
10. Ch. Gähwiller: *Mol. Cryst. Liquid Cryst.* **20**, 301 (1973)
11. A. Raviol, W. Stille and G. Strobl: *J. Chem. Phys.* **103**, 3788 (1995)
12. F.R. Schwarzl: *Polymermechanik.* Berlin, Heidelberg: Springer 1990
13. E. Castiff and T.S. Tobolsky: *J. Colloid Sci.* **10**, 375 (1955)
14. J. Heijboer: *Kolloid Z.* **148**, 36 (1956)
15. Y. Ishida, M. Matsuo and K. Yamafuji: *Kolloid Z.* **180**, 108 (1962)
16. E. Dormann, D. Hone and V. Jaccarino: *Phys. Rev. B* **7**, 2715 (1976)
17. V.J. McBrierty and K.J. Packer: *Nuclear Magnetic Resonance in Solid Polymers.* Cambridge: Cambridge University Press 1996
18. G. Burns: *Solid State Physics.* London, New York: Academic Press 1985
19. R.A. Cowley: *Phys. Rev. A* **134**, 981 (1964)
20. M.E. Lines and A.M. Glass: *Ferroelectrics and related materials.* Oxford, New York: Oxford University Press 1977
21. J.S. Kouvel and M.E. Fisher: *Phys. Rev. A,* **136**, 1626 (1964)
22. P. Weiss and R. Porrer: *Ann. Phys.* **5**, 153 (1926)
23. C. Kittel: *Introduction to Solid State Physics*, Chap. 16. New York: Wiley 1966
24. R.-J. Roe and W.-C. Zin: *Macromolecules* **13**, 1221 (1980)
25. T. Hashimoto, M. Itakura and H. Hasegawa: *J. Chem. Phys.* **85**, 6118 (1986)
26. D. Schwahn, S. Janssen and T. Springer: *J. Chem. Phys.* **97**, 8775 (1992)
27. G.R. Strobl, J.T. Bendler, R.P. Kambour and A.R. Shultz. *Macromolecules* **19**, 2683 (1986)
28. A. Lien and B. Phillips: *Phys. Rev. A* **133**, 1370 (1964)

29. D.K. McDonald and K. Mendelssohn: *Proc. Roy. Soc. A* **202**, 103 (1950)
30. A. Moore and B. Spong: *Phys. Rev.* **125**, 846 (1962)
31. H. Ibach and H. Lüth: *Festkörperphysik*, p. 223. Berlin, Heidelberg: Springer 1995
32. H.K. Onnes: *Akad. van Wetenschappen* **14**, 113 (1911)
33. N.E. Phillips: *Phys. Rev.* **134**, 385 (1964)
34. P.L. Richards and M. Tinkham: *Phys. Rev.* **119**, 575 (1960)
35. I. Giaver and K. Megerle: *Phys. Rev.* **122**, 1101 (1961)
36. H. Ibach and H. Lüth: *Festkörperphysik*, p. 249. Berlin, Heidelberg: Springer 1995
37. B.S. Deaver and W.M. Fairbank: *Phys. Rev. Lett.* **7**, 43 (1961)
38. G. Wedler: *Lehrbuch der Physikalischen Chemie*, Chap. 1.6. Weinheim: Verlag Chemie 1982
39. J.O.M. Bockris and A.K.N. Reddy: *Modern Electrochemistry*, Chap. 3. New York: Plenum Press 1970
40. J.O.M. Bockris and A.K.N. Reddy: *Modern Electrochemistry*, Chap. 4. New York: Plenum Press 1970
41. H. Ibach and H. Lüth: *Festkörperphysik*, p. 71. Berlin, Heidelberg: Springer 1995
42. R.N. Sinclair and B.N. Brockhouse: *Phys. Rev.* **120**, 1638 (1960)
43. M.D. Sturge: *Phys. Rev.* **127**, 768 (1962)
44. H. Haken and G. Strobl: *Z. Phys.* **262**, 135 (1973)
45. V. Ern, A. Suna, Y. Tomkiewicz, P. Aviakan and R.P. Groff: *Phys. Rev. B* **5**, 3222 (1972)
46. D. Boese and F. Kremer: *Macromolecules* **23**, 829 (1990)
47. G.C. Berry and T.G. Fox: *Adv. Polymer Sci.* **5**, 261 (1968)
48. T. Perkins, D.E. Smith and S. Chu: *Science* **264**, 819 (1994)
49. S.M. Shapiro, R.W. Gammon and H.Z. Cummins: *Appl. Phys. Lett.* **9**, 157 (1966)
50. I. Pelah: *Phys. Rev.* **108**, 1091 (1957)
51. R.F. Wallis: *Lattice Dynamics*. New York: Pergamon 1965

Index